T4-AHW-004

REVIEWS in MINERALOGY

(Formerly: "Short Course Notes")

Volume 6

T.M.

MARINE MINERALS

ROGER G. BURNS, Editor

The Authors:

Roger G. Burns
Virginia Mee Burns
Department of Earth & Planetary Sciences
Massachusetts Institute of Technology
Cambridge, Massachusetts 02139

Thomas M. Church
College of Marine Studies
University of Delaware
Newark, Delaware 19711

Robert A. Gulbrandsen
United States Geological Survey
Menlo Park, California 94025

John C. Hathaway
United States Geological Survey
Woods Hole, Massachusetts 02543

William T. Holser
Department of Geology
University of Oregon
Eugene, Oregon 97403

Miriam Kastner
Division of Geological Sciences
Scripps Institution of Oceanography
La Jolla, California 92093

Frank T. Manheim
United States Geological Survey
Woods Hole, Massachusetts 02543

James W. Murray
Department of Oceanography
University of Washington
Seattle, Washington 98195

Series Editor: Paul H. Ribbe
Department of Geological Sciences
Virginia Polytechnic Institute and State University
Blacksburg, Virginia 24061

MINERALOGICAL SOCIETY OF AMERICA

X

PRINTED BY

BookCrafters, Inc.
Chelsea, Michigan 48118

<u>REVIEWS IN MINERALOGY</u>

(Formerly: SHORT COURSE NOTES)

ISSN 0275-0279

Volume 6: MARINE MINERALS

ISBN 0-939950-06-5

Additional copies of this volume as well as those listed below may be obtained at moderate cost from

Mineralogical Society of America
2000 Florida Avenue, NW
Washington, D.C. 20009

Vol. 1:	SULFIDE MINERALOGY, P.H. Ribbe, Editor (1974)	284 p.
Vol. 2:	FELDSPAR MINERALOGY, P.H. Ribbe, Editor (1975; revised 1982)	∿300 p.
Vol. 3:	OXIDE MINERALS, Douglas Rumble III, Editor (1976)	502 p.
Vol. 4:	MINERALOGY and GEOLOGY of NATURAL ZEOLITES, F.A. Mumpton, Editor (1977)	232 p.
Vol. 5:	ORTHOSILICATES, P.H. Ribbe, Editor (1980)	381 p.
Vol. 6:	MARINE MINERALS, Roger G. Burns, Editor (1979)	380 p.
Vol. 7:	PYROXENES, C.T. Prewitt, Editor (1980)	525 p.
Vol. 8:	KINETICS of GEOCHEMICAL PROCESSES, A.C. Lasaga and R.J. Kirkpatrick, Editors (1981)	391 p.
Vol. 9A:	AMPHIBOLES and Other Hydrous Pyriboles - Mineralogy, D.R. Veblen, Editor (1981)	372 p.
Vol. 9B:	AMPHIBOLES: Petrology and Experimental Phase Relations, D.R. Veblen and P.H. Ribbe, Editors (1981)	∿375 p.

FOREWORD

Marine Minerals was first published in 1979 as Volume 6 of the series entitled "SHORT COURSE NOTES." In 1980 the Mineralogical Society of America changed the name of the series to "REVIEWS IN MINERALOGY," and for that reason this, the second printing of *Marine Minerals* has been reissued under the new banner. Only minor corrections have been made. The reader's attention is directed to the titles of other volumes published in the "REVIEWS" series, as listed on the opposite page.

Paul H. Ribbe
Series Editor
Blacksburg, VA
October 1981

PREFACE and ACKNOWLEDGMENTS

This volume was originated from notes prepared for a short course on Marine Minerals held in La Jolla, California, November 2-3, 1979. Sponsored by the Mineralogical Society of America, the short course was organized by the following individuals who served as lecturers and authors:

Roger G. Burns
Virginia Mee Burns
Thomas M. Church
Robert A. Gulbrandson
John C. Hathaway
William T. Holser
Miriam Kastner
Frank T. Manheim
James W. Murray

With the increased activity in marine geochemical research resulting from studies of dissolved constituents and particulate matter in seawater, diagenetic reactions in underlying sediments, submarine volcanism at spreading centers, and interactions of seawater with igneous rocks, it is timely to review the properties of some of the major mineral constituents found on the seafloor or extracted from seawater. Chapters in this volume, therefore, are devoted to marine manganese oxide, iron oxide, silica polymorphs, zeolite, clay, phosphorite, barite, evaporite, and placer minerals. Carbonates are not included; coverage of this important mineral group warrants a separate monograph. The extremely interesting sulfide and hydrothermal mineral assemblages recently discovered at oceanic spreading centers are also not discussed here.

Special thanks are due to Prof. Paul H. Ribbe, of Virginia Polytechnic Institute and State University, who did the copy editing, to Mrs. Margie Strickler, Ramonda Haycocks, and Patricia Fenton, who did the typing, and to Messrs. Brian Cooper, James Downs, and Bryan Chakoumakos, who proof-read the entire text. We gratefully acknowledge the permission of numerous publishers, editors and authors to reproduce many of the illustrations and data used in several of the chapters.

Roger G. Burns
Cambridge, MA
November 1979

TABLE of CONTENTS

Page

Chapter 9. TRACE ELEMENTS AND ISOTOPES IN EVAPORITES, continued *William T. Holser*

Chapter 10. MARINE PLACER MINERALS *Virginia Mee Burns*

Chapter 1

MANGANESE OXIDES

Roger G. Burns and Virginia Mee Burns

INTRODUCTION

Manganese oxides are ubiquitous phases in the marine environment, being major components of seafloor manganese nodules, crusts on basalts at ridge-crests and fracture zones, metalliferous sediments near spreading centers, and coatings on detritus and biogenic debris suspended in the water column and underlying sediments. The hydrous manganese oxides are frequently, but not always, intimately associated with hydrated iron oxide and oxyhydroxide phases (Murray, this volume, Ch. 2), and they generally host a variety of minor elements, some of which (e.g. Ni, Cu, Co) are enriched to economically important concentrations. Tetravalent manganese predominates in marine manganese oxides, but the presence of Mn^{2+} and Mn^{3+} ions in some phases is inferred from crystal chemistry and thermodynamic criteria. Although more than 20 manganese (IV) oxide minerals are known in continental manganese deposits, only a few of these terrestrial minerals have been positively and unambiguously identified in the marine environment. The marine manganese oxide phases are often metastable, intimately intergrown with other materials, and so poorly crystalline that attempts to identify them by conventional x-ray diffraction techniques are fraught with difficulties. The manganese oxides found in the marine environment are characterized by numerous structural defects, essential vacancies which may or may not be ordered, domain intergrowths, extensive solid solution, and cation exchange properties. These phenomena not only lead to non-stoichiometry, but detract from long-range ordering, making crystal structure determinations by x-ray diffraction methods extremely difficult. Recently, information derived from scanning electron microscopy, electron diffraction, high resolution transmission electron microscopy, infrared spectroscopy and extended x-ray absorption fine structure, has led to a better understanding of the crystal chemistry and physical and chemical properties of the intractable manganese oxide phases. These results are summarized in this chapter.

Table 1. Crystallographic Data for Selected Oxides of Manganese

Mineral or Compound	Approximate Formula	Crystal Class (Space Group)	Cell Parameters (Å)
pyrolusite (β-MnO_2)	MnO_2	tetragonal ($P4_2/mn2$)	a =4.39; c =2.87
ramsdellite	MnO_2	orthorhombic (Pbnm)	a =4.53; b =9.27; c =2.87
nsutite (γ-MnO_2)	$(Mn^{2+},Mn^{3+},Mn^{4+})(O,OH)_2$	hexagonal	a =9.65; c =4.43
ε-MnO_2	MnO_2	hexagonal	a =2.80; c =4.45
hollandite (α-MnO_2)	$(Ba,K)_{1-2}Mn_8O_{16}\cdot xH_2O$	tetragonal (I4/m) or monoclinic ($P2_1/n$)	a =9.96; c =2.86 a =10.03; b =5.76; c =9.90 β=90°42'
cryptomelane	$K_{1-2}Mn_8O_{16}\cdot xH_2O$	tetragonal (I4/m) or monoclinic (I2/m)	a =9.84; c =2.86 a =9.79; b =2.88; c =9.94 β=90°37'
romanèchite (psilomelane)	$(Ba,K,Mn^{2+},Co)_2Mn_5O_{10}\cdot xH_2O$	monoclinic (A2/m) or orthorhombic ($P2_12_12$)	a =9.56; b =2.88; c =13.85 β=92°30' a =8.254; b =13.40; c =2.864
todorokite	$(Ca,Na,K)(Mg,Mn^{2+})Mn_5O_{12}\cdot xH_2O$	monoclinic	a =9.75; b =2.849; c =9.59 β=90
buserite ("10Å manganite")	NaMn oxide hydrate	hexagonal	a =8.41; c =10.01
lithiophorite	$[Mn_5^{4+}Mn^{2+}O_{12}][Al_4Li_2(OH)_{12}]$	monoclinic (C2/m)	a =5.06; b =8.70; c =9.61 β=100°7'
Mn_5O_8 and $Cd_2Mn_3O_8$	$Mn_2^{2+}Mn_3^{4+}O_8$ $Cd_2Mn_3O_8$	monoclinic (C2/m or C2)	a =10.34; b =5.72; c =4.85 β=109°25'
chalcophanite	$Zn_2Mn_6O_{14}\cdot 6H_2O$	triclinic ($P\bar{1}$)	a =7.54; b =7.54; c =8.22 α=90°; β=117°12'; γ=120°
synthetic birnessite	$Na_4Mn_{14}O_{27}\cdot 9H_2O$	orthorhombic	a =8.54; b =15.39; c =14.26
synthetic birnessite	$Mn_7O_{13}\cdot 5H_2O$	hexagonal	a =2.84; c =7.27
birnessite	$(Na,Ca,K)(Mg,Mn)Mn_6O_{14}\cdot 5H_2O$	hexagonal	a =2.85; c =7.08-7.31
vernadite (δ-MnO_2)	$MnO_2\cdot nH_2O\cdot m(R_2O,RO,R_2O_3)$ R=Na,Ca,Co,FeMn	hexagonal	a =2.86; c =4.7
rancieite	$(Ca,Mn)Mn_4O_9\cdot 3H_2O$	hexagonal	a =2.84; c =7.07
groutite	α-MnOOH	orthorhombic (Pbnm)	a =4.56; b =10.70; c =2.85
feitknechtite	β-MnOOH	hexagonal ($P\bar{3}m1$)	a =3.32; c =4.71
manganite	γ-MnOOH	monoclinic ($B2_1/d$)	a =8.88; b =5.25; c =5.71 β=90°
hausmannite	Mn_3O_4	tetragonal ($I4_1/amd$)	a =5.76; c =9.41
pyrochroite	$Mn(OH)_2$	hexagonal ($P\bar{3}m1$)	a =3.322; c =4.734

Following a survey of the mineralogy, nomenclature and crystal structural correlations of the principal manganese (IV) oxides, thermodynamic properties, occurrences and parageneses of the marine manganese oxide minerals are described. The coverage of this chapter compliments those of recent treatises (Glasby, 1977; Varentsov, 1979) and reviews (Bonatti, 1975; Calvert, 1978; Burns and Burns, 1977a, 1979).

MINERALOGY AND NOMENCLATURE

Manganese forms a large number of oxides, ranging from refractory anhydrous phases to hydrated minerals stable at low temperatures in aqueous environments. The simpler, more refractory oxides of divalent and trivalent manganese were reviewed by Huebner (1976). So far as the higher oxides are concerned, at least 20 naturally-occurring varieties containing tetravalent manganese are currently recognized as valid mineral species (Hewett and Fleischer, 1960; Hewett *et al.*, 1963; Burns and Burns, 1977a), and several others are known as synthesis products (Wadsley, 1950a,b; McKenzie, 1970; Giovanoli, 1976; Giovanoli and Stahli, 1970; Giovanoli *et al.*, 1967, 1969, 1970a,b, 1971, 1973, 1975; Usui, 1979). Many of the manganese oxide phases pertinent to this review of marine occurrences are listed in Table 1, together with available crystallographic data and current information on chemical compositions and crystal structures. More comprehensive compendia are presented elsewhere (Burns and Burns, 1977a).

The most common method for identifying the typically opaque, cryptocrystalline manganese oxide minerals has been by x-ray diffraction analysis, preferably using Mn filtered Fe$K\alpha$ radiation. Frequently, however, the particle sizes of the crystallites are so small that coherent scattering of x-rays is reduced. Thus, many naturally-occurring manganese oxides appear to be amorphous or to give broad, diffuse lines in x-ray diffraction patterns. Samples crushed under liquid nitrogen appear to give better patterns, however (Brown, 1972). Furthermore, certain *d*-spacings are common to several minerals. Most noteworthy are the values around 2.40-2.45 A and 1.40-1.42 A, which represent diffractions of x-rays from the $(10\bar{1}0)$ and $(11\bar{2}0)$ planes of hexagonally close-packed oxygens containing manganese ions in octahedral coordination. Therefore, identification by x-ray diffraction techniques has sometimes produced ambiguous results. The added resolution of electron diffraction techniques has led to the discovery of complex intergrowths of variable lattice periodicities in some of the naturally-occurring hydrated manganese oxides (Turner and Buseck, 1979a,b; Chukhrov *et al.*, 1978a,b,c, 1979a,b,c). Such coherent and disordered intergrowths having dimensions of a few unit cells (i.e., tens of Angstroms) point to added complexities over classification and nomenclature, examples of which are now discussed.

Todorokite, Buserite and "10 A Manganite"

One of the most common manganese oxide phases identified in marine deposits is that characterized by x-ray diffraction lines at 9.5-9.8 A and 4.8-4.9 A, together with other less diagnostic *d*-spacings. Electron microprobe analyses of this phase indicate that it is a hydrated Mn-Mg-Ca-Na-Ni-Cu oxide, often with significant concentrations of Ni and Cu (see Table 2, analyses 8 and 9). Similar x-ray diffraction patterns are shown by todorokite, a Mn-Mg-Ca-Na-K oxide originally reported in a non-marine environment (Yoshimura, 1934), and by derivatives of a synthetic sodium manganese oxide hydrate (Feitknecht and Marti, 1945a,b; Wadsley, 1950a,b; McKenzie, 1970) called "10 A manganite." As a result, one of the phases occurring in manganese nodules was identified as either todorokite (Straczek *et al.*, 1960; Hewett *et al.*, 1963; Manheim, 1965) or "10 A manganite" (Buser, 1959). This dual nomenclature pervaded the literature during the 1960's (Burns and Burns, 1977a), despite the complaint (Arrhenius, 1963; Burns and Fuerstenau, 1966) that the term "10 A manganite" led to confusion with the mineral manganite (γ-$Mn^{III}OOH$). Giovanoli *et al.* (1971) proposed that the "10 A manganite" phase in marine manganese nodules be named buserite in honor of W. Buser (Hey and Embrey, 1974), and this nomenclature has been adopted by some European groups (Jeffries and Stumm, 1976; Giovanoli, 1979). It was further contended (Giovanoli *et al.*, 1971; Giovanoli and Bürki, 1975) from x-ray and electron diffraction measurements of synthetic manganese oxides that natural todorokites are a mixture of buserite and its decomposition products birnessite and manganite (γ-MnOOH), a viewpoint which has not been universally accepted.

Todorokite. Todorokite at the type locality in Japan is formed as an alteration product of the pyroxenoid, inesite (Yoshimura, 1934). The mineral has subsequently been identified in a number of continental occurrences. For example, at Charco Redondo in Cuba, todorokite is a primary mineral formed as a syngenetic deposit on the seafloor during the Eocene.

Table 2. Chemical Analyses of Todorokites and Synthetic Buserites

	1	2	3	4	5	6	7	8	9	10	11	12
MnO_2	65.39	69.57	68.46	71.75	72.98	66.3	64.3	} 71.1	78.03	80.86	74.91	67.98
MnO	12.37	11.65	10.70	8.62	6.81	7.4	7.2					7.98
CaO	3.28	2.30	2.13	1.40	1.24			1.3		1.67	1.21	5.99
SrO		0.05	0.13	0.01	0.11							
BaO	2.05	0.19	0.40	4.32	0.19			0.3		0.28	1.51	0.50
Na_2O	0.21	1.14	1.44	0.17	0.16			2.7		3.09	1.10	0.19
K_2O	0.54	0.35	0.75	0.91	0.92			0.6		0.66	1.55	0.19
MgO	1.01	3.25	3.22	2.13	2.26			3.4		3.27	3.12	1.51
CoO		0.05	0.18					0.2	0.05		0.01	0.21
NiO		0.05						6.4	6.76	0.06	0.32	
CuO		0.01	0.44		0.77		14.7	3.8	2.05	0.01	0.55	
ZnO					4.99	9.8		0.3		0.04	0.07	
Al_2O_3	0.28		0.19		0.02			0.2		0.38	0.35	
Fe_2O_3	0.20		0.07	0.02	0.59			0.1	2.22	0.35	1.64	0.08
SiO_2	0.45	0.27	0.41	0.23	0.55			0.2	1.67		0.96	0.20
H_2O	11.28	10.83	10.99	9.95	8.34	16.3	14.3	(9.4)	(9.07)	(9.51)	(12.46)	14.31
Others	1.98	0.15							0.15		0.24	0.24
TOTAL:	99.24	100.16	99.51	99.89	100.74	99.3	100.5	(100.0)	(100.0)	(100.0)	(100.0)	99.29

1. Todorokite, hydrothermal alteration product of inesite, $(Ca_2Mn_7Si_{10}O_{28}(OH)_2 \cdot 5H_2O)$, Todoroki Mine, Hokkaido, Japan (Yoshimura, 1934).
2. Todorokite, fissure or breccia filling in limestone or volcanic tuff, Charco Redondo Mine, Oriente Province, Cuba (Frondel *et al.*, 1960a).
3. Todorokite, fissure or breccia filling in limestone or volcanic tuff, Charco Redondo Mine, Oriente Province, Cuba (Straczek *et al.*, 1960). Calculated formula on the basis of 12 oxygens: $Ca_{0.27}Na_{0.30}K_{0.01}Mg_{0.57}Mn^{2+}_{0.96}Mn^{4+}_{5.01}O_{11}] \cdot 3.88H_2O$.
4. Todorokite, nodular masses, Huttenberg, Carinthia, Austria (Frondel *et al.*, 1960a).
5. Zincian todorokite, Phillipsburg, Montana (Larsen, 1962). Includes 0.55% PbO.
6. Synthetic zinc buserite (Giovanoli *et al.*, 1975; Giovanoli, 1979).
7. Synthetic copper buserite (Giovanoli *et al.*, 1975; Giovanoli, 1979).
8. Todorokite, lining of healed internal fracture, manganese nodule; north equatorial Pacific; DOMES Site A at 8°N 151°W (Burns and Burns, 1978a). Electron microprobe analysis.
9. Todorokite ("10 A Manganite"), lamination in manganese nodule; north Pacific; Challenger Station 252 at 37°52'N, 160°17'W (Stevenson and Stevenson, 1970). Electron microprobe analysis.
10. Todorokite, almost pure, very well crystallized, with slight amount of birnessite; hydrothermal mounds on south flank of the Galapagos Rift at 00°36'N, 86°04'W (Corliss *et al.*, 1978). Electron microprobe analysis.
11. Todorokite, micronodule in metalliferous sediments, Bauer Deep at 9°S, 102°W (J. Dymond and W. Eklund, unpub. data). Electron microprobe analysis. Includes $0.06Ce_2O_3$.
12. Todorokite, radiating crystallites lining cavities of intensively zeolitized and phosphatized hyaloclastic tuff, south Pacific Basin, station 6333 at 22°41'09"S, 160°50'08"W, sample 73-15 (Andrushchenko *et al.*, 1975).

Note: In analyses 8-11, all manganese is reported as wt. % MnO_2, and water is obtained by difference from 100.0%.

The deposits occur in and around centers of fumerolic hot springs which accompanied submarine volcanic activity (J. Straczek, personal communication). Pyroclastic deposits were formed at the same time and are associated with the todorokite. This todorokite was deposited either in altered basalt pyroclasts, or along pyroclastic zones intercalated in foraminiferal oozes now indurated into hard white limestone. Such a characteristic submarine volcanic paragenesis is also found in todorokite deposits in Java and the Phillipines. Subsequent weathering of todorokite near the water table in Cuba has produced some alteration to manganite, while extensive weathering above the water table has caused todorokite to be oxidized to pyrolusite. In other localities, todorokite is clearly of secondary origin (Frondel *et al.*, 1960a; Larson, 1962).

X-ray diffraction data (Burns and Burns, 1977a) show that considerable variations exist between relative intensities of comparable lines for different todorokite samples. Faulring (1962) attributed these large intensity variations, as well as the diffuseness of certain reflections, to preferred orientation of the fine fibrous crystallites of todorokite from Charco Redondo. Thus, the most intense lines at 9.65 A and 4.82 A observed when x-rays are diffracted parallel to the fiber axis were weak or undetected for x-rays diffracted perpendicular to the crystallite axis. The most intense lines in the latter orientation occur at 2.42 A and 1.42 A and, as noted earlier, are not unique to todorokite. Faulring also noted that manganite was topotactically intergrown with the Cuban todorokite she examined and suggested that the manganite either formed simultaneously with the todorokite or resulted from the alteration of todorokite during subsequent weathering. The coexistence of manganite and todorokite was cited as evidence by Giovanoli *et al.* (1971) for rejecting todorokite as a valid mineral.

Chemical analyses of todorokites in continental deposits (Table 2) show that manganese is present in two oxidation states, Mn^{II} and Mn^{IV}, with Mn^{II}/Mn^{IV} ratios falling in the range 0.15-0.23 (Levinson, 1960; Frondel *et al.*, 1960a; Straczek *et al.*, 1960; Nambu *et al.*, 1964). Significant amounts of Mg are also present, suggesting that relatively small divalent Mg^{2+} and Mn^{2+} ions are essential constituents of todorokite. The analyses also show that the larger cations Ca^{2+}, Na^{+}, and to lesser extents K^{+}, Ba^{2+}, and occasionally Ag^{+} (Radtke *et al.*, 1967) are common constituents of todorokites. Several chemical formulae have been proposed for todorokite; the one adopted here is (Ca,Na,K,Ba,Ag) (Mg,Mn^{2+}, Zn) $Mn_5^{4+}O_{12}\cdot 3H_2O$.

Most todorokites in hand specimen consist of fibrous aggregates of small acicular crystals. Electron micrographs (Straczek *et al.*, 1960; Finkelmann *et al.*, 1972, 1974; Hariya, 1961) reveal that the crystals consist of narrow lathes or blades elongated along one axis (parallel to *b*) and frequently show two perfect cleavages parallel to the (001) and (100) planes. Such morphological features are also shown by hollandite-cryptomelane and romanèchite (psilomelane) samples possessing tunnel structures, discussed later. Electron diffraction patterns of the Cuban todorokite (Straczek *et al.*, 1960; Chukhrov *et al.*, 1978c) show pseudo-hexagonal symmetry with the reciprocal translation along a^* being subdivided into four smaller reciprocal translations (Fig. 11 later). In buserite discussed later the corresponding translation is subdivided into three subcells (Giovanoli, 1979). Recently, Chukhrov *et al.*, 1978c, 1979b) identified several todorokite polymorphs from their electron diffraction patterns. The patterns reveal that, in addition to the standard Cuban todorokite with unit cell parameters a = 9.75, b = 2.84, c = 9.59 A, species with a = 14.6 and 24.38 A but having identical b (2.84 A) and c (9.59 A) parameters, occur in natural environments. Chukhrov *et al.* suggested that as many as five todorokite polymorphs might exist having a parameters which are multiples of 4.88 A.

Buserite. Buserite is synthesized by oxidation of fresh $Mn(OH)_2$ suspensions in cold aqueous NaOH by molecular oxygen (Giovanoli *et al.*, 1970a). The syntheses are more successful if solutions are colder than 10°C; there is a tendency for hausmannite (Mn_3O_4) to form at normal laboratory temperatures. Buserite consists of platelets which give an electron diffraction pattern with pseudohexagonal symmetry and a sub-cell with triple periodicity (Giovanoli, 1979). The most prominent lines in x-ray diffraction patterns are at 10.1-10.2 and 5.0-5.1 A, which are significantly larger than values measured for marine manganese oxides. A variety of buserite derivatives can be synthesized by cation exchange reactions (Giovanoli *et al.*, 1975b), the Mg^{2+}, Ni^{2+}, Cu^{2+}, etc., derivatives having smaller diagnostic *d* spacings than the parent sodium buserite. There is also considerable variation of weaker x-ray diffraction lines for the different buserites (Giovanoli, 1979). The transition-metal derivatives appear to have greater thermal stability than the Na parent.

Sodium buserite readily decomposes when exposed to air. Drying over P_4O_{10} in a vacuum leads to the formation of sodium birnessite, $Na_4Mn_{14}O_{27}\cdot 9H_2O$, which also has a platey morphology and gives an electron diffraction pattern similar to that of buserite (Giovanoli *et al.*, 1970a). Refluxing of sodium birnessite with 0.5M HNO_3 at 40°C yields sodium-free birnessite, $Mn_7O_{13}\cdot 5H_2O$ (Giovanoli *et al.*, 1970b), which on careful reduction with cinnamic alochol causes the platelets to break down to extremely thin needles presumed to be γ-MnOOH (Giovanoli *et al.*, 1971). Such observations on synthetic manganese oxides led Giovanoli *et al.* (1971, 1975, 1979) to propose that natural todorokites actually consist of platey buserite almost wholly decomposed to needles of manganite. According to this view, the small amount of buserite in todorokite accounts for the characteristic x-ray diffraction pattern, while a large amount of fibrous, nearly x-ray amorphous manganite accounts for todorokite's morphology. The proponents claim to be able to recognize, on the basis of electron microscopy observations of morphology, γ-MnOOH which is undetectable (due to small crystallite size) by x-ray diffraction. Apparently, the morphology argument was considered stronger than the diffraction evidence, as the electron diffraction patterns for the hypothetical todorokite decay products do not match those of the synthetic birnessite decay products. The relevance of the buserite break-down products to natural todorokites has been critically evaluated by Stockman and Burns (1980).

Birnessite or "7 A Manganite"

Another manganese oxide phase found in marine deposits is that characterized by x-ray diffraction lines at 7.0-7.3 A and 3.5-3.6 A. Both sets of lines must be present to be diagnostic of this phase, because minerals of the hollandite-cryptomelane, zeolite (e.g. phillipsite), and clay-mica groups also have *d*-spacings around 7 A. The phase with *d*-spacings at 7.0-7.2 and 3.5-3.6 A has been correlated with the non-marine mineral birnessite (Jones and Milne, 1956; Frondel *et al.*, 1960b) and the synthesis products manganous manganite (Feitknecht and Marti, 1945a,b; Buser *et al.*, 1954), "7 A manganite" (Buser, 1959) or sodium manganese (II,III) manganite (IV) hydrate (Giovanoli *et al.*, 1970a).

Birnessite. Jones and Milne (1956) originally reported birnessite in a fluvo-glacial deposit in Scotland. Subsequently, birnessite was identified in other terrestrial deposits (Frondel *et al*., 1960b; Hariya, 1961; Levinson, 1962; Sorem and Gunn, 1967; Brown *et al*., 1971) including desert varnish (Potter and Rossman, 1977, 1979a). Reported occurrences in marine deposits include these by Manheim (1965), Woo (1973), Glover (1977), Chukhrov *et al*. (1978b, 1979a) and Londsdale *et al*. (1979). Representative chemical analyses of birnessites are summarized in Table 3. The platey habit of birnessite observed by scanning electron microscopy of samples from continental and marine deposits is distinctive.

Synthetic Birnessite. The product obtained by synthesis, which was originally called "manganous manganite" (Feitknecht and Marti, 1945a,b; Buser *et al*., 1954), "7 A manganite" (Buser, 1959), and subsequently sodium manganese (II,III) manganite (IV) hydrate $Na_4Mn_{14}O_{27} \cdot 9H_2O$ (Giovanoli *et al*., 1970a), is formed by controlled dehydration of buserite. This sodium-rich orthorhombic phase readily undergoes a topotactic transition to the hexagonal phase manganese (III) manganate (IV) hydrate, $Mn_7O_{13} \cdot 5H_2O$ (Giovanoli *et al*., 1970b).

Vernadite or Delta-MnO_2

The phase in marine manganese deposits giving *only* two diffuse x-ray diffraction lines at 2.40-2.45 A and 1.40-1.42 A has been commonly designated as delta-MnO_2 (δ-MnO_2). Earlier literature reviewed by Burns and Burns (1977a) regarded δ-MnO_2 to be a structurally-disordered birnessite. However, since it has distinctive physical (e.g. very high specific surface areas) and chemical (e.g. significant Co, Ce and Pb contents) properties, δ-MnO_2 is conveniently regarded as a separate phase in the marine environment. Recently, however, Chukhrov *et al*. (1978a,b) proposed that δ-MnO_2 be called vernadite. Vernadite was originally named for the x-ray amorphous to slightly crystalline $MnO_2 \cdot xH_2O$ phase formed by supergene alteration of rhodonite (Betekhtin, 1937, 1940). Chukhrov *et al*. (1978a,b) claimed that the name vernadite takes precedence over delta-MnO_2 for poorly crystalline supergene hydrated manganese (IV) oxides giving reflections at 2.4 and 1.4 A in x-ray and electron diffraction patterns. The chemical analyses summarized in Table 3 show that the composition of vernadite is variable, which is reflected in its proposed formula $MnO_2 \cdot m(R_2O,RO,R_2O_3) \cdot nH_2O$, where R = Na, Ca, Co, Mn, Fe. It is doubtful whether Fe is present in the vernadite

Table 3. Chemical Analyses of Birnessites and Vernadites

	1	2	3	4	5	6	7	8	9	10	11	12
MnO_2	54.24	66.66	80.03	75.80	77.62	78.40	48.98	36.34	35.71	46.47	39.3	56.45
MnO	4.66	16.07								1.09		4.81
CaO	1.65	1.05		0.39	2.38	2.29	1.03	3.08	3.44	2.15	3.4	5.17
Na_2O	2.17	0.16	8.67	1.90	2.82	2.04				2.29	1.1	0.12
K_2O		0.09		1.80	0.76	1.68	1.44	0.36		0.60	0.24	0.23
MgO		0.23		6.20	3.06	4.60	3.48	1.16		2.62	1.9	0.28
CoO				0.14	0.05	0.13	0.25	1.53		3.41		
NiO				0.80		0.22	3.81	0.76		0.95		
CuO				0.33		0.25	0.63	0.25				
Al_2O_3	3.32	0.83					5.67	2.08	2.51	1.00	1.7	1.00
Fe_2O_3	2.88	0.86		0.62	0.12	0.24	2.86	17.16	24.74	10.47	25.6	7.00
SiO_2	18.92	2.62		0.90	0.06	1.56	1.07	5.35	11.06	0.80	4.35	1.30
TiO_2	0.28						0.03	2.00		1.50	1.5	
H_2O	10.87	10.83	10.02	(10.80)	(13.10)	(8.6)	(30.4)	(29.8)	(22.5)	25.44	(20.9)	16.53
Others		1.47		0.30			0.37	0.15		1.53		7.26
TOTAL:	98.99	100.87	98.72	(100.0)	(100.0)	(100.0)	(100.0)	(100.0)	(100.0)	100.32	(100.0)	100.15

1. Birnessite, manganese-rich hardpan in fluvio-glacial gravel deposit, Birness, Scotland (Jones and Milne, 1956). Modal analysis of the deposits: quartz 20.0%; clay 7.3%; limonite 5.2%; rutile 0.1%; birnessite 67.4%. Formulated as $(Na_{0.7}Ca_{0.3})Mn_7O_{14}\cdot 2\cdot 8H_2O$.
2. Birnessite, soft fine-grained black coating on oxidized lenses of manganiferous silicate and carbonate minerals, Cummington, Massachusetts (Frondel *et al.*, 1960b).
3. Synthetic birnessite, sodium manganese (II,III) manganate (IV) hydrate, $Na_4Mn_{14}O_{27}\cdot 9H_2O$ (Giovanoli *et al.*, 1970a).
4. Birnessite, in micronodules occurring in foraminiferal sediments, Gulf of Mexico just south of the northwest tip of Cuba at 21°32'N, 85°4.5'W (Glover, 1977). Electron microprobe analysis. Formulated as $Na_{0.46}Ca_{0.12}K_{0.25}Mg_{1.04}Fe_{0.05}Co_{0.02}Ni_{0.07}Cu_{0.03}Mn_{5.96}O_{13.6}(H_2O)_{3.8}$.
5. Birnessite, Pacific Ocean (Chukhrov *et al.*, 1978b). Electron microprobe analysis. Formulated as $(Na_{0.650}Ca_{0.303}K_{0.116})(Mg_{0.542}Co_{0.005}Zn_{0.001}Fe_{0.001}Mn^{3+}_{0.382})Mn^{4+}_{6.000}O_{14.305}(H_2O)_{5.197}$.
6. Birnessite, rods in nodules dredged from caldera of East Seamount, East Pacific Rise at 8°48.2'N, 103°53.8'W (Lonsdale *et al.*, 1979). Electron microprobe analysis.
7. Birnessite ("7 A manganite"), manganese nodule from seamount, south Pacific Ocean, at 16°29'S, 145°33'W, specimen DWHD-16 (Burns and Fuerstenau, 1966). Electron microprobe analysis.
8. Vernadite (δ-MnO_2), manganese nodule from seamount, south Pacific Ocean, at 16°29'S, 145°33'W, specimen DWHD-16 (Burns and Fuerstenau, 1966). Electron microprobe analysis.
9. Vernadite (δ-MnO_2), in hydrothermal mounds on south flank of the Galapagos Rift at 00°36'N, 86°04'W (Corliss *et al.*, 1978). Electron microprobe analysis.
10. Vernadite, hydrothermal deposit, Kurchativ fracture, Pacific Ocean (Chukhrov *et al.*, 1978a).
11. Vernadite, manganese nodule, Pacific Ocean (Chukhrov *et al.*, 1978a). Electron microprobe analyses, average of 90 analyses.
12. Vernadite, Mt. Lepkhe-Nel'm, Lovozero tundra, U.S.S.R. (Chukhrov *et al.*, 1978a).

Note: In analyses 3-9 and 11, all manganese is reported as wt. % MnO_2, and water is obtained by difference from 100.0%.

structure; instead, intimately associated or epitaxial intergrowths of feroxyhyte, δ'-FeOOH (Chukhrov *et al.*, 1976a,b; Burns and Burns, 1975) probably accounts for the high iron contents of vernadite. In electron microscope preparations, leaflets of vernadite have smaller dimensions (tens of Angstroms) than flakes of birnessite, and vernadite leaflets are often curved and folded to resemble fibers.

Psilomelane and Romanèchite

Although psilomelane was the name given to a specific mineral, the term *psilomelane* has been used to describe any black, hard, botryoidal, unidentifiable manganese oxide mineral, just as *wad* signifies soft unidentifiable samples (Wadsley, 1950b; Fleischer, 1960). Ambiguity over the use of psilomelane led to the adoption of the name *romanèchite* for the specific hydrated barium manganese oxide mineral formerly called psilomelane. Nevertheless, despite the fact that psilomelane has been discredited as a mineral name, its use in crystal chemistry lingers on due to the importance of the psilomelane structure type (Wadsley, 1953). The two mineral names psilomelane and romanèchite are used synonymously throughout this chapter. Wadsley (1953) originally formulated psilomelane (romanèchite) as $(Ba,H_2O)_2Mn_5O_{10}$ and described the crystal structure as having monoclinic symmetry. Mukherjee (1959) suggested an orthorhombic structure and a chemical formula indicating the presence of OH^- ions. The infrared spectral measurements of Potter and Rossman (1979b), however, established the presence of H_2O and not OH^- in romanèchites. The complexity of psilomelane (romanèchite) structure types, suggested originally by Mouat (1962), and confirmed recently by high resolution transmission electron microscopy (Turner and Buseck, 1979a), indicates that the psilomelane-romanèchite nomenclature may have to be revised again.

Manganite

Manganite, γ-MnOOH, is one of three oxide hydroxide polymorphs of trivalent manganese and should not be confused with synthetic "10 A manganite" (buserite) and "7 A manganite" (birnessite). Manganite commonly forms by replacement of Mn^{2+}-bearing carbonate and silicate minerals, and frequently transforms topotactically to pyrolusite (β-MnO_2) (Champness, 1971). As noted earlier, it may be synthesized by reducing acid-treated synthetic birnessite with cinnamic alcohol (Giovanoli *et al.*, 1971). Manganite also eventually forms by the aging of aerated Mn_3O_4 and γ-MnOOH suspensions precipitated from aqueous Mn^{2+} in alkaline solutions at 25°C. Such synthetic manganites consist of very small elongated needles which give poor x-ray diffraction patterns.

CRYSTAL STRUCTURES AND STRUCTURAL CORRELATIONS

Although the crystal structures of most of the manganese oxides listed in Table 1 are known, there are some important exceptions particularly for minerals widely distributed in the marine environment. Thus, crystal structures are unknown or have not been solved completely for todorokite, birnessite and vernadite because these minerals are too fine grained for single crystal x-ray structure refinements. Attempts have been made to model the structures of these cryptocrystalline minerals, and the suggested structures hinge upon structural correlations with simpler, non-marine manganese oxides.

The fundamental structural unit in manganese (IV) oxides is the $[Mn^{IV}O_6]$ octahedron, in which Mn^{4+} ions with their $[A]3d^3$ electronic configuration acquire exceptionally high crystal field stabilization energy (Burns, 1970). The basic $[MnO_6]$ octahedra in manganese oxide mineralogy are linked by corner-sharing *and* edge-sharing to give a variety of chain, tunnel and layer structures. Analogies exist in silicate mineralogy where the basic $[SiO_4]$ tetrahedra may be connected by corner-sharing only of the tetrahedral units to form chain, ring, framework, and layer structures. Edge-sharing of $[MnO_6]$ octahedra and the close proximity of Mn^{4+} ions to one another may induce structural instabilities and cryptocrystallinity of manganese oxide minerals. Added complexities in the manganese oxides arise from ordered and random vacancies in, and domain structures and non-periodic intergrowths of, the linked $[MnO_6]$ octahedral units. These and other features are highlighted in the following descriptions of the crystal structures.

Chain Structures: The Ino-Manganate (IV) Oxides

Pyrolusite. Pyrolusite (β-MnO_2) is pivotal in descriptions of the crystal structures of manganese oxides. It has the rutile (TiO_2) structure in which every metal atom is surrounded by six oxygen atoms located at the vertices of a distorted octahedron with Mn at the center. The $[MnO_6]$ octahedra share edges to form single chains running parallel to the *c* axis. These chains influence the morphology of pyrolusite which frequently has an acicular habit. All $[MnO_6]$ octahedra are equivalent, and the average Mn-O distance is 1.88 A. The unit cell *c* dimension, 2.87 A, represents the Mn^{4+}-Mn^{4+} and O-O internuclear distances, and this dimension (or multiples of it) is a common cell parameter in Mn^{IV} oxide mineralogy (Table 1). The single chains of $[MnO_6]$ octahedra in pyrolusite are cross-linked with neighboring chains by corner-sharing of oxygen atoms of adjacent octahedra. The hexagonally close-packed oxygen layers along [100] are puckered to give tetragonal symmetry to pyrolusite. The crystal structure of pyrolusite is illustrated in Figure 1. Note that the single chains of linked $[MnO_6]$ octahedra bear resemblances to the chains of linked $[SiO_4]$ in pyroxenes in silicate mineralogy.

Manganite. Manganite (γ-MnOOH) has a structure modelled on that of pyrolusite (Buerger, 1936; Dachs, 1963). Buerger showed that the manganite structure is monoclinic and chose the non-standard space group $B2_1/d$ to preserve pseudo-orthorhombic axes. The lower symmetry and doubled *a* and *c* parameters of manganite compared to pyrolusite arise from the presence of OH^- groups. In manganite, chains of edge-shared $[Mn^{III}(O,OH)_6]$ octahedra again extend along the *c* axis, but the four oxygens in the planes of edge-shared octahedra are considerably closer (Mn-O = 1.87-1.98 A) to the central Mn^{3+} ion than are the two apical oxygens (Mn-O = 2.20-2.33 A) as a consequence of the Jahn-Teller effect in Mn^{III} (Burns, 1970). Thus, the $[Mn(O,OH)_6]$ octahedra are elongated along the *b* axis relative to *a* resulting in manganite having pseudo-orthorhombic symmetry. Hydrogen bonding occurs

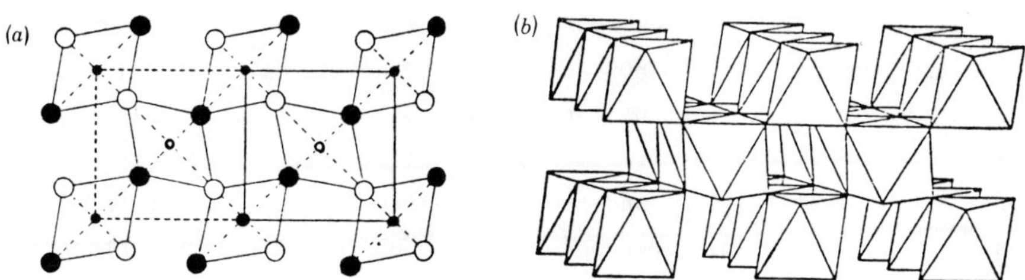

FIGURE 1. The pyrolusite structure. (*a*) Projection on to (001); closed circles, atoms at zero level; open circles, atoms at level 1/2*c* (after Byström 1949). (*b*) The single chains of edge-shared $[MnO_6]$ octahedra parallel to *c* (after Clark 1972). ○ ●, Mn; ○ ●, O.

between the OH groups in edge-shared octahedra in one chain and the apical (corner-shared) oxygens belonging to adjacent chains. The pronounced (010) and (110) cleavages resulting from hydrogen bonding cause manganite to have prismatic or acicular habits.

Ramsdellite. Ramsdellite (MnO_2) contains double chains of linked $[MnO_6]$ octahedra (Byström, 1949). The octahedra are again linked together by sharing opposite edges, producing continuous pyrolusite-like chains along the *c* axis. Two such chains are cross-linked by edge-sharing, one chain being translated by 1/2 *c* relative to the other so that an octahedron from one chain shares edges with each of two octahedra from the other chain. The double chains of linked octahedra are further cross-linked to adjacent double chains through corner-sharing of oxygen atoms to give orthorhombic symmetry and acicular habit to ramsdellite. These features are illustrated in Figure 2. All octahedra have identical configurations with an average Mn-O distance of 1.89 A.

Nsutite. An important precedent is set in Mn (IV) oxide mineralogy by the nsutite or γ-MnO_2 group, which consists of irregular structural intergrowths between pyrolusite and ramsdellite units (de Wolff, 1959; Giovanoli *et al.*, 1967). The alternating *c* axis chain segments of the basic single and double octahedral chains shown schematically in Figure 3 are random, so that no regular periodicity or superstructure is apparent. The lattice disorder causes nsutites and synthetic γ-MnO_2 phases to have extensive defects, vacancies, and non-stoichiometry. These factors, together with the small crystallite sizes of natural and synthetic phases therefore give rise to a variety of x-ray powder diffraction lines, as well as the frequently observed asymmetric and selective line broadening for nsutites. Electrolytically-deposited ε-MnO_2, which consists of a hexagonal close-packed lattice of O^{2-} ions analogous to those of pyrolusite and nsutite, differs from β-MnO_2 and γ-MnO_2 by having Mn^{4+} ions randomly distributed over the octahedral interstices (de Wolff *et al.*, 1978). The existence of domains of pyrolusite and ramsdellite structural units in natural and synthetic γ-MnO_2 and ε-MnO_2 is extremely significant in marine manganese oxide mineralogy because analogous non-periodic domain structures may occur in the todorokite and birnessite phases, as well as hollandite-romanèchite assemblages, discussed later.

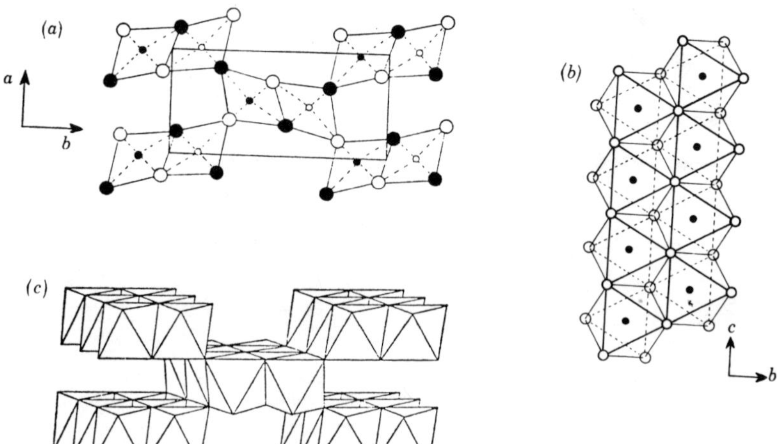

FIGURE 2. The ramsdellite structure. (*a*) Projection on to (001); open circles, atoms at level 1/4*c*; shaded circles, atoms at level 3/4*c* (after Byström 1949). (*b*) A double-chain of $[MnO_6]$ octahedra running along *c* (after Byström 1949). (*c*) The double-chains of edge-shared $[MnO_6]$ octahedra parallel to *c* (after Clark 1972). ○ ●, Mn; ○ ●, O.

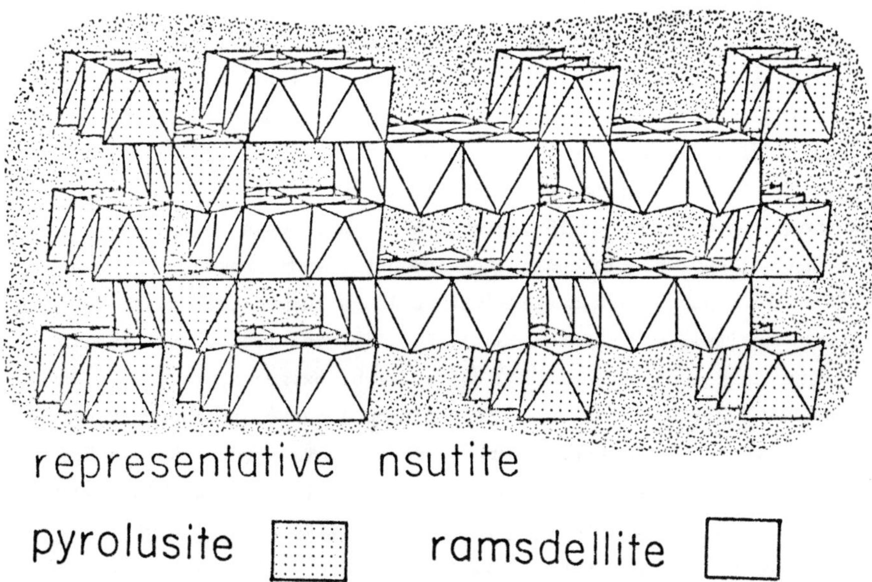

Figure 3. Idealized structure of nsutite showing irregular alternations of ramsdellite domains linked by pyrolusite domains (from Potter and Rossman, 1979b).

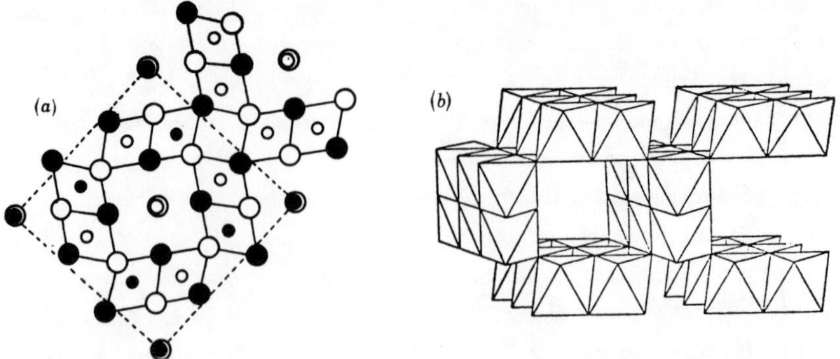

FIGURE 4. The hollandite structure-type. (*a*) The structure of hollandite projected onto (001); open circles, atoms at zero level; shaded circles, atoms at level 1/2*c* (after Byström & Byström 1950). (*b*) The framework structure of hollandite, showing tunnels between double-chains of edge-shared $[MnO_6]$ octahedra parallel to *c* (after Clark 1972). ○ ●, Mn; ◯ ⬤, O; ◎ ⊙, Ba, K, Pb, Na, or H_2O.

Ring, Framework or Tunnel Structures: The Cyclo- or Tekto-Manganate (IV) Oxides

Hollandite, etc. Minerals of the cryptomelane-hollandite or α-MnO_2 group have structures based on that of ramsdellite (Byström and Byström, 1950, 1951). The hollandite structure, which is depicted in Figure 4, again contains infinite double chains of edge-shared $[MnO_6]$ octahedra. In hollandite minerals with tetragonal symmetry, these chains extend along the *c* axis; whereas, in varieties with monoclinic symmetry the *b* axis becomes the chain direction. The octahedra of the double chains share corners with adjacent double chains to give a three-dimensional framework (Figure 4). This produces a large cavity which accommodates H_2O as well as Ba^{2+} (hollandite), K^+ (cryptomelane), Pb^{2+} (coronadite) or Na^+ (manjiroite). Each large cation is surrounded by eight oxygen atoms situated at the corners of a slightly distorted cube (in hollandite, the Ba-O distances are 2.74 A to these eight oxygens) and by four other oxygens (Ba-O = 3.31 A) at the corners of a square at the same level along the *c* axis of tetragonal varieties as the large cation. The *c* dimension, 2.86 A, again represents the Mn-Mn internuclear distance. Disordering of Ba^{2+}, K^+, H_2O, etc. occurs in the cavities which are probably less than half-filled; otherwise, unfavorable cation repulsions would occur when Ba^{2+}-Ba^{2+}, K^+-K^+, etc., pairs are as close as 2.86 A (Byström and Byström, 1951). However, significant amounts of H_2O, K^+, etc., are necessary to prevent collapse of the structure of synthetic α-MnO_2 (Butler and Thirsk, 1952); otherwise, submicroheterogeneities are formed in which regions of pyrolusite and ramsdellite are interdispersed and coexist with regions of α-MnO_2 in the same crystals. Natural cryptomelanes and hollandites, however, appear to be stable to quite high temperatures because they retain their Ba^{2+} and K^+ ions.

A significant property of the sieve-like hollandite structure is that it displays pronounced cation exchange properties, which may be an important feature noted later when considering metal uptake by minerals in marine sediments. Furthermore, in order to maintain a charge balance in the structure accommodating the large exchangeable Ba^{2+}, K^+, Pb^{2+} and Na^+ cations, the linked $[MnO_6]$ octahedra in the double chains must contain some OH groups, cation vacancies, or a proportion of the manganese in oxidation states (e.g. Mn^{II}, Mn^{III}) lower than Mn^{IV}. The latter feature is suggested by the average metal-oxygen distance of the $[MnO_6]$ octahedra being 1.98 A in the hollandite structure, which is significantly larger than the mean Mn-O distances in pyrolusite (1.88 A) and ramsdellite (1.89 A). The nonstoichiometry and local charge balancing by large cations in the tunnels and voids of vacancies or Mn^{2+}, Mn^{3+}, etc., in the $[MnO_6]$ octahedra again are significant factors which may contribute to the fractionation of minor elements in hollandite-like phases in marine sediments (Burns and Burns, 1977b).

Romanèchite. Romanèchite (psilomelane) has a structure related to that of hollandite (Wadsley, 1953). It consists of *treble* chains of $[MnO_6]$ octahedra linked by double (ramsdellite-like) chains to form a series of tunnels or tubes running along the *b* axis, as illustrated in Figure 5. The *b* dimension (2.88 A) of the psilomelane structure corresponds, therefore, to the *c* dimension of pyrolusite, ramsdellite and tetragonal hollandite-cryptomelane. The large tunnels are again occupied by Ba^{2+}, K^+, H_2O, etc., so that the psilomelane structure resembles the hollandite structure to which it breaks down at high temperatures. Psilomelane also has cation exchange properties and requires some of the manganese to be in oxidation states lower than Mn^{IV} in order to balance the charge of the exchangeable cations (Ba^{2+}, K^+). The psilomelane structure differs from those of pyrolusite, ramsdellite, and hollandite by having three distinct octahedral sites. Two of them (the *M1* and *M3* sites shown in Figure 5) each have average metal-oxygen distances of 1.91 A, which is significantly smaller than that (1.99 A) of the third site (designated *M2*), indicating that the *M2* site is the one enriched in the lower valance cations. These *M2* sites of psilomelane figure prominently in recent hypothesis on the mechanism of uptake of Ni^{2+}, Cu^{2+}, etc., into the todorokite phase in deep-sea manganese nodules (Burns and Burns, 1977b).

Note that the structures of psilomelane and hollandite show that although the minerals have tunnels of differing widths in one dimension, the tunnel widths are identical in the other dimension. This similarity of

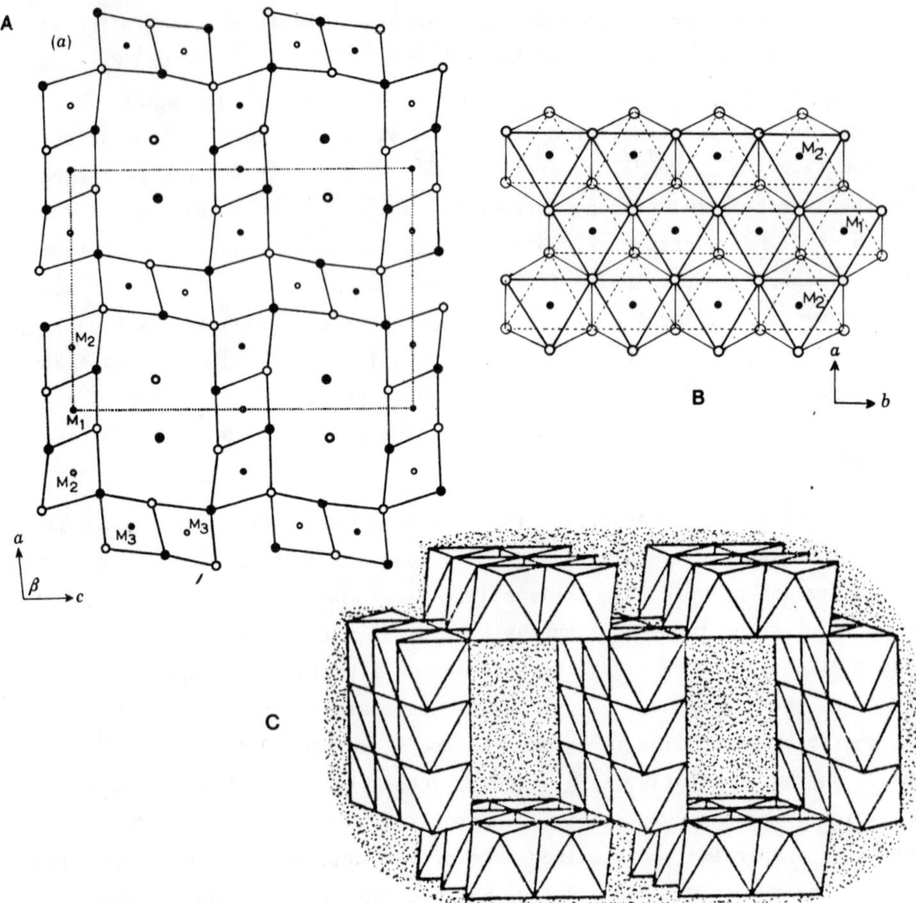

FIGURE 5. The psilomelane structure (after Wadsley 1953). (*a*) Projection on to (010), showing four linked tunnels. The unit cell is shown by broken lines. Open circles, atoms at zero level; shaded circles, atoms at level 1/2*b*. ○, ●, Mn; ○ ●, O; ◎ ⊙, Ba, K, or H_2O. (*b*) A treble-chain of [MnO_6] octahedra parallel to the *b* axis. (c) The 3 x 2 tunnels between chains of edge-shared [MnO_6] octahedra (from Potter and Rossman, 1979b).

tunnels in one dimension theoretically permits intergrowth of the two phases (Wadsley, 1964), by analogy with the pyrolusite and ramsdellite domains found in γ-MnO_2 (nsutite) and ε-MnO_2 (Figure 3). Recently, high resolution transmission electron microscopy (HRTEM) of fibrous manganese oxide minerals has revealed complex intergrowths in the tunnel structures of psilomelane and hollandite (Turner and Buseck, 1979). These intergrowths shown in Figure 6 do not appear to order periodically, and isolated structures with widths greater than the treble chains of psilomelane have been observed. These and more recent observations (Turner and Buseck, 1981) are particularly relevant to postulated structures of todorokite (Burns and Burns, 1977b; Stockman and Burns, 1980).

Layer Structures: The Phyllo-Manganate (IV) Oxides

Chalcophanite. Manganese (IV) oxides with layered structures play a prominent role in deductions of the structures of certain marine manganese oxides. One of the structural models for these phyllomanganates is chalcophanite ($Zn_2Mn_6O_{12}\cdot 6H_2O$), the structure of which is illustrated in Figure 7. The chalcophanite structure (Wadsley, 1955) consists of layers of edge-shared $[MnO_6]$ octahedra and single sheets of water molecules between which the Zn^{2+} ions are located. The stacking sequence along the *c* axis is thus: -O-Mn-O-Zn-H_2O-Zn-O-Mn-O, and the perpendicular distance between consecutive $[MnO_6]$ octahedral layers is about 7.16 A. The water molecules are grouped in open double hexagonal rings, while vacancies exist in the layers of linked $[MnO_6]$ octahedra, so that one out of every seven octahedral sites is unoccupied by manganese. Each $[MnO_6]$ octahedron shares edges with five neighboring octahedra and is adjacent to a vacancy. The ordered vacancies in the chalcophanite structure are reflected in the hexagonal net of less intense reflections observed in electron diffraction patterns of platelets corresponding to the (001) plane (Chukhrov *et al.*, 1978b, 1979a).

Of particular significance is the location of the Zn^{2+} ions which are situated above and below the vacancies in the manganese layers and are coordinated to three oxygens of each vacant octahedron in the $[MnO_6]$ octahedral layer. Each Zn^{2+} ion completes its coordination with three water molecules in the H_2O sheet so as to form an irregular six-coordination polyhedron. The chemical compositions of natural chalcophanites differ

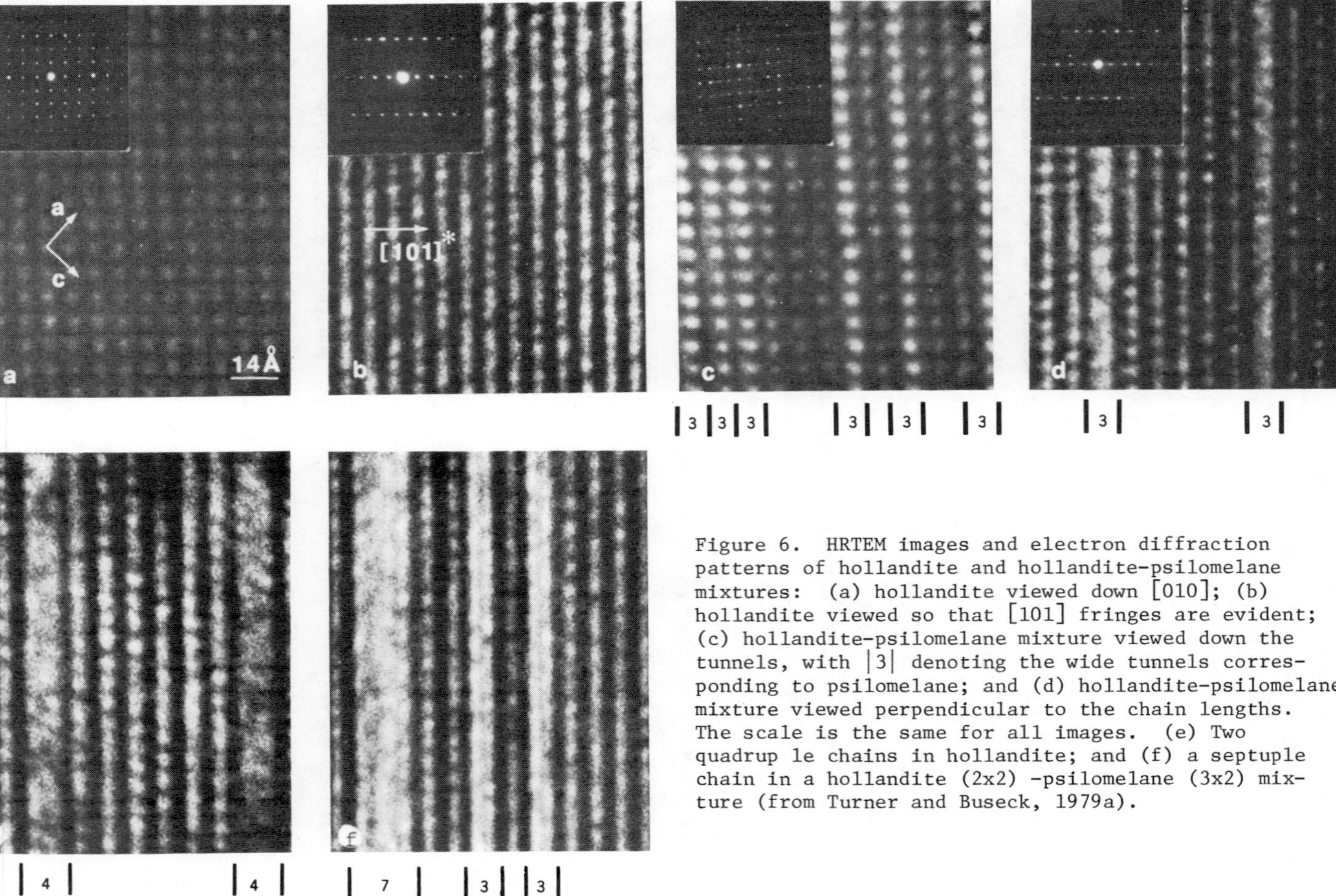

Figure 6. HRTEM images and electron diffraction patterns of hollandite and hollandite-psilomelane mixtures: (a) hollandite viewed down [010]; (b) hollandite viewed so that [101] fringes are evident; (c) hollandite-psilomelane mixture viewed down the tunnels, with |3| denoting the wide tunnels corresponding to psilomelane; and (d) hollandite-psilomelane mixture viewed perpendicular to the chain lengths. The scale is the same for all images. (e) Two quadrup le chains in hollandite; and (f) a septuple chain in a hollandite (2x2) -psilomelane (3x2) mixture (from Turner and Buseck, 1979a).

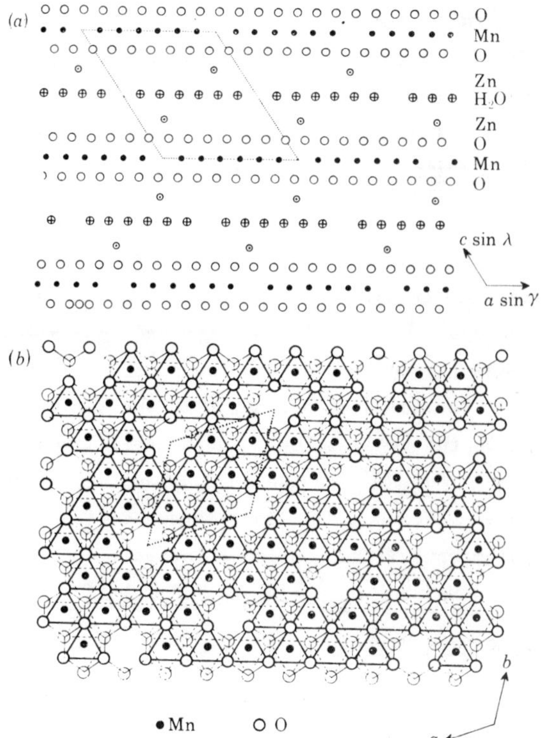

FIGURE 7. The chalcophanite structure (after Wadsley 1955). (*a*) Projection along the *b* axis. Vacancies in the Mn layers define the rhombus unit cell. Note that one out of every seven Mn positions is a vacancy. (*b*) The edge-shared [MnO_6] layer viewed normal to the basal plane. The vacant octahedral sites at the origin are at the corners of a rhombus outlining the plane of the Mn atoms. Note that each Mn atom is adjacent to a vacancy.

significantly from the ideal formula, $Zn_2^{2+}Mn_6^{4+}O_{12}\cdot6H_2O$. Not only is the water content variable, but there is a deficiency of Mn^{4+} ions and the number of cations (Zn, Mn, etc.) usually exceeds eight per formula unit. These trends indicate that some Mn^{2+} and Mn^{3+} ions replace Mn^{4+} in the linked octahedra, accounting for the larger average Mn-O distance of 1.95 A (compared with 1.88 A in pyrolusite). Additional cations also occur in interstitial positions between the H_2O sheets and oxygens of the [MnO_6] layers. Such observations have been used to model the structures of birnessite and buserite.

Synthetic $Cd_2Mn_3O_8$ *and* Mn_5O_8. Some of the important features displayed by the chalcophanite structure occur also in the synthetic phases

$Cd_2Mn_3O_8$ and $Mn_2^{2+}Mn_3^{4+}O_8$ which possess layer structures (Oswald and Wampetich, 1967). There are vacancies in the layers of edge-shared $[MnO_6]$ octahedra containing Mn^{4+} ions (average Mn^{4+}-O = 1.87 A). The Mn^{2+} and Cd^{2+} lie above and below the empty Mn^{IV} sites and are six-coordinated by oxygens forming a distorted trigonal prism (average Mn^{2+}-O = 2.21 A).

Lithiophorite. Lithiophorite also has a layer structure (Wadsley, 1952) in which layers of edge-shared $[MnO_6]$ octahedra alternate with layers of $[(Al,Li)(OH)_6]$ octahedra. The stacked sequence along the *c* axis is: -O-Mn-O-OH-(Al,Li)-OH-O-Mn-O, and two consecutive $[MnO_6]$ layers are about 9.5 A apart. Vacancies do not occur in the sheets of linked $[MnO_6]$ octahedra, nor in the $[(Al,Li)(OH)_6]$ octahedral layer. However, some substitution of Mn^{2+} for Mn^{4+} in the $[MnO_6]$ layers is required to maintain charge balance. Recent x-ray and electron diffraction studies of a synthetic stoichiometric lithiophorite (Giovanoli *et al.*, 1973) have led to the formulation $[(Mn_5^{4+}Mn^{2+}O_{12})^{2-}(Al_4Li_2(OH)_{12})^{+2}]$.

Deductions on the Birnessite Structure

Although the crystal structure of birnessite has not been determined, information has been derived from electron diffraction measurements of synthetic compounds (Giovanoli *et al.*, 1969, 1970a,b; Giovanoli and Stahli, 1970) and natural phases (Chukhrov *et al.*, 1978b, 1979a). The birnessite structure shown schematically in Figure 8 contains layers of edge-shared $[MnO_6]$ octahedra separated by about 7.2 A along the *c* axis, which enclose sheets of H_2O molecules by analogy with chalcophanite. However, in sodium birnessite, $Na_4Mn_{14}O_{27}\cdot 9H_2O$, one out of six octahedral sites in the layer of linked $[MnO_6]$ is unoccupied (compared with one out of seven for chalcophanite). The vacant Mn^{4+} sites are distributed according to different patterns for adjacent $[MnO_6]$ octahedral layers, resulting in a two-layer orthorhombic cell with periodicity *c* = 14.26 A. The Mn^{2+} and Mn^{3+} ions are arranged above and below the vacancies in the octahedral layers and are bonded with oxygens in both the $[MnO_6]$ layer and the sheet of H_2O molecules. The positions of the sodium ions is uncertain. The electron diffraction pattern of $Na_4Mn_{14}O_{27}\cdot 9H_2O$ reflects the ordered vacancies in the $[MnO_6]$ layers, for separate networks of weak and strong reflections can be distinguished.

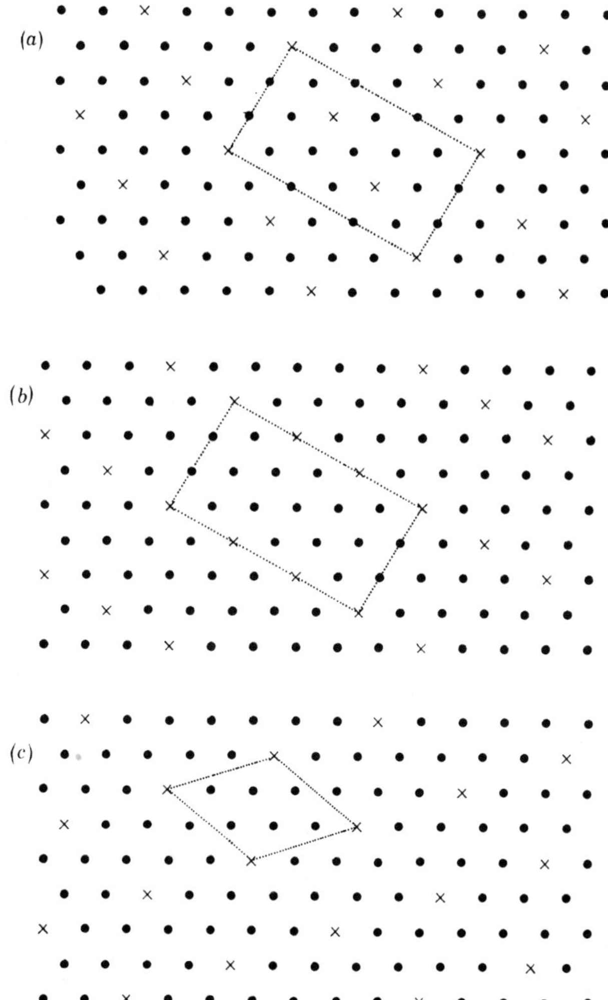

Figure 8. Projections of the structure proposed for synthetic birnessite, $Na_4Mn_{14}O_{27} \cdot 9H_2O$ (after Giovanoli *et al.*, 1970a,b). The projections show Mn atoms only in the basal planes at zero level (a) and at about 7.2 A along *c* (b). The different locations of vacancies in levels A and B necessitate doubling of the *c* parameter. A comparable projection for chalcophanite is shown for reference (c).

The structure of sodium-free birnessite, $Mn_7O_{13}\cdot 5H_2O$, is again related to those of chalcophanite and sodium birnessite (Giovanoli *et al.*, 1970b). However, vacancies are disordered in the $[MnO_6]$ layers, leading to a hexagonal cell with periodicity c = 7.27 A. The irregular positions of the vacancies in the octahedral layers and of the associated interlayer cations above and below these vacancies are responsible for the absence of a network of weak reflections in electron diffraction patterns of $Mn_7O_{13}\cdot 5H_2O$ and most natural birnessites.

Deductions on the Vernadite Structure

The structure of poorly crystalline δ-MnO_2 or vernadite has commonly been regarded to be a disordered birnessite in which there is no periodic stacking along the c axis of the edge-shared $[MnO_6]$ octahedra. However, the smaller particle size of δ-MnO_2 (indicated by its high specific surface area, typically 300 m^2gm^{-1} compared with 30-40 m^2gm^{-1} for synthetic birnessite (Buser and Graf, 1955), causes it to have different chemical properties. Chukhrov *et al.* (1978a,b) used electron microscope techniques to distinguish vernadite from birnessite. By inclining leaflets of vernadite with respect to the electron beam, reflections with d = 2.18-2.20 A, corresponding to the (101) plane, were observed, which led to an approximate c parameter of 4.7 A, differing from those of birnessite. This dimension corresponds to the width of two layers of close-packed oxygens.

The vernadite structure is thus represented as a two-layer hexagonal packing of oxygen atoms and water molecules in which the octahedra are statistically, but a little less than half, filled by tetravalent manganese. The extent of filling is apparently determined by the contents of water and other cations such as Na^+, K^+, Ca^{2+}, Mg^{2+}, Co^{3+}, etc.

Deductions on the Buserite Structure

Structural models for buserite ("10 A manganite") have been based on its crystal morphology and the proposed mechanism for its synthesis. Since buserite is synthesized by oxidation of aqueous $Mn(OH)_2$ suspensions in NaOH, it was suggested (Feitknecht and Marti, 1945a; Buser *et al.*, 1954) that buserite and $Mn(OH)_2$ have related structures. Pyrochroite ($Mn(OH)_2$) has the brucite structure and consists of layers of edge-shared $[Mn(OH)_6]$ octahedra in which each octahedron shares edges with six neighboring octahedra to form a two-dimensional layer. By analogy with lithiophorite the buserite structure was proposed to consist of oxidized layers of $[Mn^{IV}O_6]$ octahedra interspersed with layers of relatively unoxidized $[Mn^{II}(OH,H_2O)_2]$, giving the prominent basal reflections around 10 A and 5 A in x-ray powder diffraction patterns. Electron microscopy photographs of synthetic buserite also indicate that it has a layer structure like birnessite to which buserite readily dehydrates (Giovanoli *et al.*, 1970a,b, 1971, 1975). The similarity of electron diffraction patterns of (001) planes of birnessite and buserite indicate that vacancy distributions are similar in the layers of edge-shared $[MnO_6]$ octahedra. The Na^+ and Mn^{2+} ions in buserite are again postulated to be located adjacent to the vacancies, and to be exchangeable by smaller cations such as Cu^{2+}, Ni^{2+}, Co^{2+}, Zn^{2+}, Mg^{2+}, etc., leading to contraction of the structure (Giovanoli, 1979).

Deductions on the Todorokite Structure

A structural model for todorokite has also been inferred from its crystal morphology and cleavage properties (Burns and Burns, 1977b). It was noted that minerals of the hollandite-cryptomelane and psilomelane (romanèchite) groups also have fibrous or acicular habits and two perfect cleavages parallel to the fiber axis, as do a variety of Ti (IV) oxides. Since these phases possess tunnel structures consisting of double and treble chains of edge-shared $[MnO_6]$ octahedra (see Figs. 4 and 5,) todorokite was proposed to have a tunnel structure based on chains of multiple edge-shared $[MnO_6]$ octahedra extending along its *b* axis. The correlation is further borne out when comparisons are made between cell parameters of todorokite, psilomelane, and hollandite-group minerals summarized in Table 1. Furthermore, todorokite like psilomelane contains essential Mn^{2+} ions, suggesting that these and other divalent cations of comparable ionic radii (e.g. Mg^{2+},Co^{2+},Ni^{2+},Cu^{2+},Zn^{2+}) might be located in specific positions in the chains of edge-shared $[MnO_6]$ octahedra (compare the *M2* site of psilomelane, Fig. 5). The larger Ca^{2+}, Na^+ and related Group I and II cations (K^+,Ag^+,Ba^{2+},Sr^{2+}), together with H_2O molecules, were envisaged to occupy the large tunnels of a zeolite-like psilomelane-type structure proposed for todorokite. Further modelling of the structure of todorokite is discussed by Turner and Buseck (1981).

Information from Infrared Spectroscopy

Some of the problems of identification and structural correlations involving disordered and finely particulate manganese oxides have been clarified by infrared spectroscopy (Potter and Rossman, 1977, 1979a,b). In a study of numerous natural and synthetic Mn oxides, Potter and Rossman (1979b) demonstrated that it might be possible to distinguish between different $[MnO_6]$ structural groups by their spectral profiles in the mid-infrared region. They have shown, for example, that the proposed layer structure of birnessite is supported by its infrared spectrum, and have confirmed the identity of natural birnessite with its synthetic analogues. The infrared spectra also show that synthetic buserite and birnessite have analogous structures, the shift from the 10 A to 7 A spacing in x-ray diffraction patterns being the result of water loss alone rather than a structural rearrangement of the $[MnO_6]$ framework.

The infrared spectra of natural todorokites indicate that this mineral is not analogous to any synthetic phases, including buserite and its derivatives, or to any decomposition products of buserite such as birnessite and manganite. The todorokite spectra are consistent with either a layer structure of linked $[MnO_6]$ octahedra containing vacancies, or a highly polymerized chain or tunnel structure with quadruple chains (Potter and Rossman, 1979b), such as that observed in the high resolution transmission electron microscope studies of Turner and Buseck (1979, 1981).

Information from X-ray Absorption Spectroscopy

Results of extended x-ray absorption fine structure (EXAFS) measurements of synthetic manganese (IV) oxides have been described by Arrhenius *et al.* (1979) and Crane (1979). They support the structure model proposed by Giovanoli *et al.* (1970a) for synthetic birnessite based on Wadsley's (1953) chalcophanite structure. Divalent cations of Cu, Ni, Co and Zn are located directly above and below vacancies in the edge-shared $[MnO_6]$ octahedral sheets, these near-vacancy sites being labelled as R-sites. The EXAFS measurements have yielded details of the coordination environment about the R-sites, which are shown in Figure 9. The divalent cation (Cu^{2+} in Fig. 9) situated above an Mn^{4+}-site vacancy is closer to three oxygens of the $[MnO_6]$ layer than to three water molecules of the intermediate layer (*cf.* Fig. 9). The EXAFS measurements show that R-site cations are

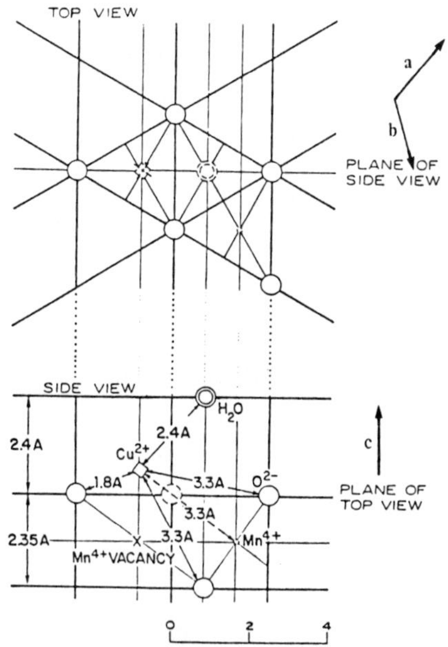

Figure 9. The birnessite structure in the vicinity of a substituent Cu^{2+} ion. The top view is a (001) projection; the side view is perpendicular to the top view (from Arrhenius *et al.*, 1979).

bound in the birnessite structure and not randomly adsorbed on external surfaces of the microcrystallites.

Similar R-type sites are found in Co^{2+} and Cu^{2+} buserites, although the nearest neighbor distances vary for each cation. Moreover, the EXAFS results indicate that buserite has in addition a larger, more regular octahedral site, designated as *A*-sites, located between sheets of water molecules of oxygen atoms. The *A*-sites are occupied preferentially by large ions such as Ca^{2+}, Na^+, H_3O^+, etc. (Crane, 1979). The *A*-sites are also accessible to smaller cations, and the EXAFS spectra show that this site deforms considerably to accommodate ions of different sizes.

THERMODYNAMIC PROPERTIES

Gibbs free energy data for the higher oxides of manganese have been assembled by Bricker (1965), Crerar and Barnes (1974), Crerar *et al.* (1979) and Hem (1978). Most of the data are for pure, well-crystallized, stoichiometric anhydrous manganese oxides not encountered in the marine environment. The occurrence of cryptocrystalline vernadite, birnessite and todorokite in seawater deposits suggests that these phases may have lower nucleation energies or lower free energies than the more crystalline oxides. However, structural features described earlier, such as defects, essential vacancies, extensive solid solution, cation exchange properties, cryptocrystallinity, and variable hydration, renders the marine manganese oxides intrinsically impure or heterogeneous phases, so that the acquisition of accurate thermodynamic data for them is virtually impossible. Nevertheless, some E_h and free energy values have been derived for synthetic birnessite or δ-MnO_2 (Bricker, 1965) and buserite (Jeffries and Stumm, 1976), from which the E_h-pH diagram illustrated in Figure 10 has been compiled (Crerar and Barnes, 1974).

The thermodynamic data suggest that Mn (IV) oxides are supersaturated relative to the oxygen content of normal seawater (Crerar *et al.*, 1979). For example, the Gibbs free energy for the reaction

$$Mn^{2+} + 1/2O_2(g) + 2OH^- \rightarrow \delta\text{-}MnO_2 + H_2O \qquad [1]$$

is -7 kcal mole^{-1} at $P_{O_2} = 10^{0.21}$ atoms, pH = 8.1, $[Mn^{2+}] = 10^{-9}$m, and T = 25°C. Since manganese must be oxidized from the Mn^{II} to Mn^{IV} state, its precipitation from seawater is E_h dependent. Using data for the half-cell reactions (Bricker, 1965; Jeffries and Stumm, 1976):

$$Mn^{2+} + 2H_2O \rightarrow \delta\text{-}MnO_2 + 4H^+ + 2e; \; E° = 1.293 \text{ v} \qquad [2]$$

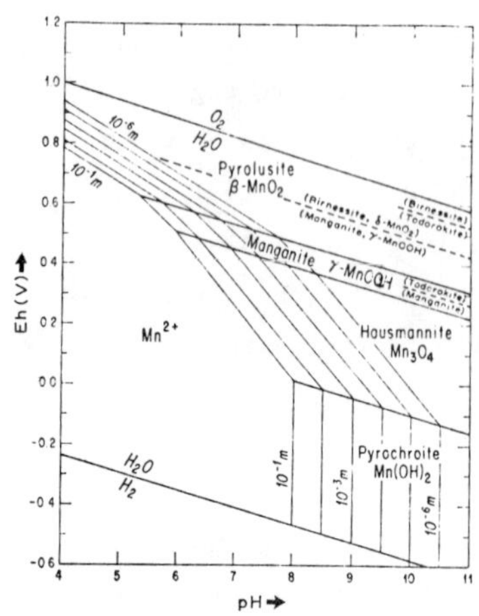

Figure 10. Eh-pH relations of the system Mn-H_2O for 25°C, 1 atmosphere, showing stability fields of Mn^{2+}, pyrolusite, manganite, hausmannite, and pyrochroite, and the metastable boundary birnessite-manganite. Metastable phases are shown in parentheses. Dissolved Mn^{2+} exceeds 0.1 m to the left of the oxide solubility contours (from Crerar and Barnes, 1974).

or

$$Mn^{2+} + 2H_2O \rightarrow \text{buserite} + 4H^+ + 2e;\ E° = 1.32\ v \qquad [3]$$

The calculated solubilities of δ-MnO_2 or buserite over the E_h range 0.20-0.45v characteristic of seawater exceeds the 1-2 ppb measured concentration of Mn by many orders of magnitude (Crerar *et al.*, 1979). Thus, Mn (IV) oxides are apparently undersaturated relative to the E_h, but supersaturated relative to the pH, of seawater.

One solution to this dilemma may be that kinetic factors also play an important role in the formation and preservation of marine oxides. The rate of oxidation of Mn^{2+} is proportional to pH, temperature, oxygen pressure, and concentrations of solid and dissolved manganese, and follows the rate law (Stumm and Morgan, 1970),

$$-\frac{d[Mn^{2+}]}{dt} = k_o [Mn^{2+}] + k\ P_{O_2}[OH^-]^2[Mn^{2+}][MnO_2] \qquad [4]$$

This equation indicates that the rate of autocatalytic oxidation increases by a factor of 100 per unit increase in pH. However, even at the pH of seawater, the oxidation of Mn^{2+} is extremely slow; equation 4 predicts that roughly 10^3 years would be required to autocatalytically oxidize 90% of the manganese dissolved in seawater. This figure, however, approximates closely the mean residence time of manganese in the oceans (Crerar *et al.*, 1979).

A three-step reaction has been proposed for the autocatalytic oxidation of manganese(Stumm and Morgan, 1970):

$$Mn^{2+} + 1/2O_2 \xrightarrow{slow} MnO_2 \qquad [5]$$

$$Mn^{2+} + MnO_2 \xrightarrow{fast} Mn(II) \cdot MnO_2 \qquad [6]$$

$$Mn(II) \cdot MnO_2 + 1/2O_2 \xrightarrow{slow} 2MnO_2 \qquad [7]$$

Here, equation 5 triggers the following reactions, equation 6 represents surface adsorption of Mn^{2+}, and equation 7 (the oxidation of adsorbed Mn^{2+}) is ultimately rate determining.

OCCURRENCES IN THE MARINE ENVIRONMENT

Todorokite, "10 A Manganite," or Buserite

The manganese oxide phase giving characteristic x-ray diffraction lines at 9.5-9.8 and 4.8-4.9 A has long been recognized as a common constituent of manganese nodules (Buser and Grütter, 1956; Grütter and Buser, 1957). The literature abounds with reported occurrences of todorokite or "10 A manganite" in nodules from both abyssal and near-shore environments (see reviews by Burns and Burns, 1977a, 1979). The todorokite polymorph has not usually been specified, but Chukhrov *et al.* (1978c, 1979b) using electron diffraction techniques have identified trillings of the polymorph having a = 14.6 A in iron-manganese concretions from the Pacific seafloor. This material is illustrated in Figure 11. Chukhrov *et al.* (1979a) have also commented on the fact that no satisfactory data exist for the natural occurrence of phases analogous to synthetic buserites.

Deep-sea manganese nodules, particularly those from the north equatorial Pacific and Indian Oceans, are strongly enriched in Ni, Cu, Co and Zn by factors of 10^4-10^6 over their average concentrations in seawater, seafloor basalts, and pelagic sediments (Mero, 1965; Margolis and Burns, 1976). The relatively high concentrations of Ni and Cu in these nodules are dramatically demonstrated by electron microprobe analyses across sectioned nodules which reveal bands of todorokite a few microns in diameter containing up to 4-5% Ni and 3-4% Cu (Burns and Burns, 1978a). The chemical analysis of one such band is listed in Table 2 (analysis 8). Such high concentrations may be attributable to atomic substitution of Ni^{2+} and Cu^{2+} for Mn^{2+} in the todorokite structure. Observations by scanning electron microscopy, SEM, such as those illustrated in Figure 12 (Burns and Burns, 1978a,b,c), suggest that post-depositional recrystallization processes lead to the formation of authigenic todorokite inside marine manganese nodules. Phillipsite and todorokite assemblages line cavities and voids adjacent to areas where extensive leaching of biogenic siliceous debris has occurred. Clusters of todorokite crystallites containing significant Ni and Cu concentrations fill fissures, line cavities, coat biogenic debris, and infill voids where biogenic debris once occurred in interiors of nodules. The source of the Ni and Cu appears to be of biological origin (Sclater *et al.*, 1976; Boyle *et al.*, 1977).

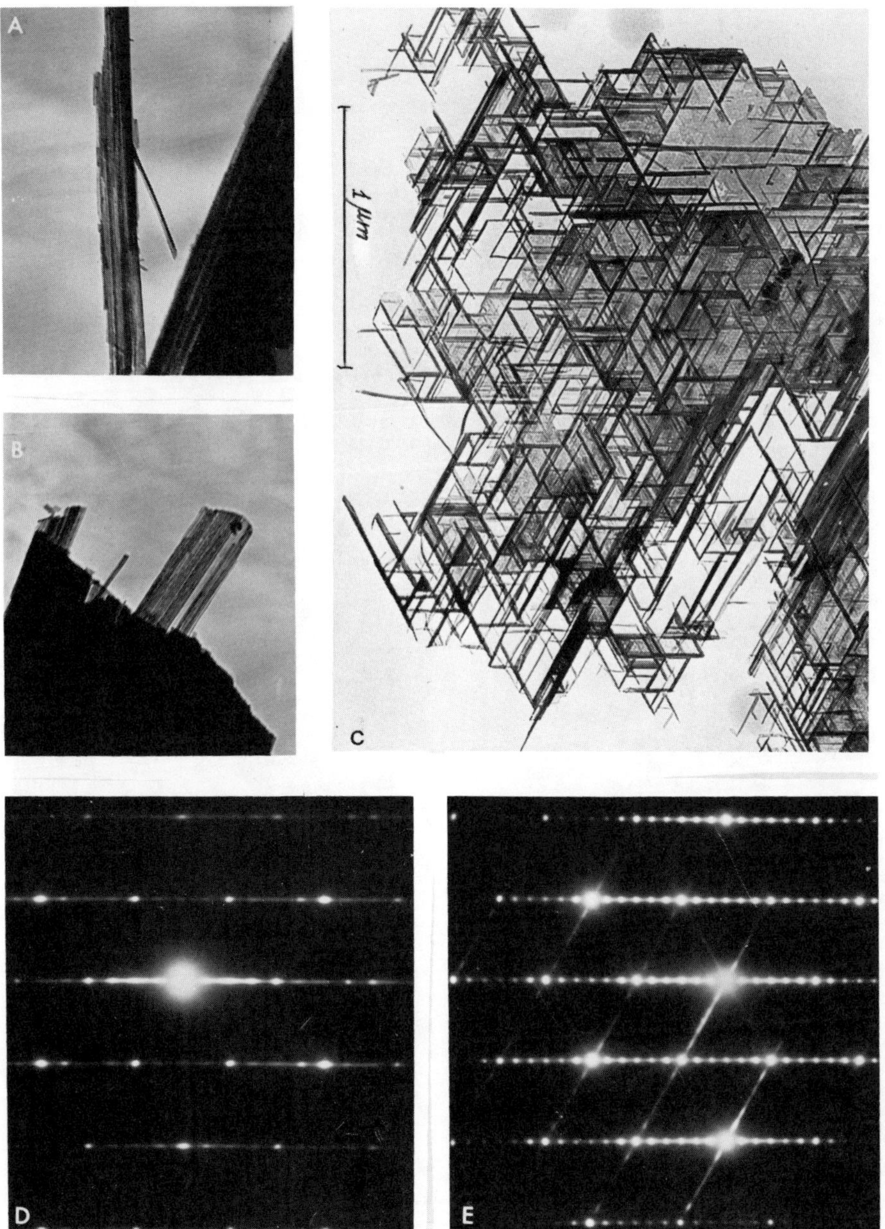

Figure 11. Electron microscope observations of todorokites (a); (b) todorokite from Charco Redondo, Cuba, with a = 9.75 A (V. M. Burns, unpublished data); (c) trillings of todorokite with a = 14.6 A from iron-manganese concretions from the Pacific sea floor; (d),(e) electron diffraction patterns of the Cuban todorokite and the treble intergrowths of the todorokite from iron-manganese concretions, respectively. [Figures c-e inclusive were provided by F. V. Chukhrov (see Chukhrov *et al.*, 1978c, 1979b).]

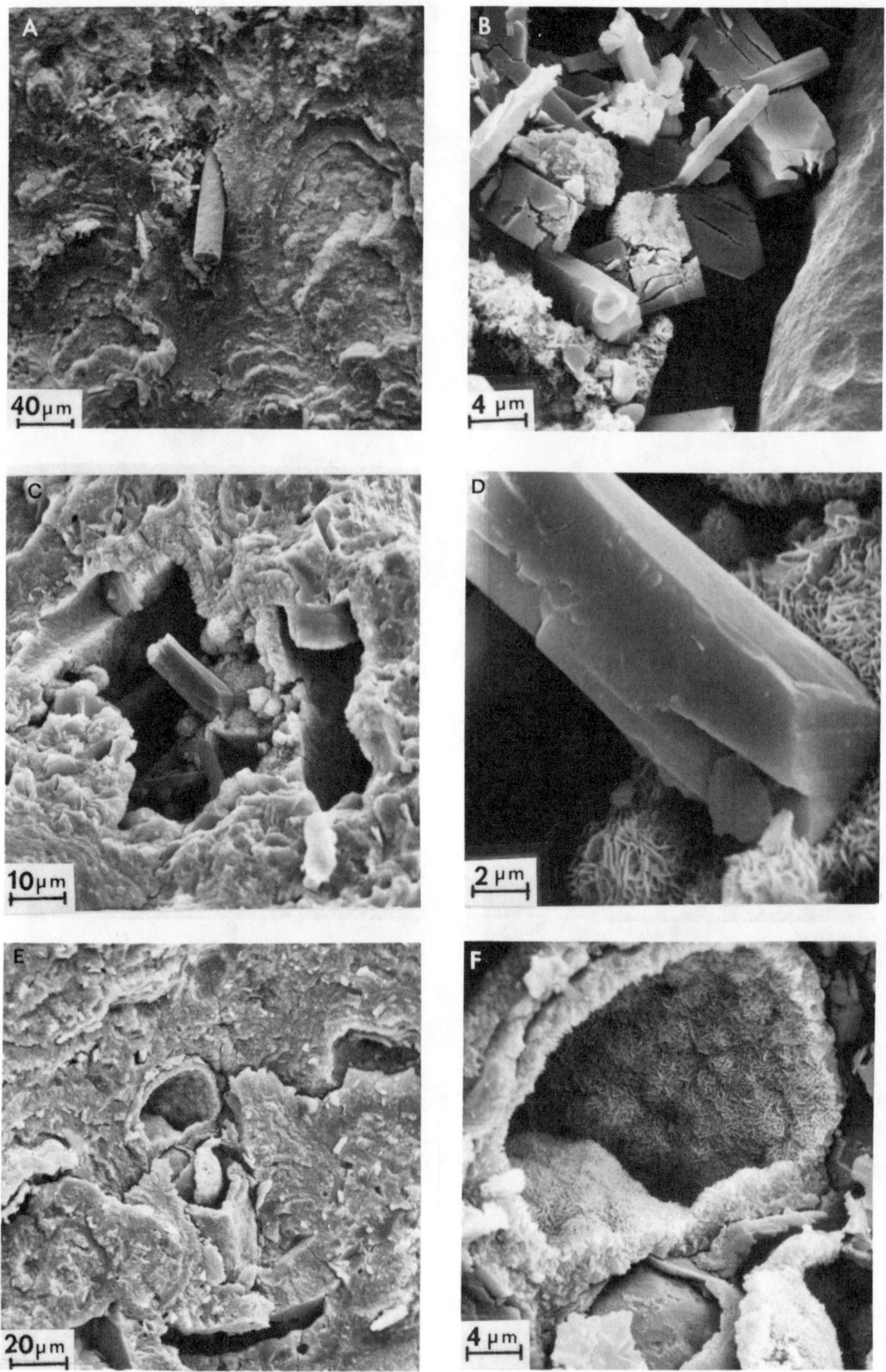
A
40µm
B
4 µm
C
10µm
D
2 µm
E
20µm
F
4 µm

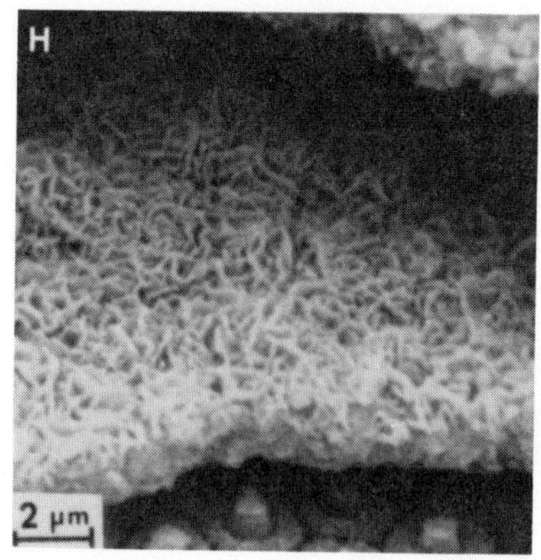

Figure 12 (this page and opposite). Authigenic todorokite and phillipsite inside manganese nodules. (a),(b) Phillipsite and todorokite crystallites lining a cavity created by dissolution of biogenous debris; (c),(d) euhedral phillipsite crystal and meshwork of todorokite crystallites inside a cavity; (e),(f) meshwork of todorokite lining a vug inside a nodule; (g),(h) todorokite crystallites coating biogenous debris entombed inside a nodule (from Burns and Burns, 1978c).

Figure 13. SEM photographs of platelets of birnessite inside a manganese nodule (from Woo, 1973). Bar scale (lower left): 1 micron.

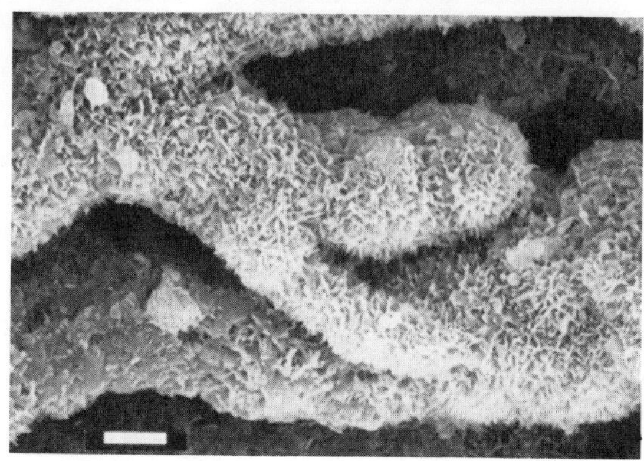

Figure 14. SEM photograph of radiating rods of birnessite crystallites from a submarine hydrothermal deposit (from V. M. Burns in Lonsdale *et al.*, 1979). Bar scale: 10 microns.

Todorokite has also been identified in ferromanganese crusts near oceanic spreading centers, adjacent to fracture zones, and in metalliferous sediments (see review by Burns and Burns, 1979). These todorokites tend to have significantly lower concentrations of Ni and Cu (compare analyses 10 and 11 in Table 2) than deep-sea manganese nodules which is attributed to the faster growth rates of the hydrothermal deposits.

Birnessite or "7 A Manganite"

The birnessite or "7 A manganite" phase, characterized by x-ray diffraction lines at 7.0-7.3 *and* 3.5-3.6 A, has also been reported in several manganese nodules (see reviews by Burns and Burns, 1977a, 1979). It appears to predominate in nodules from topographic highs, such as ridges and seamounts in the open ocean (Barnes, 1967; Lonsdale and Spiess, 1978). Particularly well-crystallized birnessite platelets have been identified in microconcretions from marine sediments (Woo, 1973; Glover, 1977; Chukhrov *et al.*, 1978b) (Figure 13). They appear to be the hexagonal polymorph. However, electron diffraction measurements of micronodules in coccolith oozes (Chukhrov *et al.*, 1979a,c) have revealed the presence of several calcium-rich birnessite polymorphs having cell parameters analogous to those of the synthetic orthorhombic sodium birnessite and hexagonal sodium-free birnessite (Giovanoli *et al.*, 1970a,b).

Birnessite is frequently a major constituent of hydrothermal ferromanganese oxide crusts adjacent to oceanic spreading centers and fracture zones, and also on flanks and calderas of submarine volcanoes (see review by Burns and Burns, 1979). The morphology of the birnessite in a hydrothermal deposit from the caldera of one such seamount on the East Pacific Rise is distinctive. The birnessite is seen by scanning electron microscopy to consist of radiating rods (Fig. 14) with crystallites coating the surface of solid rods (Lonsdale *et al.*, 1979).

Vernadite or δ-MnO_2

The fact that many marine manganese deposits frequently give featureless x-ray diffraction patterns or broad, diffuse reflections around 2.4 A and 1.4 A, is indicative that vernadite or δ-MnO_2 is an abundant constituent. Indeed, Chukhrov *et al.* (1978a) suggest that vernadite is the dominant manganese oxide phase on the seafloor. It is frequently intimately associated with the poorly crystalline iron oxide hydroxide mineral feroxyhyte, δ'-FeOOH, perhaps forming epitaxial intergrowths, and often enriches Co, Ce and Pb relative to todorokite and birnessite. The poor crystallinity and high specific surface areas of δ-MnO_2 result in this phase having high cation adsorption properties, particularly for Co (Murray, 1975a,b). The uptake of cobalt by δ-MnO_2 in manganese nodules, soils and synthesis products has been explained by atomic substitution of low-spin Co^{3+} (ionic radius 0.53 A) for Mn^{4+} (0.54 A) in the $[MnO_6]$ octahedra (Burns, 1976). The presence of trivalent cobalt on δ-MnO_2 was recently confirmed by photoelectron spectroscopy (Murray and Dillard, 1979).

Chukhrov *et al.* (1978a) suggested that the genesis of vernadite in the marine environment takes place during the rapid oxidation of dissolved Mn^{2+} ions. They believe that micro-organisms catalyse the oxidation, and they have identified vernadite in relict bacterial forms in manganese nodules. Although vernadite also forms as an oxidation product of terrestrial todorokite (Chukhrov *et al.*, 1978a), scanning electron microscopy observations (Burns and Burns, 1978a,b,c) indicate that post-depositional recrystallization of vernadite to todorokite occurs inside manganese nodules (see Fig. 12).

Other Manganese Oxide Phases

Pyrolusite, ramsdellite, nsutite, psilomelane, and rancieite have each been occasionally reported in manganese nodules and marine sediments (see reviews by Burns and Burns, 1977a, 1979). In some cases, however, the evidence is poorly documented or ambiguous.

PARAGENESIS

The origin of marine manganese oxide deposits has been debated at length, following the classic descriptions of samples collected during the *Challenger* expedition of 1872-1876 (Murray and Renard, 1884, 1891; Murray and Irvine, 1894). Although there are only two ultimate sources of manganese (namely, erosion of the continents and submarine volcanism), a variety of processes in the marine environment has generated different types of sediment. These processes include (Bonatti, 1975):

(1) hydrogenous deposits, formed by slow precipitation of metals from seawater; these may include chemical precipitates formed during reactions in seawater, or settled particulate terrigenous matter derived from the continents;

(2) diagenetic deposits, in which metals are remobilized in buried sediments as a result of redox reactions involving organic matter;

(3) hydrothermal deposits, which are associated with seafloor volcanism or hydrothermal activity;

(4) halmyrolic deposits, caused by submarine weathering of seafloor igneous rocks (especially basaltic glass); and,

(5) biogenous deposits, derived from marine organisms.

Although each one of these deposits can be recognized in marine sediments, they are inextricably associated with one another.

The processes of weathering and erosion of the continents eventually lead to the transport of dissolved species and suspended phases of manganese and other metals to the sea. Manganese probably enters the ocean as amorphous or cryptocrystalline vernadite and manganite (Stumm and Giovanoli, 1976). The particle sizes of these suspended phases probably do not exceed diameters of a few tens of Angstroms, and therefore consist of no more than a few score of $[Mn(O,OH)_6]$ coordination clusters. The small dimensions of such particles not only causes them to have large surface areas capable of adsorbing significant concentrations of other cations, but also lead to very low settling velocities making the particles susceptible to transportation by ocean currents to depositional environments far removed from the continents. The product of slow hydrogenous precipitation from seawater is mostly vernadite. However, coprecipitation of hydrated iron oxide phases creates deposits having Mn:Fe ratios of 0.5 to 2.0.

Considerably faster accretion of ferromanganese oxides occurs in the vicinity of seafloor volcanism and hydrothermal activity. The voluminous literature describing the association of ferromanganese oxide crusts and metalliferous sediments with oceanic spreading centers has been reviewed by Bonatti (1975). These deposits have formed by the interaction of hydrothermal fluids with oxygenated seawater. The manganese and iron appear to have resulted from seawater permeating through fractured seafloor basalts being heated up and leaching the metals from the volcanic rocks. Some input may also come from juvenile deep-seated mantle sources. The dissolved metals can be effectively transported as chloride complexes even when the chloride concentrations of the hydrothermal solutions are comparable to that of seawater. When these fluids emerge from submarine fracture zones or ridge crests, they come into contact with cold oxygenated seawater. The dissolved chlorides are then susceptible to hydrolysis and oxidation reactions, during which amorphous or poorly crystalline oxides (e.g. vernadite) are the most likely phases to form. Hydrothermal deposits often exhibit extreme fractionation of iron and manganese because the kinetics of oxidation of dissolved Mn^{II} complexes are slower than those of soluble Fe^{II} complexes. Incomplete oxidation of Mn^{2+} contributes to the formation of todorokite and birnessite in hydrothermal deposits.

The high valence oxides of Mn and Fe suspended in seawater and derived from terrigenous sources, submarine hydrothermal activity or weathering of seafloor basalts, ultimately become buried in pelagic sediment, together with biogenous debris and other detrital material. During burial, diagenetic processes involving redox reactions of biogenous debris with amorphous or poorly crystalline Mn^{IV}, Mn^{III} and Fe^{III} oxides occur, producing soluble Fe^{II} and Mn^{II} species. The low valence Mn and Fe cations may be transported in pore waters as soluble Cl^-, HCO_3^- or organic complexes, which enter into redox reactions near the sediment-oxygenated seawater interface.

Several mechanisms have been proposed for the oxidation of dissolved Mn^{II} species in sediment pore waters. The three-step autocatalytic oxidation of aqueous Mn^{2+} suggested by Stumm and Morgan (1970) summarized in equations 5-7, probably leads to the formation of vernadite. Such vernadite nucleates and grows epitaxially on feroxyhyte surfaces coating debris in the sediment (Burns and Burns, 1975; Chukhrov *et al.*, 1978a).

Stumm and Giovanoli (1976) suggested that oxidation of aqueous Mn^{2+}, particularly in continental waters, produces manganite. Hem (1978) proposed an alternative mechanism of Mn^{2+} oxidation in aerated pore waters which involved the disproportionation of intermediary hausmannite or manganite:

$$6Mn^{2+} + O_2(aq) + 6H_2O \rightarrow 2Mn_3O_4 + 12H^+ \qquad [8]$$

$$Mn_3O_4 + 4H^+ \rightarrow \delta\text{-}MnO_2 + 2Mn^{2+} + 2H_2O \qquad [9]$$

so that the overall reaction is:

$$Mn^{2+} + 1/2O_2(aq) + H_2O \rightarrow \delta\text{-}MnO_2 + 2H^+ \qquad [10]$$

The alternative disporportionation reaction involving manganite is:

$$2MnOOH + 2H^+ \rightarrow \delta\text{-}MnO_2 + Mn^{2+} + 2H_2O \qquad [11]$$

Chukhrov *et al.* (1978a), noting vernadite relicts of bacterial forms in iron-manganese nodules, suggested that micro-organisms catalyse the oxidation of Mn^{2+}. However, manganese nodules at the sediment-seawater interface, as well as micronodules in pelagic sediments, contain todorokite which, in continental deposits, is a well-characterized Mn^{2+}-bearing Mn (IV) oxide mineral (Table 2). Chukhrov *et al.* (1978a) also suggested that such todorokites are vulnerable to vernaditization. A solution to the dilemma of how todorokite survives in deep-sea manganese nodules has been proposed by Burns and Burns (1978a,b,c). They suggested that, since most todorokites inside manganese nodules contain significant concentrations of Ni, Cu, Zn, Mg, etc., these divalent cations are not susceptible to oxidation like Mn^{2+} ions, and therefore stabilize the todorokite structure by inhibiting reaction (3) above. Todorokites (and birnessite) deposited in hydrothermal deposits at spreading centers, however, are more susceptible to vernaditization because they have lower concentrations of stabilizing Ni^{2+}, Cu^{2+}, etc.

The inorganic mechanism of manganese oxide deposition proposed by Stumm and Morgan (1970) is modified by biological processes, particularly in manganese nodules growing in biogenic siliceous ooze sediments in the north equatorial Pacific. In a study of micronodules from Pacific deep-sea sediments, Greenslate (1974) observed that embryo concretions nucleate within microcavities inside plankton skeletal remains, and that the concretions are composed entirely of hydrated manganese oxides with negligible amounts of Fe, Ni and Cu. He postulated that a similar process is responsible for the initial uptake of Mn onto surface layers of large nodules. The formation of todorokite inside the nodules (Burns and Burns, 1978a,b,c) was interpreted to result from the oxidation of proteinaceous matter in emtombed biogenic siliceous debris by δ-MnO_2 (vernadite) or δ-FeOOH (feroxyhyte). The Fe^{2+} ions liberated in such redox reactions were suggested to be re-oxidized by aerated seawater or vernadite back to feroxyhyte which coats phillipsite crystallites and forms a substrate for the deposition of todorokite. The biogenic siliceous matter is also a possible source of Ni and Cu enriched in the manganese nodules beneath the high productivity surface waters in the equatorial Pacific Ocean.

Other diagenetic reactions leading to the fractionation of Mn and Fe have been proposed in metalliferous sediments adjacent to ridge crests, such as the Bauer Deep (Lyle *et al.*, 1977; Heath and Dymond, 1977). Hydrothermal Mn (IV) and Fe (III) oxides generated at the East Pacific Rise are transformed into Fe (III)-smectite and todorokite by reaction with biogenic silica after they were transported to the Bauer Deep. The formation of todorokite occurring in micronodules and manganese nodules in Bauer Deep sediments is attributed to the incompatibility of Mn^{IV} in the smectite phase.

SUMMARY

Todorokite, vernadite and birnessite are the predominant manganese oxide minerals found in the marine environment. These Mn (IV) oxides occur as cryptocrystalline phases in manganese nodules, ferromanganese crusts coating altered ridge-crest basalt, and microconcretions in underlying sediments. Although their crystal structures are poorly known, todorokite and birnessite appear to contain variable linkages of edge-shared $[MnO_6]$ octahedra and to be characterized by numerous structural defects, essential cation vacancies in the chains or layers of linked octahedra, domain intergrowths of mixed periodicities, extensive atomic substitution (particularly divalent Ni, Cu, Zn, Mg, etc., for Mn^{2+}, and cation exchange properties. Vernadite is a poorly crystalline to amorphous phase with high surface areas and cation adsorption properties, and concentrates cobalt (possibly by substitution of Mn^{4+} by low-spin Co^{3+}). Thermodynamic data indicate that marine Mn (IV) oxides are undersaturated in normal oxygenated seawater, suggesting that kinetic factors control the formation and preservation of each manganese oxide mineral. Micro-organisms appear to catalyse the formation of vernadite, but post-depositional recrystallization to todorokite occurs inside manganese nodules and in metalliferous sediments.

MANGANESE OXIDES: REFERENCES

Andrushchenko, N.F., *et al.* (1975) Composition and structure of metamorphosed ferromanganese nodules, new vein formations of manganese hydroxides, and the surrounding pelagic sediments in the Southern Basin of the Pacific Ocean Floor. Intern. Geol. Rev. *17*, 1375-1392.

Arrhenius, G. (1963) Pelagic sediments. *In*, M.N. Hill (ed.), *The Sea*, vol. 3 (Interscience Publ., New York), p. 655-657.

Arrhenius, G., Cheung, K., Crane, S., Fisk, M., Frazer, J., Korkisch, J., Mellin, T., Nakao, S., Tsai, A., and Wolf, G. (1979) Counterions in marine manganates. *In*, C. Lalou (ed.), *La Genèse des Nodules de Manganese*, Colloques Intern. du CNRS, no. 289, 333-356.

Barnes, S.S. (1967) Minor element composition of ferromanganese nodules. Science *157*, 63-65.

Betekhtin, A.G. (1937) New mineral species of the group of manganese hydroxides. Zap. Vses. Mineralog. Obsheh *66*, no. 4.

________ (1940) Southern Ural manganese deposits as a resource base for the Magnitogorsk Metallurgical Trust, Trudy Inst. Geol. Nauk, Akad. Nauk SSSR, no. 30.

Bonatti, E. (1975) Metallogenesis at ocean spreading centers. *In*, F.A. Donath (ed.), Ann. Rev. Earth Planet. Sci. (Annual Reviews Inc., Palo Alto, California), vol. 3, p. 401-403.

Boyle, E.A., Sclater, F.R., and Edmond, J.M. (1977) The distribution of dissolved copper in the Pacific. Earth Planet. Sci. Lett. *37*, 38-54.

Bricker, O. (1965) Some stability relations in the system $Mn\text{-}O_2\text{-}H_2O$ at 25° and one atmosphere total pressure. Am. Mineral. *50*, 1296-1354.

Brown, B.A. (1972) A low-temperature crushing technique applied to manganese nodules. Am. Mineral. *57*, 284-287.

Brown, F.H., Pabst, A., and Sawyer, D.L. (1971) Birnessite on colemanite at Boron, California. Am. Mineral. *56*, 1057-1064.

Buerger, M.J. (1936) The symmetry and crystal structure of manganite, Mn(OH)O. Z. Kristallogr. *95*, 163-174.

Burns, R.G. (1970) *Mineralogical Applications of Crystal Field Theory*. Cambridge University Press, Cambridge.

________ (1976) The uptake of cobalt into ferromanganese nodules, soils, and synthetic manganese (IV) oxides. Geochim. Cosmochim. Acta *40*, 95-102.

________, and Burns, V.M. (1975) Mechanism for nucleation and growth of manganese nodules. Nature *255*, 130-131.

_______ and ________ (1977a) Mineralogy of manganese nodules. *In*, G.P. Glasby (ed.), *Marine Manganese Deposits* (Elsevier, New York), Chapter 7, p. 185-248.

_______ and ________ (1977b) The mineralogy and crystal chemistry of deep-sea manganese nodules, a polymetallic resource of the twenty-first century. Phil. Trans. Royal Soc. London *A285*, 249-258.

Burns, R.G. and Burns, V.M. (1981) Authigenic oxides. *In*, C. Emiliani (Ed.), *The Sea, Vol. 7: The Oceanic Lithosphere*. John Wiley, New York, p. 875-914.

________, ________, Stockman, H.W., Stockman, C.T. and Benson, M.D. (1978) Depolyment of long-term seafloor mineral exposure experiments to measure changes of mineralogy and composition of manganese nodules. Marine Tech. Soc. Conf., Oceans, *78*, 662-667.

Burns, R.G. and Fuerstenau, D.W. (1966) Electron microprobe determinations of interelement relationships in manganese nodules. Am. Mineral. *54*, 895-902.

Burns, V.M. and Burns, R.G. (1978a) Post-depositional metal enrichment processes inside manganese nodules from the north equatorial Pacific. Earth Planet. Sci. Lett. *39*, 341-348.

________ and ________ (1978b) Diagenetic features observed inside deep-sea manganese nodules from the north equatorial Pacific. Scanning Electron Microscopy *1978*, 245-252.

________ and ________ (1978c) Authigenic todorokite and phillipsite inside deep-sea manganese nodules. Am. Mineral. *63*, 827-831.

Buser, W. (1959) The nature of the iron and manganese compounds in manganese nodules. *In*, M. Sears (ed.), *Internat. Oceanogr. Congr., AAAS*, p. 962-963.

Buser, W. and Graf, P. (1955) Differenzierung von Mangan (II) - manganit und δ-MnO_2 durch oberflächenmessung nach Brunauer-Emmet-Teller. Helv. Chim. Acta *38*, 830-834.

________, ________, and Feitknecht, W. (1954) Beitrag zur Kenntnis des Mangan (II)-manganite and des δ-MnO_2. Helv. Chim. Acta *37*, 2322-2333.

Buser, W. and Grütter, A. (1956) Über die Natur der Manganknollen. Schweiz. Min. Petr. Mitt. *36*, 49-62.

Butler, G. and Thirsk, H.R. (1952) Electron diffraction evidence for the existence and fine structure of a cryptomelane modification of manganese dioxide prepared in the absence of potassium. Acta Crystallogr. *5*, 288-289.

Byström, A.M. (1949) The crystal structure of ramsdellite, an orthorhombic modification of MnO_2. Acta Chem. Scand. *3*, 163-173.

Byström, A. and Byström, A.M. (1950) The crystal structure of hollandite, the related manganese oxide minerals, and δ-MnO_2, Acta Crystallogr. *3*, 146-154.

________ and ________ (1951) The positions of the barium atoms in hollandite. Acta Crystallogr. *4*, 469.

Calvert, S. E. (1978) Geochemistry of oceanic ferromanganese deposits. Phil. Trans. Royal Soc. Lond. *A290*, 43-73.

Champness, P. E. (1971) The transformation manganite → pyrolusite. Mineral. Mag. *38*, 245-248.

Chukhrov, F.V., Zvyagin, B.B., Yermilova, L.P., and Gorshkov, A.J. (1976a) Feroxyhyte, a new modification of FeO(OH). Izvest. Akad. Nauk, SSSR, ser. geol., no. 5, 5-24.

________, ________, ________, and ________ (1976b) Mineralogical criteria in the origin of marine iron-manganese nodules. Mineral Deposita *11*, 24-32.

________, Gorshkov, A.I., Rudnitskaya, E.S., Berezovskaya, V.V., and Sivtsov, A.V. (1978a) On vernadite. Izvest. Akad. Nauk SSSR, ser. geol., no. 6, 5-19.

________, ________, ________, and Sivtsov, A.V. (1978b) The characteristics of birnessite. Izvest. Akad. Nauk SSSR, ser. geol., no. 9, 67-76.

________, ________, Sivtsov, A.V., and Berezovskaya, V.V. (1978c) Structural varieties of todorokite. Izvest. Akad. Nauk SSSR. ser. geol., no. 12, 86-95.

________, ________, ________, and ________ (1979a) New mineral phases of oceanic manganese microconcretions. Izvest. Akad. Nauk SSSR, ser. geol., no. 1, 83-90.

________, ________, ________, and ________ (1979b) New data of natural todorokite. *278*, 631-632.

________, ________, ________, and ________ (1979c) A new 14 A mineral of the birnessite group in deep-sea micronodules. Nature *28* , 136-137.

Clark, G.M. (1972) *The Structures of Non-molecular Solids.* Applied Science Publications, London.

Corliss, J.B., Lyle, M., Dymond, J., and Crane, K. (1978) The chemistry of hydrothermal mounds near the Galapagos Rift. Earth Planet. Sci. Lett. *40*, 12-24.

Crane, S.E. (1979) Ion exchange in marine manganate minerals. EOS *60*, 282.

Crerar, D.A., and Barnes, H.L. (1974) Deposition of deep-sea manganese nodules. Geochim. Cosmochim. Acta *38*, 279-300.

________, Cormick, R.K., and Barnes, H.L. (1980) Geochemistry of manganese: an overview. In, I.M. Varentsov and Gy. Grasselly (Eds.) *Geology and Geochemistry of Manganese* (E. Schweizerbart'sche Verlagsbuchhandlung Publ., Stuttgart) Vol. 1, p. 293-334.

Dachs, H. (1963) Neutronen- und Röntgenuntersuchungen am Manganit, MnOOH, Z. Kristallogr. *118*, 303-326.

Faulring, G.M. (1962) A study of Cuban todorokite. *In*, M. Mueller (ed.), *Advances in X-ray Analysis*, vol. 5, p. 117-126.

Feitknecht, W. and Marti, W. (1945a) Über die oxydation von mangan (II) hydroxid mit molekularem sauerstoff. Helv. Chim. Acta *28*, 129-148.

________ and ________ (1945b) Über manganite und künstlichen braunstein. Helv. Chim. Acta *28*, 149-157.

Finkelman, R.B., Matzko, J.J., Woo, C.C., White, J.S., Jr., and Brown, W.R. (1972) A scanning electron microscopy study of minerals in geodes from Chihuahua, Mexico. Mineral. Rec. *3*, 205-212.

________, Evans, H.T., Jr., and Matzko, J.J. (1974) Manganese minerals in geodes from Chihuahua, Mexico. Mineral. Mag. *39*, 549-558.

Fleischer, M. (1960) Studies of the manganese oxide minerals. III Psilomelane, Am. Mineral. *45*, 176-187.

Frondel, C., Marvin, U.B., and Ito, J. (1960a) New occurrences of todorokite. Am. Mineral. *45*, 1167-1173.

________, ________, and ________ (1960b) New data on birnessite and hollandite. Am. Mineral. *45*, 871-875.

Giovanoli, R. (1969) A simplified scheme for polymorphism in the manganese dioxides. Chimia *23*, 470-472.

________ (1976) Vom hexaquo-mangan zum mangan-sediment reaktionssequenzen feinteiliger fester manganoxidhydroxide. Chimia *30*, 102-103.

________ (1980) On natural and synthetic manganese nodules. *In*, I.M. Varentsov and Gy. Grasselly (Eds.) *Geology and Geochemistry of Manganese* (E. Schweizerbart'sche Verlagsbuchhandlung Publ., Stuttgart), Vol. 1, p. 159-202.

________, and Bürki, P. (1975) Comparison of x-ray evidence of marine manganese nodules and non-marine manganese ore deposits. Chimia *29*, 266-269.

________, Bühler, H., and Sokolowska, K. (1973) Synthetic lithiophorite: electron microscopy and x-ray diffraction. J. Microsc. *18*, 271-284.

________, Bürki, P., Giuffredi, M., and Stumm, W. (1975) Layer structured manganese oxide hydroxides. IV: The buserite group; structure stabilized by transition elements. Chimia *29*, 517-520.

________, Feitknecht, W., and Fischer, F. (1971) Über oxidhydroxide des vierwertigen mangans mit schichtengitter. 3. Reduction von mangan (III) - Manganat (IV) mit zimtalkohol. Helv. Chim. Acta *54*, 1112-1124.

________, Maurer, R., and Feitknecht, W. (1967) Zur struktur des γ-MnO_2. Helv. Chim. Acta *50*, 1073-1080.

________, and Stahli, E. (1970) Oxide and Oxidhydroxide des drei - und vierwertigen mangans. Chimia *24*, 49-61.

________, ________, and Feitknecht, W. (1969) Über struktur und reaktivität von mangan (IV) oxiden. Chimia *23*, 264-266.

________, ________, and ________ (1970a) Über oxidhydroxide des vierwertigen mangans mit schichtengitter. 1. Natrium-mangan (II,III) manganat (IV). Helv. Chim. Acta *53*, 209-220.

________, ________, and ________ (1970b) Über oxidhydroxide des vierwertigen mangans mit schichtengitter. 2. Mangan (III)-manganat (IV). Helv. Chim. Acta *53*, 453-464.

Glasby, G.P. (ed.) (1977) *Marine Manganese Deposits*, Elsevier, Amsterdam, The Netherlands.

Glover, E.D. (1977) Characterization of a marine birnessite. Am. Mineral. *62*, 278-285.

Greenslate, J. (1974) Microorganisms participate in the construction of manganese nodules. Nature *249*, 181-183.

Grütter, A., and Buser, W. (1957) Untersuchungen an mangansedimenten. Chimia *11*, 132-133.

Hariya, Y. (1961) Mineralogical studies on todorokite and birnessite from the Todoroki mine, Hokkaido. Jap. J. Assoc. Min. Petr. Econ. Geol. *45*, 219-230.

Heath, G.R., and Dymond, J. (1977) Genesis and transformation of metalliferous sediments from the East Pacific Rise, Bauer Deep, and Central Basin, Northwest Nazca plate. Geol. Soc. Am. Bull. *88*, 723-733.

Hem, J.D. (1978) Redox processes at surfaces of manganese oxide and their effects on aqueous metal ions. Chem. Geol. *21*, 199-218.

Hewett, D.F. and Fleischer, M. (1960) Deposits of the manganese oxides. Econ. Geol. *55*, 1-55.

________, ________, and Conklin, N. (1963) Deposits of the manganese oxides: supplement. Econ. Geol. *58*, 1-51.

Hey, M.H. and Embrey, P.G. (1974) Twenty-eighth list of new mineral names. Mineral. Mag. *39*, 903-932.

Huebner, J.S. (1976) The manganese oxides--a bibliographic commentary. *In*, D. Rumble, III (ed.), *Oxide Minerals*. Reviews in Mineralogy 3, Chapter 7.

Jeffries, D.A. and Stumm, W. (1976) The metal-adsorption chemistry of buserite. Can. Mineral. *14*, 16-22.

Jones, L.H.P. and Milne, A.A. (1956) Birnessite, a new manganese oxide mineral from Aberdeenshire, Scotland. Mineral. Mag. *31*, 283-288.

Larson, L.T. (1962) Zinc-bearing todorokite from Philipsburg, Montana. Am. Mineral. *47*, 59-66.

Levinson, A.A. (1960) Second occurrence of todorokite. Am. Mineral. *45*, 802-807.

________ (1962) Birnessite from Mexico. Am. Mineral. *47*, 790-791.

Lonsdale, P. and Spiess, F.N. (1979) A pair of young cratered volcanoes on the East Pacific Rise. J. Geol. *87*, 157-174.

________, Burns, V. M., and Fisk, M. (1980) Nodules of hydrothermal birnessite in the caldera of a young seamount. J. Geol. 88, 611-618.

Lyle, M., Dymond, J., and Heath, G.R. (1977) Copper-nickel-enriched ferromanganese nodules and associated crusts from the Bauer Basin, Northwest Nazca Plate. Earth Planet. Sci. Lett. *35*, 55-64.

Manheim, F.T. (1965) Manganese-iron accumulations in the shallow marine environment. *In*, D.R. Schink and J.T. Corless (eds.), *Symposium on Marine Chemistry*, Occas. Publ. Univ. Rhode Island, *3*, 217-276.

Margolis, S.V. and Burns, R.G. (1976) Pacific deep-sea manganese nodules: their distribution, composition, and origin. Ann. Rev. Earth Planet. Sci. *4*, 229-263.

McKenzie, R.M. (1970) The synthesis of birnessite, cryptomelane, and some other oxides and hydroxides of manganese. Mineral. Mag. *28*, 493-502.

Mero, J.L. (1965) *The Mineral Resources of the Sea*. Elsevier Press, Amsterdam, The Netherlands, Ch. 6, p. 103-241.

Mouat, M.M. (1962) Manganese oxides from the Artillery Mountains area, Arizona. Am. Mineral. *47*, 744-757.

Mukherjee, B. (1959) X-ray study of psilomelane and cryptomelane. Mineral. Mag. *32*, 166-171.

Murray, J. and Irvine, R. (1894) On the manganese oxides and manganese nodules in marine deposits. Trans. Roy Soc. Edinburgh *37*, 712-742.

Murray, J. and Renard, A.F. (1884) On the nomenclature, origin and distribution of deep-sea deposits. Proc. Royal Soc. Edinburgh *12*, 474-495.

________ and ________ (1891) Report on deep-sea deposits based on specimens collected during the voyage of H. M. S. "Challenger" in the years 1872-1876. *In*, C. Wyville Thomson (ed.), *Report on the Scientific Results of the Voyage of H.M.S. "Challenger"* (Eyre and Spottiswoode, London), *5*, p. 1-525.

Murray, J.W. (1975a) The interaction of metal ions at the manganese dioxide-solution interface. Geochim. Cosmochim. Acta *39*, 505-519.

________ (1975b) The interaction of cobalt with hydrous manganese dioxide. Geochim. Cosmochim. Acta *39*, 635-647.

________ and Dillard, J.G. (1979) The oxidation of cobalt (II) adsorbed on manganese dioxide. Geochim. Cosmochim. Acta *43*, 781-787.

Nambu, M., Okada, K., and Tanida, K. (1964) Chemical composition of todorokite. Jap. J. Assoc. Mineral. Petr. Econ. Geol. *51*, 30-38.

Oswald, H.R. and Wampetich, M.J. (1967) Die kristallstrukturen von Mn_5O_8 und $Cd_2Mn_3O_8$. Helv. Chim. Acta *50*, 2023-2034.

Potter, R.M. and Rossman, G.R. (1977) Desert varnish: the importance of clay minerals. Science *196*, 1446-1448.

________ and ________ (1979a) The manganese and iron oxide mineralogy of desert varnish. Chem. Geol. *25*, 79-94.

________ and ________ (1979b) The tetravalent manganese oxides: identification, hydration, and structural relationships by infrared spectroscopy. Am. Mineral. 64, 1199-1218.

Radtke, A.S., Taylor, C.M., and Hewett, D.R. (1967) Aurorite, argentian todorokite, and hydrous silver-bearing lead manganese oxide. Econ. Geol. *62*, 186-206.

Sclater, F.R., Boyle, E., and Edmond, J.M. (1976) On the marine geochemistry of nickel. Earth Planet. Sci. Lett. *31*, 119-128.

Scott, R.B., Rona, R.A., Butler, L.W., Nalwalk, A.J., and Scott, M.R. (1972) Manganese crusts of the Atlantis Fracture zone. Nature Phys. Sci. *239*, 77-79.

Sorem, R.K. and Gunn, D.W. (1967) Mineralogy of manganese deposits, Olympic Peninsula, Washington. Econ. Geol. *62*, 22-56.

Stevenson, J.S. and Stevenson, L.S. (1970) Manganese nodules from the Challenger Expedition at Redpath Museum. Can. Mineral. *10*, 599-615.

Stockman, H.W. and Burns, R.G. (1980) The todorokite-buserite problem: implications to the mineralogy of marine manganese nodules. Am. Mineral. (submitted).

Straczek, J.A., Horen, A., Ross, M., and Warshaw, C.M. (1960) Studies of the manganese oxides. IV. Todorokite. Am. Mineral. *45*, 1174-1184.

Stumm, W. and Giovanoli, R. (1976) On the nature of particulate manganese in simulated lake waters. Chimia *30*, 423-425.

Stumm, W. and Morgan, J.J. (1970) *Aquatic Chemistry* (Wiley-Interscience, New York), p. 1-583.

Turner, S. and Buseck, P.R. (1979) Manganese oxide tunnel structures and their intergrowths. Science *203*, 456-458.

________ and ________ (1981) Todorokites: a new family of naturally occurrin manganese oxides. Science *212*, 1024-1027.

Usui, A. (1979) Nickel and copper accumulations as essential elements in 10-A manganite of deep-sea manganese nodules. Nature *279*, 411-413.

Varentsov, I. (ed.) (1979) *Geology and Geochemistry of Manganese* (Hungarian Acad. Sci., Budapest), in press.

Wadsley, A.D. (1950a) A hydrous manganese oxide with exchange properties. J. Am. Chem. Soc. *72*, 1782-1784.

________ (1950b) Synthesis of some hydrated manganese minerals. Am. Mineral. *35*, 458-499.

________ (1952) The structure of lithiophorite, $(Al,Li)MnO_2(OH)_2$. Acta Crystallogr. *5*, 676-680.

________ (1953) The crystal structure of psilomelane, $(Ba,H_2O)_2Mn_5O_{10}$. Acta Crystallogr. *6*, 433-438.

________ (1955) The crystal structure of chalcophanite, $ZnMn_3O_7 \cdot 3H_2O$. Acta Crystallogr. *8*, 165-172.

________ (1964) Inorganic non-stoichiometric compounds. *In*, L. Mandelcorn (ed.), *Non-Stoichiometric Compounds* (Academic Press), p. 98-209.

de Wolff, P.M. (1959) Interpretation of some γ-MnO_2 diffraction patterns. Acta Crystallogr. *12*, 341-345.

________, Visser, J.W., Giovanoli, R., and Brütsch, R. (1978) Über ε-mangandioxid. Chimia *32*, 257-259.

Woo, C.C. (1973) Scanning electron micrographs of marine manganese micronodules, marine pebble-sized nodules, and fresh water manganese nodules. *In*, M. Morgenstein (ed.), *Papers on the Origin and Distribution of Manganese Nodules in the Pacific and Prospects for Exploration*, Honolulu, Hawaii, p. 165-171.

Yoshimura, T. (1934) Todorokite, a new manganese mineral from the Todoroki Mine, Hokkaido, Japan. J. Fac. Sci. Hokkaido Univ. Sapporo, ser. 4, *2*, 289-297.

Chapter 2

IRON OXIDES

James W. Murray

INTRODUCTION

Iron and manganese oxide minerals are important constituents of marine sediments. They occur together as major components in potentially economically-important ferromanganese nodules. They compose a major fraction of hydrothermal sediments produced near the crests of spreading ridge systems, and together they are thought to be the most important removal mechanism of many trace metals from seawater. Geophysicists consider them to impart the magnetic signature to marine sediments.

In this chapter I will first review the basic mineralogy and crystal chemistry of the forms of iron oxides and iron oxide hydroxides that might be found in the marine environment. I will then review the basic free energy data necessary for determining the equilibrium phases. The relative stability of goethite and hematite will be discussed in some detail as a case study because of the geochemical importance of these two phases. I will then review what we know about the pathways of formation of the various iron oxide minerals. Some minerals turn out to be excellent indicators of certain conditions or certain pathways of formation. Finally, I will briefly review the marine geochemical occurrence of these minerals.

STRUCTURAL BASICS

The mineralogy and crystal chemistry of the oxides, oxide hydroxides, oxyhydroxides and hydrated oxide phases of iron have been recently reviewed by Burns and Burns (1977). Fasiska (1967) has also published a summary of crystallographic information for some of these iron phases. The structures of the cubic (spinel group) and rhombohedral (e.g., hematite) oxides have been thoroughly reviewed by Lindsley (1976). The magnetic structures are described in Stacey and Banerjee (1974). A summary of crystallographic data is given in Table 1 (Burns and Burns, 1977) and Table 2 (Fasiska, 1967).

The basic structures of many of the iron oxides consist of close-packed oxygens containing Fe^{+3} and/or Fe^{+2} ions in various octahedral

Table 1. Nomenclature and crystallographic data for the oxides, oxyhydroxides and hydrated oxides of iron. (Primarily from Burns and Burns, 1977, in press, and Misawa et al., 1974)

Compound	Mineral	Formula	Crystal system, space group	Unit cell parameters (A)	Structure type
α-FeOOH	goethite	FeOOH	orthorhombic	a = 4.65 b = 10.02 c = 3.04	ramsdellite (groutite), hexagonal close-packed oxygen
β-FeOOH	akaganéite	$(Cl,OH,H_2O)_{1-2}$ $Fe_8(O,OH)_{16}$	tetragonal	a = 10.48 c = 4.53	hollandite (α-MnO_2), body-centered cubic oxygen
γ-FeOOH	lepidocrocite	FeOOH	orthorhombic	a = 3.88 b = 12.54 c = 3.07	boehmite, cubic close-packed oxygen
δ-FeOOH	synthetic	FeOOH	hexagonal	a = 2.95 c = 4.53	CdI_2
δ'-FeOOH	feroxyhyte	FeOOH	hexagonal	a = 2.95 c = 4.53	disordered CdI_2
ε-FeOOH	synthetic	FeOOH	monoclinic, $P2_1/c$	a = 8.721 b = 5.164 c = 5.680 β = 90°	
Hydrated ferric oxyhydroxide polymer	synthetic	$Fe_5HO_8 \cdot 4H_2O$ (Towe and Bradley)	hexagonal	a = 5.08 c = 9.4	
		$Fe_2O_3 \cdot 1.2H_2O$ (Van der Giesen)	cubic	a = 8.37	
	ferrihydrite	$5Fe_2O_3 \cdot 9H_2O$	hexagonal	a = 5.08 c = 9.4	related to hematite
α-Fe_2O_3	hematite	Fe_2O_3	hexagonal, $R\bar{3}c$	a = 5.04 c = 13.77 Z = 6	corundum
Fe_3O_4	magnetite	Fe_3O_4	cubic	a = 8.391	inverse spinel
γ-Fe_2O_3	maghemite	Fe_2O_3	cubic or tetragonal	a = 8.32 a = 8.338 c = 25.014	defect spinel
$Fe(OH)_2$	amakinite	$(Fe,Mg,Mn)(OH)_2$	hexagonal, $P\bar{3}m1$	a = 3.465 c = 4.85 Z = 2	brucite
Green rust I	synthetic from solutions	$(Fe^{+2}Fe^{+3})_3(OH,O)_8 \cdot Fe^{+3}(O_2anion^-)_2$	hexagonal	a = 3.198 c = 24.2	9 layers of oxygen (-ABCBCACAB-)
Green rust II	synthetic from SO_4 solutions		hexagonal	a = 3.23 c = 22.5	4 layers of oxygen (-ABAC-)

interstices which can be viewed as different assemblages of $[FeO_6]$ octahedra. The Fe-Fe interatomic distances across these edge-shared octahedra range from 2.95-3.05 A and the $(10\bar{1}0)$ and $(11\bar{2}0)$ planes of the hexagonal close-packed system have spacings of 2.50-2.56 A and 1.48-1.54 A, respectively. These dimensions figure prominently in x-ray powder diffraction and cell parameter data of the iron oxide phases. These spacings are somewhat larger than those for the manganese (IV) oxides because of the larger ionic radius of Fe^{+3} (0.64 A) compared to Mn^{+4} (0.60 A). X-ray powder diffraction data based on the Fink Indexing System are summarized in Table 3.

The nomenclature followed here is similar to that proposed by Burns and Burns (1977; 1981). The five polymorphs of "FeOOH" are called iron (III) oxide hydroxides according to the International Union of Pure and Applied Chemistry (IUPAC). The x-ray amorphous hydrous oxide hydroxide phase often referred to as amorphous $Fe(OH)_3$, iron (III) oxide hydrate gel, colloidal ferric hydroxide, hydrated ferric oxide polymer will be called ferrihydrite even though the author has some reservations about whether this material is a new mineral.

Table 2. Crystallographic data for some iron oxides and oxyhydrates (from Fasiska, 1967).

Compound	System	Space group and atom position	Atomic fractional co-ordinates	Cell constants (A)
Fe (wüstite) (Fe : O ratio = 0·83 → 0·95)	Cubic (NaCl structure)	Fm3m (No. 225) O–4b Fe–4a	O x = y = z = 0·5* Fe x = y = z = 0	a = 4·28 → 4·31 ± 0·01†
Fe_3O_4 (magnetite) (Fe : O ratio = 0·744 → 0·750)	Cubic (fcc-inverse spinel)	Fd3m (No. 227) Fe^{3+}–8a Fe^{2+}, Fe^{3+}–16d O–32e	O x = 0·379 ± 0·001	a = 8·397 → 8·394 ± 0·005
γ-Fe_2O_3 (maghemite) (Fe : O ratio = 0·67 → 0·72)	Cubic (spinel)	Fd3m (No. 227) (21–⅓) Fe^{3+}–8a, 16d (32) O–32e	O x = 0·379 ± 0·001	a = 8·33 → 8·38 ± 0·01†
γ-FeO(OH) (lepidocrocite)	Orthorhombic	Cmcm (No. 63) Fe–4c O–4c OH–4c	Fe y = 0·178 ± 0·002† O y = 0·21 ± 0·01 OH y = 0·425 ± 0·002	a = 3·06 ± 0·01† b = 12·4 ± 0·1 c = 3·87 ± 0·01
α-FeO(OH) (goethite)	Orthorhombic	Pnma (No. 62) Fe–4c O–4c OH–4c	Fe x = 0·146, z = − 0·045 ± 0·002† O x = − 0·047, z = − 0·200 ± 0·002 OH x = − 0·200, z = 0·310 ± 0·002	a = 10·0 ± 0·1† b = 3·03 ± 0·01 c = 4·64 ± 0·01
α-Fe_2O_3 (hematite)	Trigonal (hexagonal unit cell)	R$\bar{3}$c (No. 167) Fe–12c O–18e	O x = 0·300 ± 0·001† Fe z = 0·355 ± 0·001	a = 5·035 ± 0·005† c = 13·72 ± 0·01

*Special positions, i.e. positions of iron and oxygen atoms on 4-fold axes.
†Estimated uncertainty.

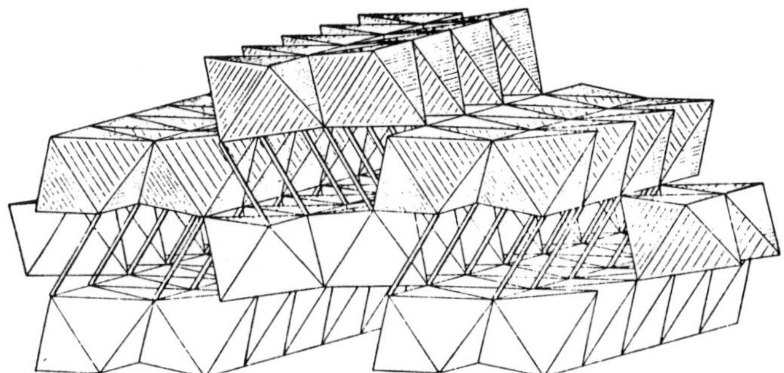

Figure 1. A pictorial representation of the diaspore structure with the hydrogen bonds shown as tubes (from Ewing, 1935). Diaspore and goethite have analogous structures.

Table 3. X-ray diffraction data for oxide compounds of iron (Fink Classification System).

Mineral or compound	The eight most intense d-spacings (intensity ⩾ 10%)								JCPDS card or reference
Goethite (α-FeOOH)	4.18 (100)	2.69 (30)	2.490 (16)	2.452 (25)	2.192 (20)	1.721 (20)	1.564 (16)		17—536
Akaganéite (β-FeOOH)	7.40 (100)	5.25 (40)	3.311 (100)	2.616 (40)	2.543 (80)	1.944 (60)	1.635 (100)	1.438 (80)	13—157
Lepidocrocite (γ-FeOOH)	6.26 (100)	3.29 (90)	2.47 (80)	1.937 (70)	1.732 (40)	1.524 (40)	1.367 (30)	1.075 (40)	8—98
δ-FeOOH	4.61 (20)	2.545 (100)	2.255 (100b)	1.685 (100vb)	1.471 (100)	1.271 (20)	1.223 (20)		13—87
ϵ-FeOOH	3.314 (100)	2.567 (50)	2.489 (80)	2.223 (50)	1.755 (70)	1.682 (80)	1.658 (50)	1.447 (50)	(1)
Hydrated ferric oxy-hydroxide polymer	2.52 (sss)		2.25 (s)	1.97 (s)	1.72 (s)		1.48 (sss)		(2), (3)
Idem	2.54 (s)	2.47 (m)	2.24 (ms)	1.98 (m-b)	1.725 (w)	1.515 (m)	1.47 (s)		(4)
Idem	2.54					1.52			(5)
Ferrihydrate	2.50 (m)	2.21 (m-w)	1.96 (w)	1.72 (vw)	1.51 (m-w)	1.48 (m)			(6)
Hydrogoethite	4.16 (90)	2.67 (70)	2.44 (100)	1.716 (80)	1.564 (40)	1.510 (50)	1.454 (40)	1.318 (40)	(7)
Hematite (α-Fe_2O_3)	3.66 (25)	2.69 (100)	2.51 (50)	2.201 (30)	1.838 (40)	1.690 (60)	1.484 (35)	1.452 (35)	13—534
Maghemite (γ-Fe_2O_3)	5.95 (60)	3.75 (100)	3.42 (65)	2.950 (100)s	2.521 (100)s	2.089 (100)s	1.702 (100)s	1.608 (100)s	15—615
Magnetite (Fe_3O_4)	4.85 (8)	2.967 (30)	2.532 (100)	2.424 (8)	2.096 (20)	1.715 (10)	1.616 (30)	1.485 (40)	19—629 11—614
Green rust (xFe$(OH)_2$·yFeOCl $z H_2O$)	8.02 (100)	4.01 (80)	2.701 (60)	2.408 (60)	2.037 (30)	1.598 (40)	1.567 (40)		13—88
Amakinite (Fe,Mg,Mn)$(OH)_2$	5.49 (70)	2.80 (80)	2.30 (100)	1.728 (90)	1.551 (70)	1.530 (80)	1.386 (70)	1.265 (70)	15—125

References:

1. N. A. Bendeliani, M.I. Baneyeva and D. S. Poryvkin (1972). Geochem. Int., 9: 589—590.
2. A. A. van der Giessen (1966). J. Inorg. Nucl. Chem., 28: 2155—2159.
3. S. Okamoto, H. Sekizawa and S. I. Okamoto (1972). In: *Reactivity of Solids*. Chapman and Hall, pp. 341—350.
4. K. M. Towe and W. F. Bradley (1967). J. Colloid Interface Sci., 24: 384—392.
5. W. Feitknecht, R. Giovanoli, W. Michaelis and M. Müller (1973). Helv. Chim. Acta, 56: 2847—2856.
6. F. V. Chukhrov, B. B. Zvyagin, A. I. Gorshkov, L. P. Yermilova and V. V. Balashova (1973). Proc. Acad. Sci., USSR, 4: 23—33.
7. P. F. Andrushchenko and N. S. Skornyakova (1965). In: Manganese Deposits of the Soviet Union. Israel Progr. Sci. Transl., Jerusalem, 1970, pp. 101—124.

Burns and Burns (1977) have drawn an analogy between the structures of the silicate minerals and the structure of the Fe and Mn oxides. The silicate minerals are composed of different linkages of corner-shared $[SiO_4]$ tetrahedra. The Fe and Mn oxides are composed of $[FeO_6]$ and $[MnO_6]$ octahedra edge-shared in different arrangements. As a result isostructural comparisons can be made between some of the iron and manganese oxides. These comparisons probably have great importance for the geochemistries of these groups of minerals because of the possibilities for solid solution and for epitaxial intergrowths. The structural types and isostructural compounds are also shown in Table 1 (Burns and Burns, 1981).

Goethite (α-FeOOH)

Goethite is the most common form of the Fe (III) oxyhydroxides and is the polymorph to which most other FeOOH phases eventually revert to upon aging. It is isostructural with the Mn minerals ramsdellite (MnO_2) and groutite (α-MnOOH) and consists of double chains of linked $[Fe(O,OH)_6]$ octahedra in which hydrogen-bonding plays an important role. Diaspore (α-AlOOH) has a similar structure, and a representation is shown in Figure 1.

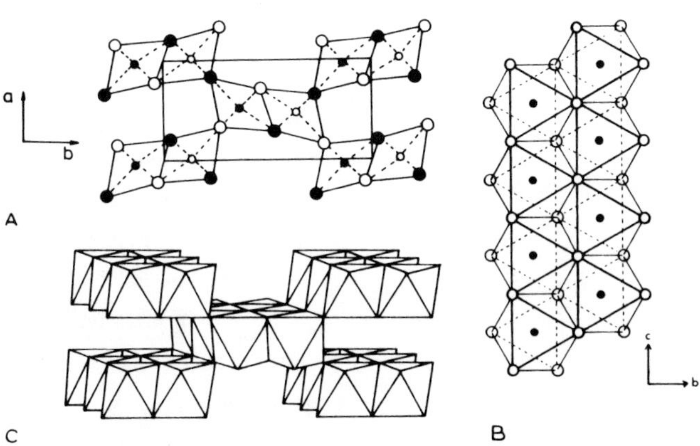

Figure 2. The ramsdellite structure. (A) Projection onto (001); white circles, atoms at level 1/4*c*; black circles, atoms at level 3*c*/4 (after Byström, 1949). (B) Double chain of $[MnO_6]$ octahedra running along *c* (after Byström, 1949). (C) The double chains of edge-shared $[MnO_6]$ octahedra parallel to *c* (after Clark, 1972). Small circles Mn, large circles oxygen (from Burns and Burns, 1977).

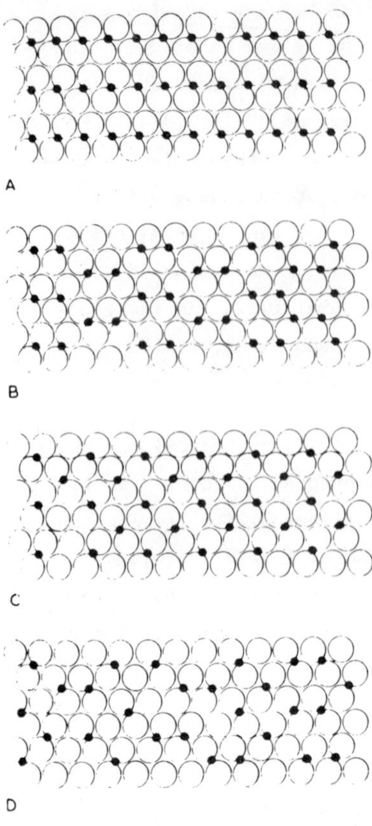

Figure 3. Hexagonal close-packed oxygen arrays, showing anion arrangements for: (A) amakinite-$Fe(OH)_2$; (B) goethite-α-FeOOH; (C) δ-FeOOH; (D) feroxyhyte.

The model structure is built up of alternating double chains of linked $[FeO_6]$ octahedra (thus resembling amphiboles) (Fig. 2). The octahedra are linked together by sharing opposite edges, thus producing continuous pyrolusite-like chains along the *c* axis. Two such chains are crosslinked by edge sharing with one chain being displaced *c*/2 with respect to the other so that an octahedron from one chain shares an edge with two octahedra from the other chain. The double chains of linked octahedra are further crosslinked to adjacent double chains through corner sharing of oxygen atoms to produce orthorhombic symmetry. This structure results in an acicular habit for goethite. Besides acicular needles 0.1 to 1.0 μm long, goethite may form twinned crystals under certain conditions (Atkinson *et al.*, 1968). Goethite is responsible for the yellow-brown color of some recent sediments and weathering outcrops (Berner, 1971).

The structure can also be viewed as a hexagonal close-packed lattice of oxygen and hydroxyl anions. The stacking of the oxygen planes is along the [001] direction with an -AB-AB-AB- sequence. The Fe^{+3} cations are distributed in an ordered array among the octahedral sites (Fig. 3). The iron atoms occupy only octahedral positions.

As will be discussed later, goethite can be synthesized by many different pathways including both hydrolysis and precipitation of Fe (III) in the absence of Cl^- as well as oxidation of Fe (II) through lepidocrocite. Thus its occurrence is not diagnostic of the geochemical history of the sample.

Goethite is an antiferromagnetic mineral which means it remains magnetized even when a field is removed. The magnetization is not reversible. Ideally, goethite should have perfectly compensated spins and hence no net spontaneous moment. However, it has been reported that both natural and synthetic goethite are capable of acquiring a thermoremnant magnetization (TRM) when cooled through the Néel point ($\sim$120°C). The reason for this is due to the fact that there are oxygen-ion vacancies in goethite which result in broken links in the continuous antiferromagnetic iron-oxygen-iron chains. Since the vacancies occur at random, some of the broken links can contain an odd number of anti-parallel iron atoms and hence such links can carry spontaneous magnetic moments (Strangeway *et al.*, 1968; Stacey and Banerjee, 1974).

Lepidocrocite (γ-FeOOH)

The structure of lepidocrocite is based on a cubic close-packed oxygen lattice (as opposed to hexagonal close packing for goethite).[1] It does not have any structural analogues among the manganese oxide or hydroxide phases. The structure may be described as a stacking of oxygen-hydroxyl planes with an -ABC-ABC- sequence along the [051] direction. The [051] direction of the orthorhombic cell corresponds to the [111] direction of a distorted cubic cell (Fasiska, 1967). As in goethite, the iron atoms occupy only octahedral positions. A projection of the γ-FeOOH structure is shown in Figure 4. Lepidocrocite is an orange-colored form of FeOOH and characteristically forms lath-shaped crystals ranging from 0.5 to 1.0 μm long.

Lepidocrocite is neither antiferromangetic nor ferrimagnetic at temperatures above 77°K, and so it can carry no magnetic remnance. On heating it is transformed to maghemite (γ-Fe_2O_3) at 250° to 300°C (Strangeway, 1970).

----

[1]A cubic close-packed framework of 32 oxygens contains a total of 64 tetrahedral sites and 32 octahedral sites. Not all of both sets can be filled, however, for this would require face-sharing between octahedra and tetrahedra which is energetically unstable because of electrostatic repulsion of the cations. The filling of eight tetrahedral sites precludes occupancy of more than 16 octahedral sites; conversely, the filling of 16 octahedral sites permits occupancy of only eight of the 64 possible tetrahedral sites.

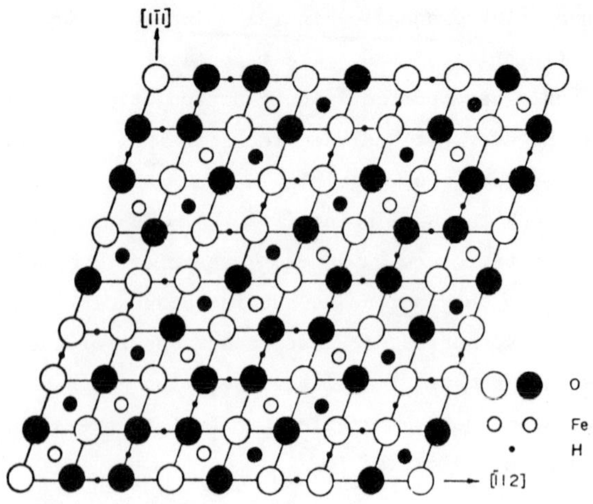

Figure 4. Projection of the γ-FeOOH crystal. The structure of γ-FeOOH is orthorhombic, but is based on a face-centered cubic framework of oxygen. Accordingly, this approximate figure is illustrated by an fcc structure. The plane of the figure is (110) of the fcc structure. Open and solid circles represent atoms at different elevations (from Misawa *et al.*, 1974).

Akaganéite (β-FeOOH)

The crystal chemistry of akaganéite is markedly different from other iron oxides. Burns and Burns (1981) have sought to compare β-FeOOH with the Mn mineral todorokite. Akaganéite (β-FeOOH) is uniquely characterized by tunnels that run parallel to the *c* axis. It is isostructural with the hollandite group (α-MnO_2) of the manganese oxides with OH^- ions replacing half of the O^{-2}. The basic building blocks, the [FeO_6] octahedra, share edges and form double chains along the *c* axis. The octahedra of the double chains share corners with adjacent double chains to give a three-dimensional framework with a body-centered tetragonal cell (a = 10.48 A, c = 3.023 A) (MacKay, 1960; see Fig. 5). This results in an open structure which contains one tunnel per unit cell. The tunnels are square, 5 A per side, bounded by two rows of octahedra (Gallagher and Phillips, 1969), and run the length of the crystals. The tunnels can accommodate H_2O molecules as well as OH^-, Cl^-, F^-, SO_4^{-2}, and NO_3^- ions. Large anions (especially Cl) are essential for the formation of this tunnelled structure. Gallagher and Phillips (1969) claimed that the Cl content could be reduced to very low levels by prolonged washing with distilled water. However, the experiments of Ellis *et al.* (1976) show that Cl is specifically adsorbed on β-FeOOH and cannot be fully eluted by washing or by exchange with inert (NO_3^-) or specifically interacting (F) anions. Cl is not only vital for the formation of β-FeOOH, but it also seems to stabilize the structure. For the analogous manganese minerals of the hollandite family it is the cations Ba^{+2}, Pb^{+2}, K^+ and Na^+ that are essential to the structure. The role the anions play during hydrolysis of Fe (III) will be discussed more later; however, it is important to note now that anions play an important role in determining iron oxyhydroxide mineralogy, and cations play an analogous role in manganese mineralogy. This may be due to the surface charge characteristics of the different groups of minerals. The pH (pzc) of the iron oxyhydroxides are between 7 and 8, whereas the hydrous manganese phases are between 2 and 3.

Tritium exchange studies (Gallagher and Phillips, 1969) indicate that the activation energy for tritium exchange between β-FeOOH and water (97 joule $mole^{-1}$) is considerably higher than for α-FeOOH (34.3 joule $mole^{-1}$) and γ-FeOOH (58.7 joule $mole^{-1}$). The shortest hydrogen bond length in β-FeOOH is 2.85 A, which is longer than the hydrogen bonds in ice and water (2.76 A). Such a long bond indicates a very weak interaction or nearly free OH groups.

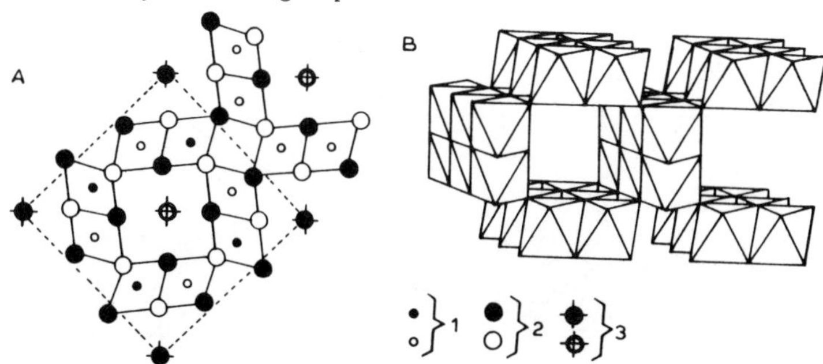

Figure 5. Akaganéite (β-FeOOH) has an open structure similar to hollandite. This structure is shown here (from Burns and Burns, 1977). (A) The structure of hollandite (or α-MnO_2) projected onto (001); white circles, atoms at zero level; black circles, atoms at level *c*/2 (after Byström and Byström, 1950). (B) The framework structure of hollandite showing tunnels between double chains of edge-shared [MnO_6] octahedra parallel to *c* (after Clark, 1972). 1 = Mn; 2 = oxygen; 3 = Ba, K, Pb, Na, or H_2O.

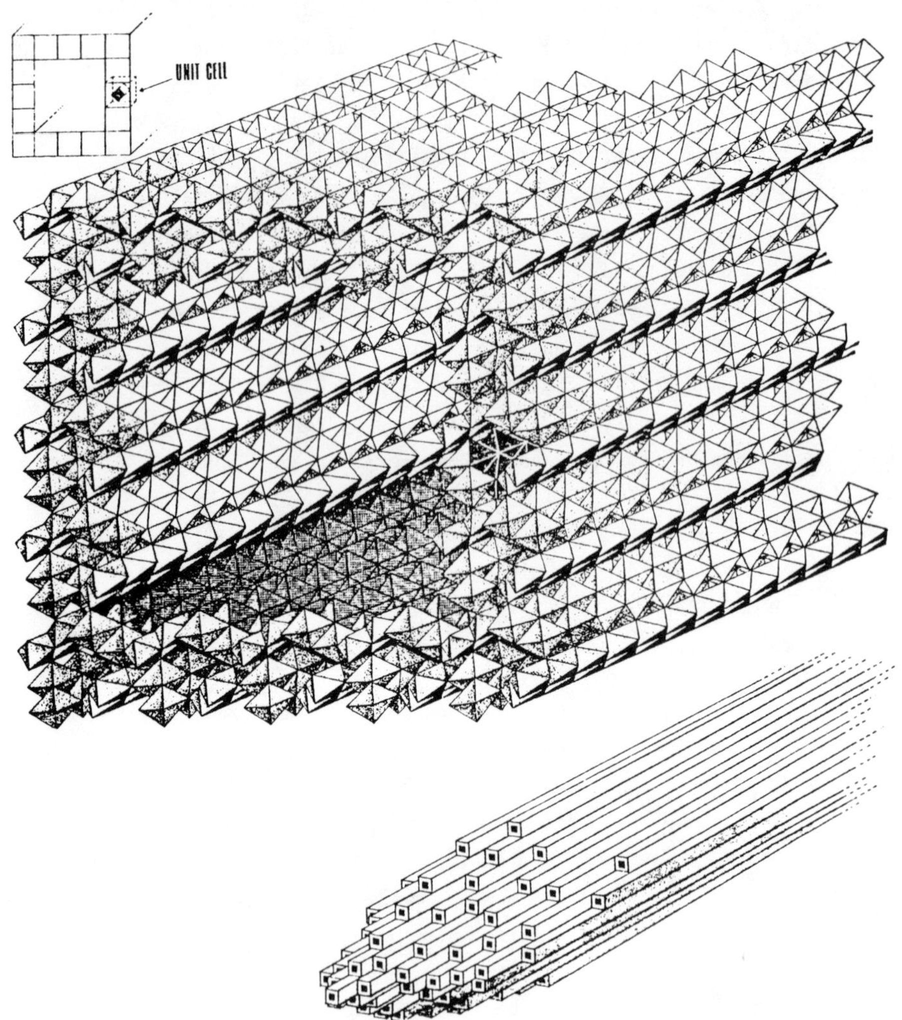

Figure 6. (a) Proposed atomic superstructure of a single tubular subcrystal of β-FeOOH (from Gallagher, 1970). Each octahedron represents an Fe^{+3} ion (at the center) octahedrally coordinated by O^{-2} or OH^{-} (at the corners). Both corner and edge sharing take place. Walls one unit cell thick form large square channels of side 31.4 A. A unit cell is shown in the inset and contains a small tunnel which contains essential Cl^{-} ions. (b) Parallel tubular rods or subcrystals such as shown in part (a) are packed to form cigar-shaped crystals or somatoids of β-FeOOH (from Gallagher, 1970).

Synthetic spindle-shaped (somatoid) crystals of β-FeOOH appear to be built up of bundles of parallel needles which under close observation appear to be an orthogonal array of closely-packed rods (Watson *et al.*, 1962; see Fig. 6). The crystals have tapered or irregular ends. The edges of these rods are built up of five unit cells which give an outer dimension of about 5 x *a* or 52.4 A. If the wall of the tube is one unit cell thick, this would give a square interior channel of about 3 x *a* or 31 A per side (Gallagher, 1970). This agrees well with the value of 30 A found by Watson *et al.* (1962) using electron microscopy and the value of 28.4 A by Gallagher and Phillips (1969) using N_2 adsorption isotherms. The difference between the model value of 31 A and that determined by N_2 adsorption (28.4 A) may correspond to a layer of water on the inner surface of the pores which would probably remain even after drying. This model must still be considered tentative, however, because Paterson and Tait (1977) have used surface area measurements and electron micrographs to conclude that the orthogonally-packed rods are solid. Thus akaganéite appears to contain two different sizes of cavities. These are small square tunnels (5 A per side) that are inherent to the hollandite type structure and are stabilized by Cl and large tunnels (30 A per side) that result from packing of the rods.

β-FeOOH has been prepared by a number of different methods, and the only essential requirement appears to be that Cl^- (or F^-) must be present during the hydrolysis and precipitation process (Feitknecht *et al.*, 1973). Regardless of the method of preparation, the crystals of akaganéite are small. Those shown by Ellis *et al.* (1976) and most other references are about 0.5 μm long, and that is also about the size of the crystals we have made in our laboratory. Because of its small particle size and structural cavities, β-FeOOH presents an unusually large proportion of its ions to the surface and thus should be especially effective as an ion exchanger.

Howe and Gallagher (1975) have studied the Mössbauer spectra and magnetic properties of β-FeOOH and β-Fe_2O_3 (prepared by dehydration of β-FeOOH). The Néel temperature for the antiferromagnetic β-FeOOH is about 295°K. The evidence supports the model that all Fe^{+3} ions are in octahedral coordination.

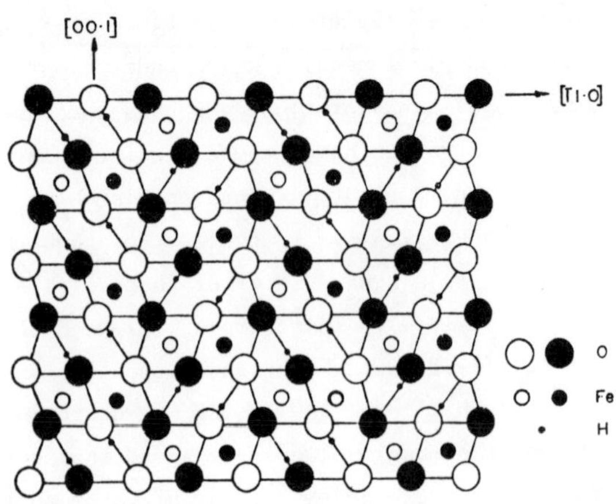

Figure 7. A projection of the $Fe(OH)_2$ structure (amakinite) which is thought to oxidize with little change in structure to form δ-FeOOH. The plane of figure is (110). The open and solid circles represent atoms at different elevations (from Misawa *et al.*, 1974).

δ-FeOOH

The brown-colored iron (III) oxide hydroxide with the formula δ-FeOOH has no mineral name because it has not been observed in nature. A modification (δ'-FeOOH) has been reported to exist in natural samples and the name feroxyhyte has been proposed (Chukhrov *et al.*, 1976; see next section).

The structure consists of a hexagonal close-packed array of O^{-2} and OH^- ions (a = 2.941 A, c = 4.49 A). Fe^{+3} has an ordered distribution among the octahedral sites; however, there is disorder of the O^{-2} and OH^- ions (Fig. 3). The white solid $Fe(OH)_2$ (amakinite) has a CDI_2 structure. A projection of $Fe(OH)_2(s)$ is shown in Figure 7 (Misawa *et al.*, 1974). δ-FeOOH is thought to have a disordered CdI_2 structure. Feitknecht (1959) obtained δ-FeOOH by vigorous bubbling of oxygen, air or H_2O_2 through alkaline solutions with solid $Fe(OH)_2$. He came to the conclusion that δ-FeOOH is formed by the topotactic replacement of $Fe(OH)_2(s)$ with preservation of the form of its platy particles and inheritance of its structural ordering. Large ordered crystals of δ-FeOOH are strongly magnetic. A relatively fast oxidation of $Fe(OH)_2(s)$ is required to produce δ-FeOOH, whereas δ'-FeOOH is produced under conditions favoring a somewhat slower oxidation (Chukhrov *et al.*, 1976). δ-FeOOH begins to transform into hematite after 10 days at 70°C in pH 4 and 8 solutions.

According to Kulgawczuk *et al.* (1968) and Chukhrov *et al.* (1976), δ-FeOOH is ferrimagnetic at room temperature. Particles 1 μm and larger show the largest degree of magnetic order. With decreasing particle size the magnetic saturation decreases and the residual magnetization disappears.

Feroxyhyte (δ'-FeOOH)

Both δ'-FeOOH and δ-FeOOH are formed topotactically by the oxidation of $Fe(OH)_2$. The arrangement of the oxygen anions in each of these phases is the same in principle, corresponding to hexagonal close packing (-AB-Ab-AB-). They differ only in the arrangement of the iron atoms. δ'-FeOOH is obtained under conditions of relatively slow oxidation (slightly acid or alkaline solutions and weakly aerated). Like δ-FeOOH, δ'-FeOOH is believed to contain Fe^{+3} ions in the octahedral sites of a hexagonal close-packed oxygen system. The Fe^{+3} ions, while almost randomly distributed in the octahedral sites, retain a slight degree of order among them (Fig. 3). In this respect δ'-FeOOH resembles ε-MnO_2 (Burns and Burns, this volume, Ch. 1).

δ'-FeOOH is thought to be a magnetically disordered form of δ-FeOOH (Chukhrov *et al.*, 1976). The basic difference in the electron diffraction patterns of δ-FeOOH and δ'-FeOOH is that the latter has no reflections characterizing an ordered distribution of Fe^{+3} in the octahedral sites. Feroxyhyte is unstable and transforms in air to goethite. Chukhrov *et al.* (1976) have proposed that δ-FeOOH occurs in nature as the mineral feroxyhyte. This claim needs further verification and should be viewed with some degree of skepticism because of the pathway of formation (topotactic oxidation of $Fe(OH)_2(s)$; see later discussion).

Hematite (α-Fe_2O_3)

The basic structure of hematite is a hexagonal close-packed array of oxygen anions in which 2/3 of the octahedral sites in every layer are occupied by Fe^{+3} (Fig. 8). The oxygen planes are stacked in an -AB-AB-AB- sequence along [001]. Hematite typically appears as small thick crystals of hexagonal outline. It may occur intergrowh with goethite (Atkinson *et al.*, 1968). The stoichiometric number of iron atoms occupy octahedral

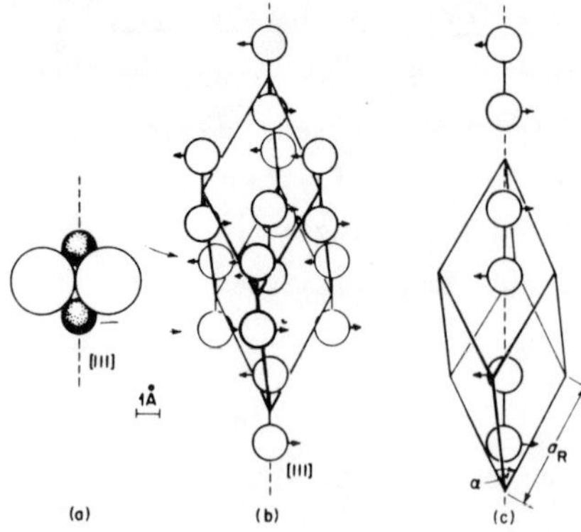

Figure 8. The structure of hematite: rhombohedral aspects. (a) Fe-O_3-Fe "pseudo-molecule." The "triplets" of three oxygens (large spheres) are essentially close packed and lie in the (III) plane [= (0001) plane, hexagonal axes]. (b) The rhombohedral cell of hematite according to the Pauling-Hendricks origin. Each corner and the center of the unit rhombohedron is occupied by the center of a FeO_3-Fe unit; the oxygen "triplets" are omitted for clarity. Only four Fe^{3+} ions lie within the cell. The magnetic structure is indicated schematically by the arrows: all Fe^{3+} ions in a given (111) layer have parallel spins but are antiparallel to those of the adjacent layers. (c) The rhombohedral cell of hematite with the origin shifted by (-1/4,-1/4,-1/4) (Barth-Posnjak origin) (from Lindsley, 1976).

interstices between the oxygen monolayers. An excellent description of the structure of hematite was given in Volume 3 of these notes (Lindsley, 1976). Hematite imparts the red coloration to rocks, giving rise to the term red beds.

The iron atoms in hematite are all octahedrally coordinated Fe^{+3} in the high spin state. Thus each ion has a magnetic moment of 5 Bohr megnetons (μ_B). The magnetic moments of the ions within a specific *c*-plane are ferromagnetically coupled (parallel to each other). Antiferromagnetic coupling between the planes couples them in parallel pairs, but with alternating polarity of the pairs of planes. The spin alignment is thus antiferromagnetic but the antiparallelism is imperfect. Above -10°C the spin moments are oriented in the *c*-plane, but instead of being precisely antiparallel they are slightly canted, resulting in a weak spontaneous magnetization (I_s = 0.4 emu/g) within the *c*-plane but normal to the spin-axis. Below -10°C there is no spin canting so that $I_s = 0$, i.e., at low T hematite is perfectly antiferromagnetic. The -10°C transition is called the Morin Transition (Stacey and Banerjee, 1974). Lattice defects may also cause ferrimagnetic regions.

<u>Ferrihydrite</u> ($5Fe_2O_3 \cdot 9H_2O$)

There has been a considerable amount of research conducted on the yellow-brown to dark brown x-ray amorphous hydrous iron oxides that are first to form during the hydrolysis and precipitation of dissolved iron from solution. Chukhrov *et al.* (1973) summarized the previous work and their own observations on field samples and proposed the name ferrihydrite for the mineral of composition $5Fe_2O_3 \cdot 9H_2O$. This name has been accepted by the nomenclature commission of the International Mineralogical Association. The studies by van der Giessen (1966) and Towe and Bradley (1967) on synthetic samples were especially important in piecing together the structural picture of ferrihydrite.

Van der Giessen (1966) was the first to obtain distinct x-ray diffraction patterns for amorphous iron oxide. He determined that the light-brown iron oxide gel prepared from NO_3^- solution at 20°C consists of 30 A diameter particles of irregular size and shape. These gel particles were proposed to have a cubic structure (a = 8.37 A) and a density of 3.8 $\pm$ 0.1 gm cm^{-3}.

Towe and Bradley (1967) used x-ray powder diffraction results on a similar product (from NO_3^- solution at 85°C) to suggest a structure based on the hexagonal close-packing of oxygens. In this sense the structure is similar to that of hematite (α-Fe_2O_3) except that the two strongest spacings (3.67 A and 2.69 A) derived from the ordered rhombohedral arrangement of Fe^{+3} ions in hematite are missing. The hematite-like model proposed by Towe and Bradley (1967) is shown in Figure 9 in projection on the plane $(11\bar{2}0)$. Chukhrov *et al.* (1973) have accepted this structure for the mineral ferrihydrite. The diffraction patterns from various investigators attributed to ferrihydrite by Chukhrov *et al.* (1973) are shown in Table 4. The indexing refers to a four-layer hexagonal cell of close packing of O^{-2}, OH^- and H_2O. The basic distinguishing features of ferrihydrite are the listed sequences of five reflections.

In contrast to hematite the position of the iron is partly vacant and the period of repeat is not six but four layers of hexagonal anionic packing. In four successive layers the period of repeat of octahedral sites populated by Fe^{+3} should be 2/3, 1/3, 1/3, 1/3. Ferrihydrite and hematite are characterized by unit cell parameters of a = 5.08 A, c = 9.4 A and a = 5.04 A, c = 13.77 A, respectively. Thus the structure proposed by Towe and Bradley (1967) is related to hematite but differs by containing some H_2O and OH^- instead of O^{-2} and less Fe in the octahedral portions. This leads to a much lower Fe/O ratio.

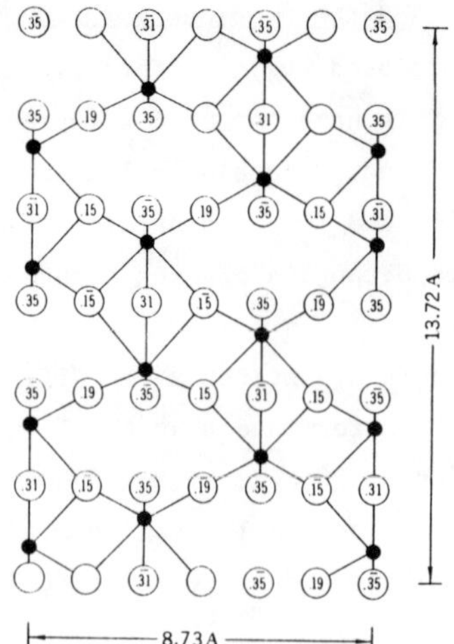

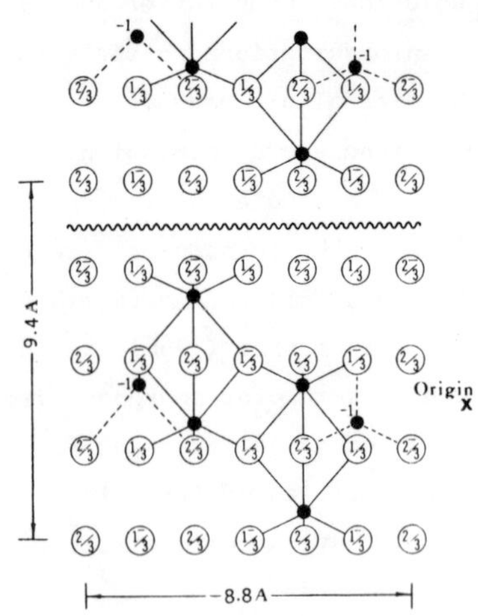

Figure 9. Ferrihydrite is proposed to have a similar structure to hematite: (a) Projection of the hematite structure on (11$\bar{2}$0). Full circles are Fe in the (11$\bar{2}$0) plane. (b) Projection of the proposed structure of ferrihydrite on (11$\bar{2}$0). Idealized detail of 4Fe octahedra are shown. Full circles are Fe in (11$\bar{2}$0). The wavey line indicates a level at which lateral displacements are optional. The c-period consists of four oxygen levels (from Towe and Bradley, 1967).

Table 4. Diffraction characteristics of synthetic and natural ferrihydrite.

hkl	Giessen, 1966		Towe,Bradley,1967		Chukhrov et al., 1971		Jackson,Keller, 1970		Towe,Lowenstam, 1967		Miyake, 1939	
	d	I	d	I	d	I	d	I	d	I	d	I
110	2.52	vs	2.54	s	2.50	m	2.52	m-s	2.54-2.47	s-m	2.56	s
112	2.25	s	2.24	m-s	2.21	m-w	2.23	ms	2.24	s	2.28	v.w.
113	1.97	s	1.98	m	1.96	w	1.98	v.w.	1.98	m	2.09	v.w.
114	1.72	s	1.725	w	1.72	vw	1.70	w-vw	1.725	w		
115			1.515	m	1.51	m-w						
300	1.48	vs	1.47	s	1.48	m	1.47	m	1.47	s	1.48	

van der Giessen, A. A. (1966) J. Inorg. Nucl. Chem *28*.
Towe, K. M. and W. F. Bradley (1967) J. Colloid Interface Sci. *24*.
Chukhrov, F. V. *et al.* (1971) AN SSSR Izvestiya, ser. geol. no. 1.
Jackson, T.A. and W. D. Keller (1970) Am. J. Sci. *269*.
Towe, K. M. and L. Lowenstam (1967) Ultrastructure Res. *17*.
Miyake, S.L. (1939) Sci. Papers Inst. Phys. Chem. Research (Japan) *36*.

Chukhrov *et al.* (1973) admit that the original x-ray diffraction data for ferrihydrite are limited and the structural model should be considered as tentative. Atkinson *et al.* (1968), for example, criticized the cubic structure proposed by van der Giessen (1966), and thus Towe and Bradley (1967) and Chukhrov *et al.* (1973) as well. Atkinson *et al.* (1968) suggested that the principal lines at d = 2.52 A, d = 1.48 A and d = 1.27 A could arise from a combination of goethite $d(021)$ = 2.58 A and $d(040)$ = 2.49 A; $d(002)$ = 1.51 A and $d(061)$ = 1.45 A; and $d(042)$ = 1.29 A and $d(080)$ = 1.25 A. The other lines attributed to ferrihydrite (Table 4) also occur in the goethite pattern. As noted earlier, the spacings at 2.55 A and 1.48 A are characteristic of a hexagonal close-packed system and are common in x-ray diffraction data of most iron oxyhydroxide phases (Table 3). Chukhrov *et al.* (1973) propose that the spacings 2.5 A and 1.5 A are characteristic of a precursor of ferrihydrite called protoferrihydrite. It cannot be unequivocally ruled out that the mineral referred to as ferrihydrite is merely a poorly ordered arrangement of [FeO_6] octahedra (Burns and Burns, 1977).

There is also some disagreement regarding the product ferrihydrite converts to on aging. Chukhrov *et al.* (1973) reported that ferrihydrite transforms into hematite in moderately acid to moderately alkaline solutions. At 80°C this transition occurs in a few hours, and at 40°C it takes 10 to 14 days. Transformation into goethite only occurred in strongly alkaline (pH > 11) or acid (pH < 3) solutions.

Schwertmann and Fischer (1973), however, could not get ferrihydrite to convert into hematite at pH 6 and 70°C even after two weeks. Goethite was obtained after aging in 1-M KOH (two weeks, 70°C) and in 0.02-M $FeSO_4$ solution at pH 6. They concluded that ferrihydrite is not converted to hematite by a topotactic dehydration reaction. It is more likely that ferrihydrite serves as a source of dissolved iron for the crystallization of goethite from the solution phase. This will be discussed more later.

Magnetite (Fe_3O_4)

Magnetite is in the group of oxide minerals with the inverse spinel structure. It forms a geochemically important solid solution series with ulvöspinel (Fe_2TiO_4). Lindsley (1976) has previously prepared an excellent set of notes on spinels and magnetite, and the brief summary given here comes from his work. The unit cell is face-centered, cubic and contains

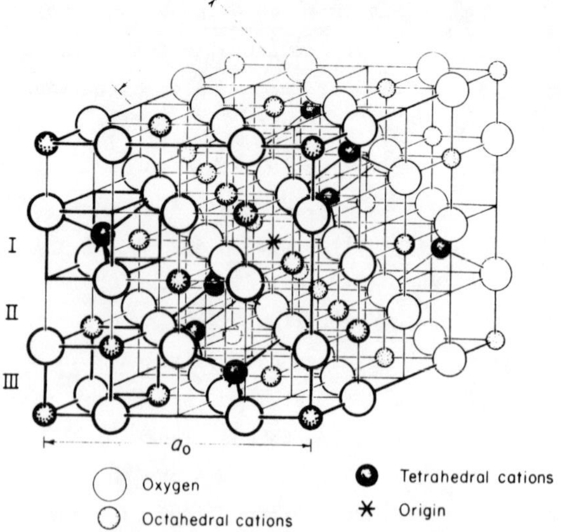

Figure 10. The spinel unit cell, oriented so as to emphasize the (111) planes. Atoms are not drawn to scale; the circles simply represent the centers of atoms. The origin in this diagram lies at the center of symmetry, as recommended by the International Tables (1952); it differs by (1/8,1/8,1/8) from the origin used in much of the literature.

32 oxygens, which form a nearly cubic close-packed framework as viewed along the cubic diagonals [111] (Fig. 10). The cations occupy some of the interstices within the oxygen framework. By convention the set of compatible tetrahedral and octahedral sites occupied in spinels is 16*d* (octahedral) and 8*a* (tetrahedral). The general chemical formula for ideal spinels is XY_2O_4 (or $X_8Y_{16}O_{32}$ per unit cell), where X and Y are cations of different valence. In magnetite X = Fe^{+2} and Y = Fe^{+3}; in ulvöspinel X = Ti^{+4} and Y = Fe^{+2}. When eight X cations occupy tetrahedral sites the spinel has a "normal" structure. It may be designated as $X[Y_2]O_4$ where brackets denote cations in octahedral sites. Other spinels have 8Y cations in the octahedral and tetrahedral sites and 8X cations in the octahedral sites to produce an "inverse" spinel with the formula $Y[YX]O_4$. Some spinel end members are shown in Table 5. Magnetite ($Fe^{+3}[Fe^{+2}Fe^{+3}]O_4$) has the inverse spinel structure with Fe^{+3} ions equally distributed between tetrahedral and octahedral sites. The magnetic properties of magnetite have been very instrumental in determining the ordering of the cations among the sites. See Stacey and Banerjee (1974) for a review.

Table 5. Spinel end members (from Lindsley, 1976).

Mineral name	Formula	Cell edge $\underline{a}$, in Å	Oxygen parameter, μ	Structure
Magnetite	$Fe^{3+}[Fe^{2+}Fe^{3+}]O_4$	8.396	0.2548	I
Magnesioferrite	$Fe^{3+}[Mg^{2+}Fe^{3+}]O_4$	8.383	0.257	I
Jacobsite	$Fe^{3+}[Mn^{2+}Fe^{3+}]O_4$	8.51		I
Chromite	$Fe^{2+}[Cr_2{}^{3+}]O_4$	8.378		N
Magnesiochromite	$Mg^{2+}[Cr_2{}^{3+}]O_4$	8.334	0.260	N
Spinel	$Mg^{2+}Al_2{}^{3+}O_4$	8.103	0.262	7/8 I
Hercynite	$Fe^{2+}[Al_2{}^{3+}]O_4$	8.135		N
Ulvöspinel	$Fe^{2+}[Fe^{2+}Ti^{4+}]O_4$	8.536	0.261	I

N, "normal" cation distribution, $X[Y_2]O_4$, where [ ] indicate octahedral cations.
I, "inverse distribution, $Y[XY]O_4$. Many spinels are probably intermediate between these extremes. Data from Burns (1970, p. 110).

Table 6. The partial molal Gibbs free energy of formation ($\Delta\bar{G}_f^o$) of iron solid phases and aqueous solution species at 25°C(298.15°K) and 1 atm total pressure.

Substance	ΔG_f^o(cal mole^{-1})	Source
Fe(II)		
Fe^{+2}(aq)	- 21,800 ± 500	1
$FeOH^+$ (aq)	- 67,168 ± 570	1
$Fe(OH)_2^o$ (aq)	-111,173 ± 600	1
$Fe(OH)_3^-$ (aq)	-148,249 ± 600	1
Fe(III)		
Fe^{+3} (aq)	- 4,020 ± 500	1
$FeOH^{+2}$ (aq)	- 57,358 ± 550	1
$Fe(OH)_2^+$ (aq)	-107,635 ± 600	1
$Fe(OH)_3^o$ (aq)	-155,595 ± 600	1
$Fe(OH)_4^-$ (aq)	-200,920 ± 550	1
Solids		
$Fe(OH)_2$(s)	-117,584 ± 570	1
$FeCO_3$ (siderite)	-162,390 ± 510	1
Fe_3O_4 (magnetite)	-243,094 ± 510	2
αFe_2O_3 (hematite)	≥-177,728 ± 310	2
γFe_2O_3 (maghemite)	≤-169,764 ± 500	1
αFeOOH (goethite)	≥-115,280 ± 160 to -116,375 ± 160	1
γFeOOH (lepidocrocite)	≤-169,614 ± 500	1
βFeOOH (akaganéite)	-179,700	3
$FeOOH \cdot XH_2O$ (hydrated ferric oxyhydroxide polymer)	≤-167,460 ± 600 to ≤-169,040 ± 1,100	1
Miscellaneous		
H_2O (ℓ)	- 56,687	4
OH^-	- 37,594	4
O_2	+ 3,900	4
$SO_4^=$	-177,970	4
H_2S	- 6,660	4
HS^-	+ 2,880	4
$S^=$	+ 20,500	4
CO_2(s)	- 94,254	4
H_2CO_3	-148,940	4
HCO_3^-	-140,260	4
$CO_3^=$	-126,170	4
Cl^-	- 31,370	4

1) Langmuir (1969), 2) Robie and Waldbaum (1968), 3) Biedermann and Chow (1966), 4) Wagman et al. (1968)

Magnetite has been classically thought to be a detrital component from the continents and submarine volcanism. Later we will summarize evidence that suggests that magnetite can also form during the oxidation of Fe (II) from solution such as might occur in the interstitial waters of some marine sediments.

Maghemite (γ-Fe_2O_3)

Maghemite is so named because its chemical composition is identical to hematite, although unlike hematite it is strongly magnetic and has a crystal structure similar to magnetite. When synthesized by oxidation of Fe (II), maghemite forms dark-brown sub-rounded particles (0.02 μm to 0.06 μm) (Taylor and Schwertmann, 1974b). In maghemite 1/9 of the metal atom sites in the Fe_3O_4 spinel structure are vacant. These vacant metal sites are distributed in one of two ways; either distributed randomly throughout the tetrahedral (8*a*) and octahedral (16*d*) sites or confined only to the (16*d*) octahedral positions (Lindsley, 1976).

Maghemite is formed by the low-temperature oxidation of magnetite (which in turn may either be detrital or authigenic). The oxidation is achieved by a topotactic process, so called because the oxygen structure is left unchanged. The Fe^{+2} ions in magnetite diffuse to the surface of a grain and oxidize to Fe^{+3} leaving lattice vacancies. If the diffusion can proceed unhindered, one microcrystal of magnetite can be completely transformed to maghemite without an additional new crystal of maghemite being formed. For each Fe^{+2} diffusing out of the magnetite structure, two more are converted to Fe^{+3} to maintain constant total cationic change within the structure. Thus $Fe_2^{+3}Fe^{+2}O_4$ becomes $Fe^{+3}_{8/3}\square_{1/3}O_4$, i.e., 1/3 of the original Fe^{+2} sites become vacancies represented by $\square$. The vacancies are in the octahedral positions. This way of representing the chemical formula also indicates that the crystal structure is similar to the structure of magnetite (Fe_3O_4). The structure is not exactly that of a spinel because the vacancies are ordered along a particular [100] axis, the repeat distance being three times the cubic cell edge. The crystal structure is, therefore, tetragonal due to the vacancy superlattice. Although the cell edge of magnetite (8.39 A) is similar to the cell edges of pseudo cubic maghemite (8.34 A), it is possible to distinugish maghemite from hematite by the presence of tetragonal diffraction peaks which are forbidden in magnetite.

FREE ENERGIES OF FORMATION

The basic data needed for equilibrium calculations and for ascertaining the direction of spontaneous reaction are the partial molar or molar-free energies of formation of substances under well-defined conditions. The partial molar-free energy of formation $\bar{G}_f^o$ is the free energy change accompanying the formation of a substance from the elements in their standard states. The partial molar free energy at a given temperature is equal to the standard chemical potential at the same temperature and 1 atm pressure:

$$\mu_i^o \text{ (T, 1 atm)} = \bar{G}_{f,i}^o \text{ (T, 1 atm)} \qquad [1]$$

Most recent geochemistry textbooks (e.g., Garrels and Christ, 1965; Berner, 1971) and literature have relied on the original data from Latimer (1952), Sillén and Martell (1964) and Wagman *et al.* (1969) for the Gibbs free energy data for iron-containing solid phases and aqueous solution species. Hem and Cropper (1959) provided one of the earliest compilations of free energy data which, in turn, was based largely on Latimer's compilation. These compilations were updated and improved by Langmuir (1969). The free energies of formation as tabulated by Langmuir (1969) are summarized in Table 6. The major difference between this table and those in earlier references is that it incorporates new values for Fe^{+2}(aq), Fe^{+3}(aq), and goethite. Latimer (1952) gave $\bar{G}_f^o$ (Fe^{+2}) = -20.3 kcal mole^{-1} and $\bar{G}_f^o$ (Fe^{+3}) = -2.53 kcal mole^{-1}. Larson *et al.* (1968) redetermined $\bar{G}_f^o$ (Fe^{+2}) = -21.8 kcal mole^{-1}. Combining this value with Latimer's (1952) value of E^o = -0.771 V for $Fe^{+2} = Fe^{+3} + e^-$ results in $\bar{G}_f^o$ (Fe^{+3}) = -4.02 kcal mole^{-1}. Incorporation of revised values for $\bar{G}_f^o$ (Fe^{+2}) and $\bar{G}_f^o$ (Fe^{+3}) results in slightly different values for the solid phases which, in turn, are based on solubility measurements. In most cases the changes are slight relative to earlier reported values.

A value for $\bar{G}_f^o$ for β-FeOOH (akaganéite) has not been previously reported. The value in Table 6 was estimated from a solubility study of ferric ion in 0.5 M sodium chloride (Biedermann and Chow, 1966). The solid phase had the composition $Fe(OH)_{2.7}Cl_{0.3}$(s) and was identified by x-ray diffraction and electron microscopy as akaganéite (Söderquist and Jansson, 1966). The constant given by Biedermann and Chow (1966) for the reaction

$$Fe^{+3} + 2.70\ H_2O + 0.30\ Cl^- = Fe(OH)_{2.70}Cl_{0.30}(s) + 2.7\ H^+$$

$$\log K = -3.05 \pm 0.10 \qquad [2]$$

was corrected for ionic strength and the free energy of formation for the solid phase was calculated to be -179.7 kcal $mole^{-1}$. The approximate nature of this calculation should be emphasized.

The values for $Fe(OH)_2^o(aq)$ and $Fe(OH)_3^o(aq)$ should be considered judiciously. Early experimental work supported the importance of $Fe(OH)_3^o$ (e.g., Gayer and Woontner, 1956); however, Lengweiler *et al.* (1961) concluded from ultracentrifugation and electron microscopy experiments that the species presumed to be molecular $Fe(OH)_3^o$ was actually colloidal-sized ferric oxyhydroxides. More recently Byrne and Kester (1976) have conducted extensive seawater solubility experiments for freshly precipitated hydrous ferric oxyhydroxide polymer. In some of these experiments the solubility was independent of pH from pH 7 to 9 at a concentration of 10^{-8} M which indicated the importance of $Fe(OH)_3^o(aq)$.

The free energy values in Table 6 can be used to calculate the relative stabilities of the different modifications of iron oxides. Diagrams can be drawn to demonstrate the relationships of the stabilities to various parameters. Iron exhibits two main oxidation states, Fe (II) and Fe (III), thus pE (= $-\log(e^-)$) or Eh are frequent parameters used. Numerous other parameters, such as pH, pS and P_{CO_2}, are also used depending on the relationships to be described. Diagrams of this type are only valid for the species considered. Garrels and Christ (1965) and Stumm and Morgan (1970) are excellent references for step-by-step instructions on how to construct such diagrams.

The literature is full of examples of stability diagrams, and two will be shown here to illustrate some basic features. An equilibrium Eh-pH diagram for iron (10^{-4} and 10^{-6} M) at 25°C and 1 atm is shown in Figure 11 (Garrels and Christ, 1965). Additional conditions are total dissolved sulfur = 10^{-6} M and total dissolved carbonate = 1 M. This diagram shows at a glance the effect of several variables. Under oxidizing conditions hematite (α-Fe_2O_3) is the most stable form of iron oxide, while under reducing conditions magnetite (Fe_3O_4) has a stability field in the alkaline pH range. Both siderite ($FeCO_3$) and pyrite (FeS_2) have stable domains in reducing solutions at the expense of magnetite. In spite of the extremely

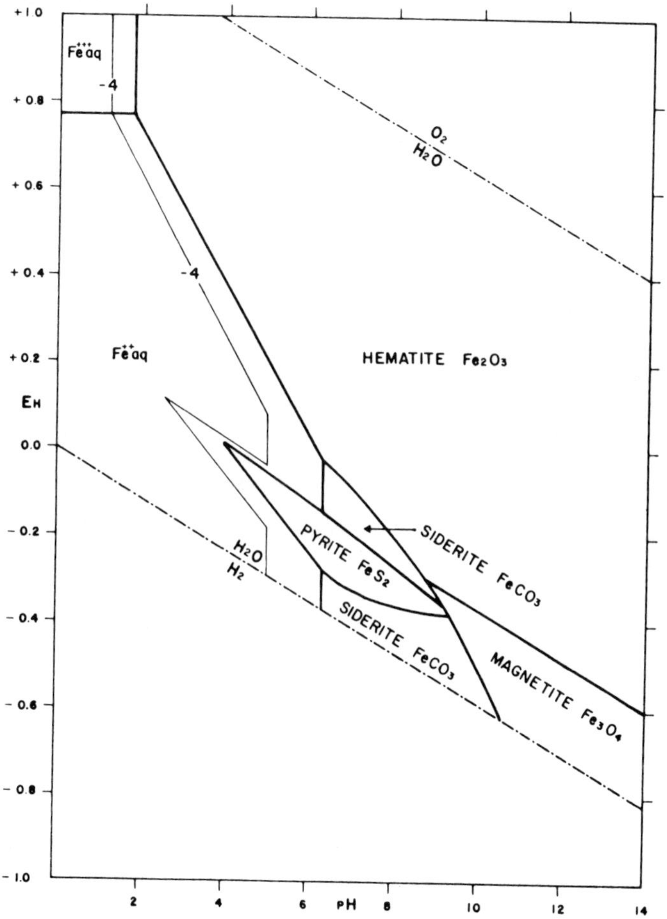

Figure 11. Stability relations of iron oxides, sulfides, and carbonate in water at 25°C and 1 atmosphere total pressure. Total dissolved sulfur = 10^{-6}. Total dissolved carbonate = 10°. Note elimination of FeS field by $FeCO_3$ under strongly reducing conditions, and remarkable stability of pyrite in presence of small amount dissolved sulfur (from Garrels and Christ, 1965).

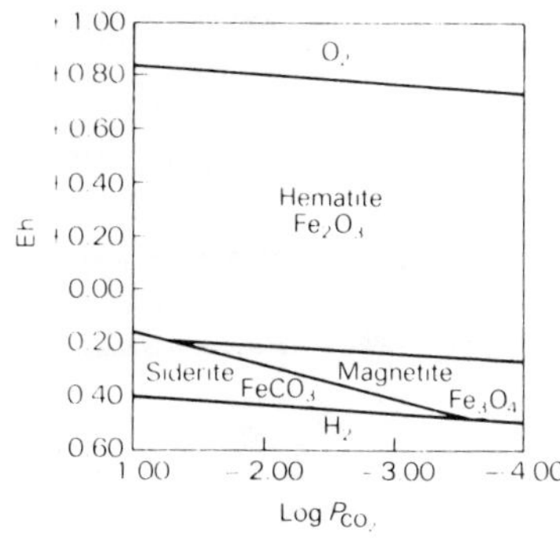

Figure 12. Eh-log P_{CO_2} diagram for hematite, magnetite, and siderite in marine sediments. T = 25°C, P_{total} = 1 atm, $a_{C3++} = 10^{-2.58}$ equilibrium with calcite assumed. O_2 and H_2 represent areas where water is thermodynamically unstable relative to the respective gases. The value for pS^{--} is assumed to be so high that pyrite and pyrrhotite do not plot stably (from Berner, 1971).

small amount of dissolved sulfur relative to carbon, pyrite still has a significant stability field.

Another approach is to assume that pH is controlled by the equilibrium $CaCO_3(s)-CO_2$ system and plot Eh *vs* P_{CO_2} (Fig. 12) (Berner, 1971). Again, we see that hematite is the stable oxide form under oxidizing conditions.

Both of these examples indicate that hematite is the thermodynamically stable phase in oxygenated seawater. There has been considerable discussion in the literature, however, about the relative stabilities of goethite and hematite, and the equilibrium is made complicated by the consideration of the effects of surface area and particle size.

The free energies of formation for the solid phases in Table 6 refer to macroscopic crystals. The change in free energy involved in subdividing a coarse solid into a finely divided one of molar surface area S is

$$\Delta G = 2/3\ \bar{\gamma}S \qquad [3]$$

where $\bar{\gamma}$ is the mean free surface energy (interfacial tension) of the solid-liquid interface (in erg cm^{-2}).

At constant T and P and for particles of uniform size (Enüstün and Turkevich, 1960; Stumm and Morgan, 1970)

$$dG = \mu_o dn + \bar{\gamma} ds \qquad [4]$$

or

$$\mu = \mu_o + \frac{\bar{\gamma} ds}{dn} \qquad [5]$$

or

$$\mu = \mu_o + \frac{M}{\rho}\bar{\gamma}\frac{ds}{dv} \qquad [6]$$

where M = formula weight, ρ = density, n = number of moles, and s = surface area of a single particle.

If the surface area and volume of a specific single crystal can be approximated by $s = kd^2$ and $\nu = \ell d^3$ where d is any characteristic dimension and k and ℓ are geometric constants, then:

$$\frac{ds}{dv} = \frac{2s}{3\nu} \qquad [7]$$

If there are N particles per mole, then the molar surface area is S = Ns and the molar volume $V = N\nu = M\rho^{-1}$.

Thus

$$\mu = \mu_o + 2/3\ \bar{\gamma}S \qquad [8]$$

The solubility product is then a function of surface area

$$\frac{d\ln K_{SO}}{dS} = \frac{2}{3}\frac{\bar{\gamma}}{RT} \qquad [9]$$

and

$$\log K_{SO}(s) = \log K_{SO}(S{=}0) + \frac{2/3\ \bar{\gamma}}{2.3\ RT}\ S \qquad [10]$$

The molar surface area S can also be expressed as

$$S = \frac{M\alpha}{\rho d} \qquad [11]$$

where $\alpha = k\ell^{-1}$ is a geometric factor which depends on the shape of the crystals. α equals six for cubes of edge length d and three for spheres of diameter d.

The effect of particle size is especially important for the iron oxyhydroxides. Goethite can range from as small as 50 A (van der Kraan and Medema, 1969) to 1000 A or greater (Feitknecht and Michaelis, 1962). Akaganéite particles are normally less than 500 A (e.g., Soderquist and Jansson, 1966; Ellis *et al.*, 1976). Mössbauer studies also suggest small particle sizes ($\leq$ 900 A) for goethite in manganese nodules (e.g., Carpenter and Wakeham, 1973).

Case Study: Goethite/Hematite Stability

Goethite is one of the forms of iron oxide most commonly identified in modern sediments and ferromanganese nodules. However, it is rare in ancient sedimentary rocks (Fischer, 1963) where hematite is a common constituent. Hematite often imparts a red coloration to sedimentary rocks giving rise to the term red beds. This suggests that goethite is the first form of hydrous iron oxide to be deposited, and that with time it ages into hematite.

Thermodynamic Calculations. There has been considerable argument about the relative stabilities of goethite and hematite as described by:

$$2\ \alpha\text{-FeOOH} = \alpha\text{-Fe}_2O_3 + H_2O \qquad [12]$$

This reaction has received a considerable amount of attention by geochemists interested in planetary atmospheres. The mineral goethite (or limonite, which is a mixture of α-FeOOH and α-Fe_2O_3) closely matches the photometric and polarimetric properties of the surface of Mars (see Pollack *et al.*, 1970).

Adamcik (1963) proposed that the atmospheric water vapor on Mars could be controlled by this reaction. Pollack *et al.* (1970) reviewed the thermodynamic data and concluded that goethite is apparently unstable at the surface of Mars during the daytime but may be stable deeper in the soil where the temperature should be more constant. Their calculations suggest that the dehydration-rehydration cycle may be balanced and, if so, reaction 12 could buffer the time averaged content of Martian atmospheric water vapor. Fish (1966), O'Connor (1968) and Berner (1969), however, have used some of the same thermodynamic data to show that goethite is probably unstable on Mars.

Bischoff (1969) undertook similar calculations in order to explain the distribution of goethite and hematite in the sediments in and around the Red Sea hot brine deposits. He concluded that goethite is the stable phase under normal submarine environments (i.e., the majority of oceanic sediments) The stability field of goethite expands with increasing pressure and decreasing temperature (below the equilibrium temperature of about 130°C; Posjnack and Merwin, 1922; Smith and Kidd, 1949). In one of the cores from the Atlantis II deep in the Red Sea deposits, hematite was predominant, suggesting the proximity of high temperatures. In yet another assessment relative to marine sediments Berner (1969) calculated a maximum value of $\Delta G^o = -0.15$ kcal mole^{-1} for reaction 12 at 25°C and 1 atm.

Most of these early attempts at thermodynamic calculations were fruitless exercises, however, because they did not have access to the measured heat capacity and entropy data for goethite published by King and Weller (1970).

Experimental Studies. The experimental approach has not yielded clearer results. Smith and Kidd (1949) reviewed much of the earlier work (especially Posnjak and Merwin, 1922) and conducted new experiments which showed that goethite decomposes to hematite in neutral solutions above 125 $\pm$ 15°C and in alkaline solutions (0.1 N base) above 165 $\pm$ 5°C at the vapor pressures of water at those temperatures. The effect of pressure on the transition temperature was small (less than 5°C per 1000 atmospheres). Unfortunately, they used color rather than x-ray diffraction techniques to identify the solid phases in most of their experiments. Though they stated that the reaction could be reversed and hematite converted into goethite (x-ray identification) after two years at room temperature, the result was not well documented. Berner (1969) criticized Smith and Kidd (1949) for reasons I have just stated and undertook his own experiments. Berner's approach was to measure the solubility of goethite and hematite in strong acid (0.01 M HCl) according to:

$$3H^{+} + FeOOH(s) = Fe^{+3} + 2H_2O \quad [13]$$

$$3H^{+} + 1/2Fe_2O_3 = Fe^{+3} + 3/2H_2O \quad [14]$$

where the equilibrium constants for both reactions are:

$$K = \frac{(Fe^{+3})}{(H^{+})^3} \quad [15]$$

When the two reactions are examined at the same pH

$$\frac{K_{FeOOH}}{K_{Fe_2O_3}} = \frac{(Fe^{+3})_{FeOOH}}{(Fe^{+3})_{Fe_2O_3}} \quad [16]$$

and thus ΔG_r^o for reaction 12 equals ΔG^o (reaction 13) - ΔG^o (reaction 14) or:

$$\Delta G^o = RT \ln \frac{(Fe^{+3})_{FeOOH}}{(Fe^{+3})_{Fe_2O_3}} \quad [17]$$

The reactions (in 0.01 M HCl) were very slow at 25°C and equilibrium was not reached after 200 days. Thus the experiments were conducted at 85°C. The solubilities of fine-grained goethite and hematite, starting from undersaturation, were followed with time. In addition, experiments were run in which ferric chloride was added to hematite suspensions to examine the approach to equilibrium starting from supersaturation. The results of Berner's (1969) experiments are shown in Figure 13. The dissolution results show that fine-grained goethite is more soluble than hematite at 85°C. The precipitation and dissolution curves for hematite suggest that there are kinetic problems that prevent reaching a truly reversible equilibrium. Apparently, hematite crystallization is very slow.

Assuming that the steady state iron concentrations from the dissolution experiments represent equilibrium solubilities, the approximate free energy change for reaction 12 at 85°C is ≤ -1.20 kcal mole^{-1}. The free energy could be even more negative because goethite solubility was approached only from undersaturation. This is equivalent to $\leq -0.33 \pm 0.08$ kcal mole^{-1}

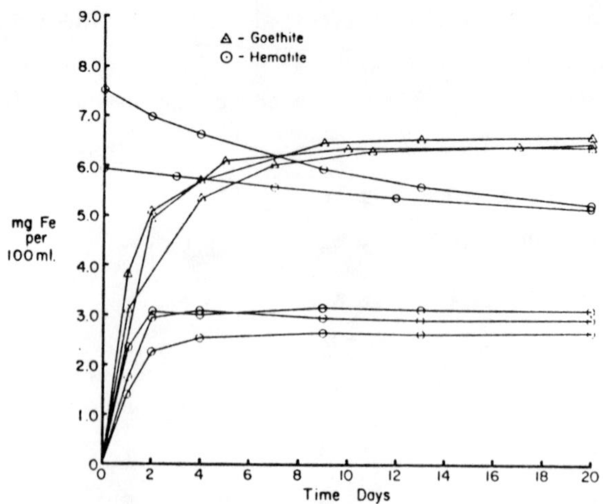

Figure 13. Dissolution and precipitation curves for goethite and hematite in 0.12 M HCl at 85 ± 1°C. For two runs ferrichloride was added to the original solution in order to study the approach to equilibrium with hematite from supersaturation (from Berner, 1969).

at 25°C and agrees well with Berner's (1969) calculated value of -0.15 kcal $mole^{-1}$ at 25°C. Thus Berner concluded that fine-grained goethite (and the other polymorphs of FeOOH as well) is unstable relative to hematite plus water over virtually all geological conditions.

Effect of Particle Size. A lot of the controversy about the goethite-hematite stability may have been solved by Langmuir (1971) who included new basic thermodynamic data for goethite as well as including the effect of particle size on reaction 12. Ferrier (1966) measured the heats of solution of goethite and hematite in HCl as a function of surface area at 70°C. From his study:

$$\Delta H^{o}_{343}\ (\text{cal mole}^{-1}) = 3440 \pm 500 - \frac{72.3}{x} + \frac{34.2}{y} \qquad [18]$$

where x and y are the edge length in microns of hypothetical cubes of goethite and hematite, respectively. These results are equivalent to surface enthalpies of hematite and goethite of 770 ergs cm^{-2} and 1250 erg cm^{-2} at 70°C. Using recently published values for the heat capacity and entropy of goethite, Langmuir calculated the following values for reaction 12 at 70°C.

$$\Delta C_{P343} = 6.48 \pm 0.2 \text{ cal deg}^{-1}\text{mole}^{-1} \qquad [19]$$

$$\Delta S^o_{343} = 9.70 \pm 0.4 \text{ cal deg}^{-1}\text{mole}^{-1} \qquad [20]$$

Thus at 70°C and 1 atmosphere total pressure

$$\Delta G^o_{343} = H^o_{343} - 343\ S^o_{343} \qquad [21]$$

or

$$\Delta G^o_{343} = 112 \pm 520 - \frac{72.3}{x} + \frac{34.2}{y} \qquad [22]$$

If we assume that the heat capacity is constant and that neither heat capacity nor entropy is a function of particle size, then the free energy for temperatures near 343°K and 1 atm can be calculated from:

$$\Delta G^o_T = \Delta G^o_{343} + \Delta H_{343}(1 - \frac{T}{343}) - \Delta C_{P,343}(343 - T + 2.30\ T \log \frac{T}{343}) \qquad [23]$$

Including particle size effects gives:

$$\Delta G^o_T(\text{cal mole}^{-1}) = 1328 - 3.54\ T - 14.92 \log \frac{T}{343}$$
$$= \frac{1}{x}\ (144.6 - 0.211\ T) + \frac{1}{y}\ (68.6 - 0.0997\ T) \qquad [24]$$

At 25°C

$$\Delta G^o_{298} = 545 \pm 600 - \frac{81.7}{x} + \frac{38.9}{y} \qquad [25]$$

The free energy of the goethite-hematite transition is a function of particle size. This equation is plotted in Figure 14 for three different cases.

For particles of goethite and hematite of equal size (curve II) and larger than 1 μm, the surface effects are negligible and $\Delta G^o_{298} = 545 \pm 600$ kcal mole^{-1}. For particles of goethite and hematite progressively smaller than 1 μm, goethite is favored for particles greater than 0.08 μm while hematite is favored for particles smaller than 0.08 μm. When the hematite is larger than 1 μm and only goethite varies in particle size, goethite is more stable than hematite down to goethite particle sizes of about 0.15 μm (curve III). Curve I represents the case for goethite particles greater than 1 μm and a variable hematite particle size. Goethite is favored for all sizes of hematite for this case. The area between curves II and III is probably most relevant for geochemical problems because goethite particles are usually smaller than those of hematite when they are both present. Furthermore, hematite crystals tend to be equant while goethite is needle-like which further enhances the particle size effect on goethite.

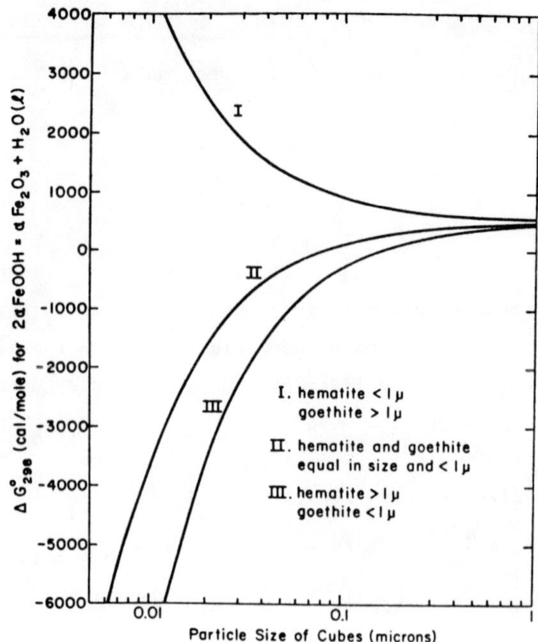

Figure 14. Effect of particle size on the Gibbs free energy of the goethite → hematite + water reaction. The special cases are shown (from Langmuir, 1971).

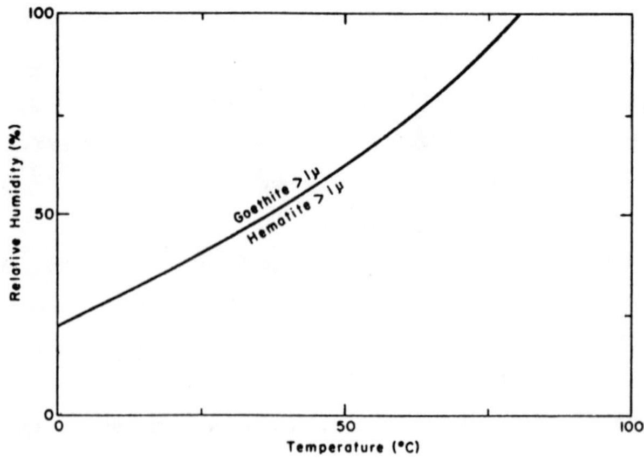

Figure 15. Relative humidity for equilibrium between coarse-grained goethite and coarse-grained hematite from 0° to 100°C at 1 bar total pressure (from Langmuir, 1971).

The free energy for reaction 12 from 0° to 100°C at 1 atmosphere can also be expressed as:

$$\Delta G^o_T = -RT \ \ell n \frac{f_{H_2O}}{f^o_{H_2O}} \qquad [26]$$

where f_{H_2O} is the fugacity of water at goethite-hematite equilibrium and $f^o_{H_2O}$ is the fugacity of water on the liquid-vapor curve. The goethite-hematite stability as a function of temperature is shown in Figure 15 where percent relative humidity is the fugacity ratio times 100. This curve shows that coarse-grained goethite is stable relative to coarse-grained hematite in liquid water (100% relative humidity) up to 80°C. At 25°C goethite is favored for relative humidity values greater than 40 percent. These results are consistent with Bischoff's (1969) conclusion that the presence of hematite in some of the sediments in the Red Sea deposits requires a local source of high temperature.

Langmuir's (1971) calculated results for goethite-hematite equilibrium as a function of temperature and pressure are shown in Figure 16. The curve

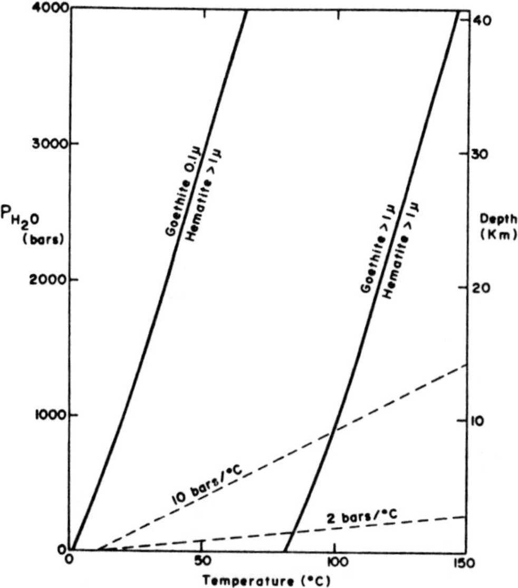

Figure 16. P_{H_2O}-T curves for equilibrium between coarse-grained goethite and coarse-grained hematite and between 0.1 μm goethite cubes and coarse-grained hematite (from Langmuir, 1971).

for fine-grained goethite is about 80° lower than the curve for coarse-grained goethite. Clearly, fine-grained goethite is unstable relative to coarse-grained hematite for the ranges of temperature and pressure applicable to marine geochemistry.

Earlier we reviewed Berner's (1969) experimental study that suggested hematite is stable relative to goethite (Fig. 13). Those results correspond to $\Delta G^o_{358} \leq -1200$ cal mole^{-1} for reaction 12. Substituting this free energy value into equation 24 and assuming the hematite is coarse grained leads to an estimated particle size of 0.062 µm for hypothetical cubes of goethite. Electron micrographs of the goethite actually used in Berner's experiments revealed lath-like crystals averaging 0.050 µm in width and 0.284 µm in length. The dimensions are equivalent in area to hypothetical goethite cubes 0.030 µm on a side. While the agreement is not perfect, it suggests that the conclusions of Langmuir and Berner are not incompatible. Possible explanations for the disagreement between calculated and observed particle sizes include failure to include the effect of surface area on the entropy of reaction and the effects of crystal corner and edge contributions to dissolution.

Kinetics. Kinetics play an important role in the goethite-hematite transition. In principle, coarse-grained hematite can crystallize from seawater dissolving fine-grained goethite or by dehydration of fine-grained goethite; however, the rates of conversion of goethite to hematite are extremely slow at low temperatures and pressures. Hematite appears in laboratory studies only after months to years of aging at room temperature. In addition, there is no convincing evidence that the reverse reaction occurs. Coarse-grained goethite is stable relative to hematite; however, hematite persists in sediments of Cambrian age in the presence of water.

To further complicate the picture are laboratory studies that suggest that goethite and hematite are not interrelated genetically by the simple hydration-dehydration reaction (reaction 12) but are formed by separate pathways from a common source (Feitknecht and Michaelis, 1962; Fischer and Schwertmann, 1975). Environmental conditions play a decisive role in determining which of the two forms are produced.

Fischer and Schwertmann (1975) formed ferrihydrite from $Fe(NO_3)_3$. The initial solid consisted of fine particles with an average diameter of 50 to 100 A. Ferrihydrite has the same oxygen arrangement as hematite with

with partial replacement of O by OH and H_2O and with random distribution and efficiency of Fe atoms.

Hematite is formed by "internal dehydration" of aggregates of these amorphous particles. Thus aggregation of the small ferrihydrite particles followed by nucleation and crystallization of hematite appear to be the essential steps of the process. Conditions which enhance aggregation of the ferrihydrite particles favor hematite formation. These are the normal conditions that favor rapid kinetics of coagulation of particles: higher temperature, increasing suspension concentration, precipitation and aging of the hydroxide in the neighborhood of its isoelectric point (possibly pH 7 to 8). As the aggregates form they gradually assume the hexagonal shape of hematite.

Goethite, on the other hand, occurs when the kinetics of aggregation of the ferrihydrite particles are slow. Goethite grows from the dissolving ferrihydrite particles. This dissolution and reprecipitation pathway was called reconstructive transformation by Mackay (1960). The conditions that would favor goethite formation include low temperature, low suspension concentration and pH values away from the isoelectron point for ferrihydrite. This is supported by Schwertmann and Fischer's (1973) observation that goethite was formed by aging ferrihydrite in 1 M KOH.

Implications. Thermodynamic calculations indicate that coarse-grained (>1 μ) goethite is stable relative to coarse-grained hematite in water up to about 80°C at one atmosphere pressure. However, because goethite in marine sediments is fine grained, it is probably unstable relative to coarse-grained hematite for most geological conditions. Particle size effects represent a major control on the relative stability. For kinetic reasons the conversion of goethite to hematite is slow so that goethite is more common than hematite in most recent low-temperature geochemical deposits. Hematite, once formed, does not appear to rehydrate to form coarse-grained goethite.

PATHWAYS OF FORMATION

During the past few years the slow hydrolysis and precipitation of iron has received considerable attention in laboratory studies. While these experiments are frequently conducted under conditions not observed in nature, the results are highly instructive for understanding the pathways of formation of the different iron oxides.

Hydrolysis of Iron (III) Solutions

The slow hydolysis of iron (III) basically follows the same pathway in all solutions except those containing Cl^-. In all cases (except Cl^-) the end product is goethite (α-FeOOH). When Cl^- is present akaganéite (β-FeOOH) is formed.

NO_3^- and ClO_3^- Solutions. When less than stoichiometric amounts of base are added to ferric iron solutions of NO_3^-, Cl^- or ClO_4^-, the first entities to form are spherical polycations of 15-30 A diameter (Murphy *et al.*, 1976a,b,c). These polycations have the same structure regardless of the solution and appear to be similar to the material called ferrihydrite by Chukhrov *et al.* (1973). These polycations grow and age into easily recognizable α-FeOOH or β-FeOOH. The divergency in the aging behavior is due primarily to the presence or absence of Cl^-.

In NO_3^- (Murphy *et al.*, 1976a) and ClO_4^- (Murphy *et al.*, 1976c) solutions the spherical polycations form over a wide range of OH-added/Fe-added ratios and in all cases form a goethite precipitate. The polycations were observed after four hours. As the solutions aged, rods appeared that were composed of two to five linked spheres. The proportion of isolation polycation spheres decreased as more of them are tied up into rods. With time these rods form rafts; however, the rods remain the same length. The spheres comprising the rods tend to become less distinct with age as some sort of structural rearrangement occurs in the rods. The individual rods and spheres gave no electron diffraction pattern, but in the rafts where coalescence had occurred goethite was identified. There is a clear progression from spherical polycations to rods to rafts in these solutions. These changes take place as the polycation charge decreases due to the slow hydrolysis of water. Under some conditions, laths of lepidocrocite (γ-FeOOH) may form during the early stages of hydrolysis of NO_3^- and ClO_4^- solutions. These laths appear to form during the early stages directly from unpolymerized ferric species ($Fe(OH)^{+2}$ and $Fe_2(OH)_2^{+4}$). Once the polycations dominate the system no more lepidocrocite forms and goethite forms from the polycations. The lepidocrocite formed appears to dissolve and contribute iron to the polycations.

Dousma and De Bruyn (1976) studied Fe (III) hydrolysis in NO_3^- solutions by a base titration approach and also found α-FeOOH to form at 24° and 55°C. All precipitations at 90°C yielded hematite (α-Fe_2O_3) as the end product. Again, small polymeric species served as intermediates (ferrihydrite?). They proposed a hydrolysis-precipitation process which incorporates olation (OH^- addition) as well as oxolation as represented below.

$$Fe(OH)^{2+} \rightleftarrows Fe\langle{}^{OH}_{OH}\rangle Fe \overset{OH^-}{\rightleftarrows} \left[Fe\langle{}^{OH}_{OH}\rangle Fe\langle{}^{OH}_{OH}\rangle Fe\langle{}^{OH}_{OH}\rangle Fe\right]^{6+}$$

$$\updownarrow OH^-$$

$$\left[Fe\langle{}^{OH}_{O}\rangle Fe\langle{}^{OH}_{O}\right]_{n/2} \rightleftarrows \left[Fe\langle{}^{OH}_{OH}\rangle Fe\langle{}^{OH}_{OH}\right]^{n+}_{n/2}$$

Matizevic and Scheiner (1978) have also found that α-Fe_2O_3 and α-FeOOH result from hydrolysis of $Fe(NO_3)_3$ and $Fe(ClO_4)_3$ solutions and aging at 100°C.

Cl^- Solutions. As for NO_3^- and ClO_4^- solutions the initial products of hydrolysis of Fe (III) in Cl^- solutions were 15 to 30 A spherical polycations. These spheres then link together to form short rods two to four units long. These rods continued to grow in length (to 200 A) and thickness (to 300 A) apparently by addition of unpolymerized ferric species rather than by addition of spheres. Eventually raft-like arrays of these rods formed and were identified as β-FeOOH (Murphy *et al.*, 1976b). Some of the smaller polycations dissolve in order to provide a source of unpolymerized ferric species. The Cl^- ion must inhibit the addition of spheres to the rods while allowing growth using the unpolymerized ferric species. Cl^- can even change the course of the hydrolysis of polycations made in NO_3^- solutions if it is added before one day has elapsed (Murphy *et al.*, 1976d).

Many authors have verified that β-FeOOH is the product of hydrolysis in Cl^- and F^- solutions (see earlier discussion of the structure). We have also found that β-FeOOH is the solid phase that forms when Fe (III) is precipitated in seawater at 25°C (Murray, 1978). Matizevic and Scheiner (1978) also found β-FeOOH to form at 100°C; however, with extended aging at that temperature it converted into hematite. A schematic diagram of the two different hydrolysis pathways of Fe (III) in different anionic solutions is shown below:

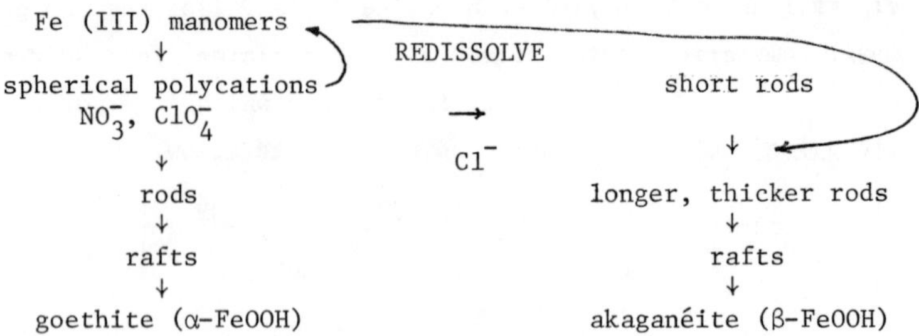

One question that is not clear at this time is the relationship of the spherical polycations of Murphy *et al.* (1976d) to the ferrihydrite crystals of Towe and Bradley (1967) and Chukhrov *et al.* (1973). The size and shape appear to be similar, yet Murphy *et al.* (1976d) could obtain no indication of structure in the polycations by electron diffraction. Moreover, the pathway of conversion of ferrihydrite to goethite is thought to occur by dissolution and reprecipitation (Fischer and Schwertmann, 1975) rather than the rods to rafts mechanism described above. It appears that goethite can form from either ferrihydrite or from spherical polycations by different pathways.

Oxidation of Fe (II) Solutions

Whereas goethite (α-FeOOH), akaganéite (β-FeOOH) and hematite (α-Fe_2O_3) are the forms of iron oxide hydroxides that result from hydrolysis and precipitation of Fe (III) from aqueous solution, a wide range of different compounds are formed from the oxidation of Fe (II) solutions. A summary of these pathways is shown schematically in Figure 17.

Ferrous iron will exist in solution as Fe^{+2}, $FeOH^+$ (colorless solutions) or precipitate as $Fe(OH)_2$ (amakinite) (white) depending on the pH and Fe (II) concentration. The solubility of $Fe(OH)_2(s)$ is $pK_{SO} = 14.7$ moles$^3 \ell^{-3}$ (Stumm and Morgan, 1970). At pH 8 the equilibrium iron concentration is $Fe^{+2} = 10^{-2.7}$M. In seawater or interstitial waters this concentration is never reached because saturation with $FeCO_3(s)$ is always reached at a lower concentration of Fe^{+2}. For example, if $[CO_3^=] = 10^{-4}$M, the Fe^{+2} in equilibrium with $FeCO_3(s)$ ($pK_{SO} = 10.4$) is $Fe^{+2} = 10^{-6.4}$.

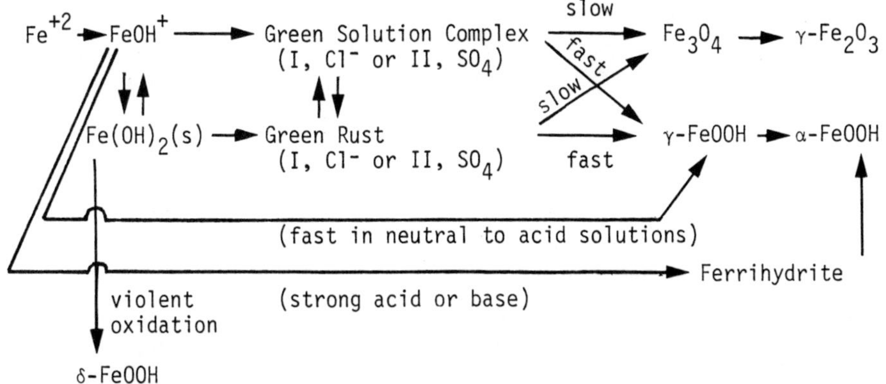

Figure 17. Schematic pathways for the oxidation of Fe(II).

As Fe (II) is oxidized in neutral or slightly basic solutions it passes through green intermediate products. Either green solution complexes or solid green rusts may form depending on the conditions. Further oxidation at pH values around neutrality is governed by their existence. The oxidation of Fe (II) to Fe (III) by dissolved O_2 in neutral and slightly acidic solutions follows the rate law:

$$\frac{-d\ Fe(II)}{dt} = k\ Fe(II)(O_2)(OH^-)^2$$

(Stumm and Lee, 1960; Singer and Stumm, 1970; Tamura *et al.*, 1976; Kester *et al.*, 1975; Murray and Gill, 1978). Thus the relative amounts of O_2, Fe (II) and OH^- will influence the rate of oxidation and consequently the end product that forms.

The green solution complexes are composed of both Fe (II) and Fe (III) held together by ol- and oxo-bridges formed during the consumption of OH^-. Green complex I forms from Cl^- solutions while green complex II forms from $SO_4^=$ solutions. The Fe(II)/Fe(III) ratio is 2:1 in green complex I (Kiyama, 1969), while in green complex II it is 1:1 (Misawa *et al.*, 1973). Accordingly, the formulas are estimated to be:

$$\text{G.C. I:}\quad [Fe(II)_2Fe(III)\ O_x(OH)_y]^{(7-2x-y)+}$$

$$\text{G.C. II:}\quad [Fe(II)_1Fe(III)_1O_x(OH)_y]^{(5-2x-y)+}$$

Misawa *et al.* (1973) have also observed that in ClO_4^- solutions the Fe(II)/

Fe(III) ratios in the respective green rusts is the same as the green complexes (Misawa *et al.*, 1974). Schwertmann and Thalmann (1976) describe green rust as a greenish-blue crystalline compound of alternating (Fe^{+2}, Fe^{+3})$_3$(OH,O)$_8$ and Fe^{+3}(O_2, anion$^-$)$_2$ layers (Feitknecht, 1959). According to this formula the Fe(II)/Fe(III) ratio may vary from 3.6:1 to 0.8:1 without changing the principal structure (Feitknecht and Keller, 1950). The included anions are integral parts of the structure. The green rusts are stable at a much lower pH than $Fe(OH)_2$(s). Bernal *et al.* (1959) use the formula $Fe(OH)_8$ or $Fe_3O_4 \cdot 4H_2O$ for green rust and estimate the solubility product as:

$$K_{SO} = (Fe^{+2})(Fe^{+3})^2(OH^-)^8 < 10^{-86}$$

Though a specific Fe(II)/Fe(III) ratio is indicated, Bernal *et al.* (1959) did not actually make this measurement. Both green rust I and II have predominantly a hexagonal structure (Table 1).

Lepidocrocite (γ-FeOOH) or magnetite (Fe_3O_4) form by the further oxidation of either green solution complexes or green rusts. The pathway depends on the initial concentration of ferrous iron. At low Fe (II) concentrations the green complex pathway is favored.

Magnetite (Fe_3O_4) can form by the slow oxidation of dissolved Fe (II) in neutral and slightly basic solutions. Misawa *et al.* (1974) argue that the green rusts and solution complexes must be intermediates in this process. It would be difficult for $Fe(OH)_2$(s) which has a hexagonal close-packed oxygen structure to convert directly into Fe_3O_4 which has a cubic close-packed structure. Consequently, Misawa *et al.* (1974) suggest that the green rusts have both hexagonal and cubic layers of close-packed oxygen. For Fe_3O_4 (inverse spinel) to form from $Fe(OH)_2$, two of three $Fe(OH)_2$ molecules must be oxidized as well as dehydration, deprotonation and reorganization. Hence, Fe_3O_4 is only possible during slow oxidation.

The magnetite that forms by slow oxidation of Fe (II) eventually may convert into maghemite (γ-Fe_2O_3) (Taylor and Schwertman, 1974a,b). It has been suggested that the formation of γ-Fe_2O_3 by slow oxidation of Fe (II) (through the intermediates green rust and Fe_3O_4) is the major source of maghemite in Australian soils (Taylor and Schwertmann, 1974a). Taylor and Schwertmann (1974b) were able to make synthetic maghemite under conditions similar to those expected in soils. The Fe (II) content of the synthetic maghemites varied from 2-7% of the total iron. This is on the low side of the possible range of the maghemite-magnetite series (0-33.3%). γ-Fe_2O_3 formation was favored by lower oxidation rates, higher total Fe concentrations, higher temperature and higher pH (7 *vs* 6). Maghemite will not form if the pH is less than 6 (Schwertmann and Thalmann, 1976).

Lepidocrocite (γ-FeOOH) formation appears restricted to the cases of rapid oxidation of Fe (II), either through green rusts or green complexes as intermediates. Because of the rapid oxidation and incomplete deprotonation of green rusts γ-FeOOH has a disordered cubic structure of oxygen and hydroxyl groups. Another view is that on fast oxidation the Fe (III) proportion in green rust will increase rapidly leading to complete oxidation before a dehydrated cubic phase can be formed (Taylor and Schwertmann, 1974b). In spite of this general knowledge, a large number of details on the mode and conditions of lepidocrocite formation are still obscure. In a qualitative sense the oxidation rate of Fe (II) depends on the O_2 supply in relation to the amount of green rust formed in the system. For a low O_2/green rust ratio maghemite is formed. As the ratio increases lepidocrocite is the only oxide to be formed. At still lower ratios (especially at low pH and low Fe (II)) ferrihydrite forms in place of lepidocrocite.

As shown in the schematic pathway diagram (Fig. 17), γ-FeOOH can also form directly during oxidation of Fe (II) solutions without the presence of the intermediate green compounds (Misawa *et al.*, 1974). This polymerization process may be as follows:

$$FeOH^+ \rightarrow Fe(OH)_2^+ \rightarrow \begin{matrix} HO \\ HO \end{matrix}\!\!>\!Fe\text{-}O\text{-}Fe\text{-}OH^+ \xrightarrow{\text{precipitation}} \gamma\text{-}FeOOH$$

Although lepidocrocite may occasionally form from pure Fe (III) systems (e.g., Murphy *et al.*, 1976a), it is generally agreed that its formation in most geochemical environments requires Fe (II). Lepidocrocite is less stable than its polymorph goethite. Laboratory studies indicate that this transformation is through the solution phase rather than topotactic (Schwertmann and Taylor, 1972a,b). The rate-controlling step can be either (1) the dissolution of lepidocrocite or (2) the formation of goethite nuclei and subsequent growth. The kinetics are faster at higher temperatures and in more alkaline solutions. In strong acid or base oxidation of Fe (II) can lead to the formation of small spherical (40 A) ferrihydrite particles which link together in linear arrays which age into goethite (Murphy *et al.*, 1976a; Schwertmann and Thalmann, 1976).

The final pathway to be discussed from Figure 17 is the violent oxidation of white $Fe(OH)_2(s)$ and green complexes and green rusts by H_2O_2 or O_2 in strongly alkaline solutions to form brown, ferrimagnetic δ-FeOOH (Misawa *et al.*, 1974). The oxidation is thought to be a topotactic or solid-solid transformation between two solid phases of hexagonal close-packed structures of oxygen.

$$Fe(OH)_2 + 1/2O_2 \rightarrow \delta\text{-}FeOOH + 1/2H_2O$$

$Fe(OH)_2$ has a CdI_2-type structure while δ-FeOOH has a more disordered CdI_2 structure (Bernal *et al.*, 1959). δ-FeOOH has not been observed in nature. Chukhrov *et al.* (1976), however, reported finding a disordered form which they called δ'-FeOOH and for which they proposed the mineral name feroxyhyte. This proposal should be viewed with caution, because it is highly unlikely that the necessary precursor $Fe(OH)_2(s)$ would form in a natural environment.

From the preceding discussion certain trends emerge. Some minerals can serve rather confidently as indicator minerals for certain pathways or certain conditions:

Akaganéite (β-FeOOH) is formed only from Cl^-rich solutions. A simple prediction would be that hydrolysis of iron (II) in seawater should produce primarily akaganéite.

Lepidocrocite (γ-FeOOH) appears to be a reliable indicator of relatively rapid oxidation of Fe (II) solutions.

Magnetite (Fe_3O_4) and *maghemite* ($\gamma\text{-}Fe_2O_3$) result from the relatively slow oxidation of Fe (II) solutions.

δ-*FeOOH* would result from the violent oxidation of $Fe(OH)_2(s)$.

Goethite (α-FeOOH) is a poor indicator mineral as it is the end product of both hydrolysis and oxidation pathways. According to the results summarized here, goethite should be an indicator of the absence of Cl^- which inhibits its formation through the solution phase. Thus, the widely reported occurrence of goethite in marine sediments and ferromanganese nodules could be considered anomalous.

OCCURRENCES IN THE MARINE ENVIRONMENT

The literature reviewed herein has led to the hypothesis that the occurrence of some iron minerals in marine deposits may provide valuable clues to the pathways of formation and the geochemical environment. The strength and generality of these clues needs further testing by detailed studies of the field occurrence of specific minerals.

Most of the iron phases discussed have been suggested or positively identified in marine sediments or ferromanganese nodules. Identification is difficult. The iron oxide/oxyhydroxide minerals are fine grained and poorly crystalline so that attempts to identify them by conventional x-ray diffraction techniques have often been unsuccessful. Electron diffraction techniques have found increased use. Mössbauer spectroscopy is a frequent approach; however, because of the symmetry of the d^5 electron shell in the ferric ion, it is not always possible to get a unique spectrum from each of the different minerals. In addition, in small particles a significant fraction of the Fe^{+3} ions are at or near the surface of the "crystal" resulting in significant line broadening. Some investigators have attempted to interpret this in terms of more than one iron oxide phase. Table 7 summarizes some observations. For more review see Chester and Aston (1976), Elderfield (1976), Cronan (1976), Burns and Burns (1977; in press) and Glasby (1972).

There are some occurrences or lack thereof that require special comment.

(A) Akaganéite (β-FeOOH) is the form of iron that precipitates from Cl^- solutions such as seawater. Biedermann and Chow (1966) first made this observation at seawater concentrations of Cl^-, and we have verified the finding in our own experiments. Some confusion still exists because Riley and Chester (1971) incorrectly stated that the solid Biederman and Chow precipitate was goethite (α-FeOOH).

It is possible that the failure to identify β-FeOOH is because it rapidly converts to the more stable goethite under seawater conditions. We have found in our experiments, however, that β-FeOOH is stable for up to two years at pressures up to 1000 atmospheres. It may be that akaganéite is simply too small to be identified by classical x-ray diffraction techniques.

Another possibility is that in some cases it may have been misidentified as phillipsite. Many studies of iron-rich hydrothermal deposits

Table 7. Some reported occurrences of iron oxide minerals.

Mineral	Formula	Occurrence	Reference	Comments
ferrihydrite	$5\ Fe_2O_3 \cdot 9\ H_2O$	Freshwater springs in N. England	Coey & Readman (1973)	No X-ray diffraction pattern. Shapeless particles less than 100Å that convert to hematite in electron beam.
		Freshwater springs and Red Sea hydrothermal deposits	Chukhrov et al. (1973)	Identified in conditions favoring rapid hydrolysis or oxidation of iron. Frequently associated with iron bacteria (Gallionella, Leptathrix, Toxathrix)
		Red Sea hydrothermal deposits	Bischoff (1969a)	In association with goethite as 1-30μm spherical particles
		Marine nodules	Calvert (1978)	Postulated to occur
goethite	α-FeOOH	Marine Mn-Fe nodules and concretions	Buser & Grütter (1956) Grütter & Buser (1957)	Identified in residuals of acid leached nodules
			Johnson & Glasby (1969)	Mössbauer
			Manheim (1975) Glasby (1972)	
		Pacific sediments	Hein et al. (1976)	Reported as opaque fines in sediments of DOMES Site C
		Gulf of Mexico	Watson & Angino (1969)	Iron-rich layers in sediments give weak goethite X-ray patterns
		Nazca Plate sediments	Heath & Dymond (1977)	Traces of goethite and iron-rich smectite
		Marine gastropods	Lowenstam (1962)	X-ray identification in teeth of several species
		Red Sea hydrothermal brine deposits	Bischoff (1969a,b)	Extensive beds of orange to yellow goethite + amorphous material (ferrihydrite?)
akaganéite	β-FeOOH	New Zealand soils	Logan et al. (1976)	Proposed to be a major form of iron in soils. Evidence based on fitting multiple peaks to Mossbauer spectra
		Marine and fossil manganese nodules	Johnston & Glasby (1978)	Based on Mossbauer curve fitting. No X-ray evidence
		Manganese nodules	Goncharov et al. (1973)	Mössbauer identification
		Bauer Deep Sediments	Chukhrov et al. (1978)	
lepidocrocite	γ-FeOOH	Manganese nodules	Chukhrov et al. (1978) Goodell et al. (1971) Glasby (1972) Okada et al. (1972)	There have been many reports especially in association with other minerals
		Sediments of Indian Ocean	Harrison & Peterson (1965)	Reported as a minor constituent associated with goethite and maghemite in a core from 23°56'S, 73°53'E
		Red Sea hydrothermal deposits	Bischoff (1969a) Strangway et al. (1969)	Identified by X-ray diffraction in 2-62μm fraction
		Soils	Schwertmann & Thalmann (1976)	Occurs in hydromorphic soils from many regions. Appears to form by oxidation of Fe(II).

Table 7. (cont.)

Mineral	Formula	Occurrence	Reference	Comments
		Marine chiton	Lowenstam (1967)	In mature denticles of three species of marine chiton. Always found between apatite and magnetite layers. In tropical sediments may be 10 mg/M^2
feroxyhyte	δ'-FeOOH	Manganese nodules	Chukhrov et al.(1976)	Impossible to separate pure δ'FeOOH. Found as admixtures with clay minerals and/or goethite. Present as yellow-brown, 0.1-0.4 μ plates. Diffraction lines similar to layer silicates but converts to wüstite under electron beam
hematite	α-Fe_2O_3	Red Sea hydrothermal sediments	Bischoff (1969a,b) Strangway et al. (1969)	In two cores from Atlantis II deep. Proposed to form by dehydration of goethite
		N.E. Equatorial Pacific	Hein et al. (1976)	observed in DOMES site C
magnetite	Fe_3O_4	Manganese nodules (fresh water and sea water)	Carpenter et al. (1972)	Thermomagnetic properties suggest presence of magnetite and maghemite. Not more than 0.1-0.6%
		Pacific sediments	Kobayashi & Nomura (1974)	Observed in 10-50 μ size range in extracted ferromagnetic minerals. Angular shape suggests detrital origin
		Red Sea hydrothermal sediments	Hackett & Bischoff (1973)	Identified in one core from Atlantis II deep. In coexistence with hematite.
		Marine chitons	Lowenstam (1962)	Found on denticles of the radular teeth of marine chitons (Polyplacophora)
		Marine bacteria and fresh water	Blakemore (1975) Frankel et al.(1979)	Certain bacteria contain chains of magnetite. Bacteria may use magnetic field to find anoxic sediments.
		Marine sediments	Haggerty (1970)	Studied 110 samples from 22 pelagic cores. Observed oxides in magnetite-ulvospinel and hematite-ilmenite solid solution series. High T origin.
maghemite	γ-Fe_2O_3	Manganese nodules	Goodell et al.(1971)	Found in five nodules from South Pacific, Scotia Sea and Drake Passage
		Crusts from Jan Basin Pacific	Bertine (1974)	In dredge samples of tholeiitic basalts. Apparently from weathered magnetite.
		Sediments	Harrison & Peterson (1965)	In top of two cores from Indian Ocean. X-ray identification. Spacings suggest incompletely oxidized magnetite
		Soils (Britain)	Oades & Townsend (1963)	Maghemite present in most soils and formed by pedologic processes from action of organic matter.

identify phillipsite as a component based on x-ray diffraction patterns (e.g., Dymond *et al.*, 1973; Lyle *et al.*, 1977). In some cases this is supported by visual verification in the coarse size fractions. Consideration of Figure 18 demonstrates the problem. This is a differential x-ray diffractogram of iron-rich sediments from DSDP cores (Dymond *et al.*, 1973). It was obtained by subtracting the diffractogram after a reducing-acid soluble leach from a diffractogram taken before the leach. Thus it represents the diffractogram of the leached material. Peaks labeled G and P were identified as goethite and phillipsite, respectively. The solid lines drawn on the figure represent the major lines of akaganéite precipitated from seawater. A combination of goethite and akaganéite may be an acceptable alternative explanation.

(B) The Red Sea metalliferous brine deposits are an exciting location for mineral discoveries. Goethite, lepidocrocite, hematite, magnetite and maghemite have been observed (Bischoff, 1969a,b; Strangway *et al.*, 1969; Hackett and Bischoff, 1973). The pathway of genesis proposed by Bischoff and co-workers is that iron hydroxides (limonite) precipitate at the interface between the 56°C and 44°C brines. The precipitate settles to the bottom where it dehydrates to goethite. Further dehydration results in hematite. Magnetite may subsequently form from hematite according to:

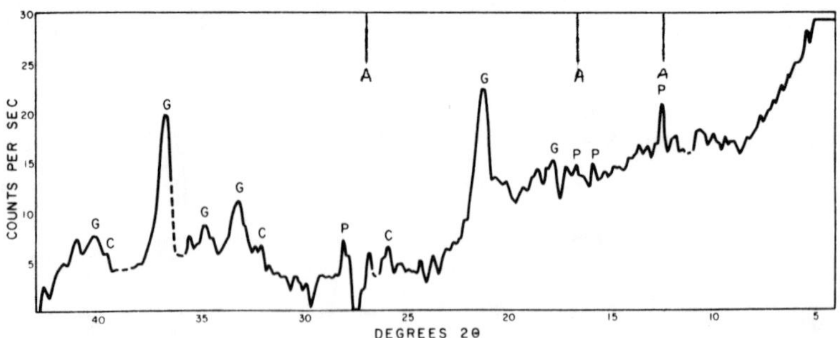

Figure 18. Differential x-ray diffractogram of the reducing-acid soluble fraction of DSDP sample 39-2-6 (from Dymond *et al.*, 1973). Dymond *et al.* (1973) concluded that phillipsite (P), carbonate fluorapatite (phosphorite) (C) and goethite (G) were present in this sample. The major lines for akaganéite precipitated from seawater are shown as (A). In this case phillipsite was confirmed by usual observation of the 2-62 μm size fraction, but the diffractogram demonstrates the difficulty that might be encountered in relying on x-ray data alone.

$$Fe^{+2} + Fe_2O_3 + H_2O = Fe_3O_4 + 2H^+$$

Thermodynamic arguments support each step in this reaction sequence.

An alternative that comes from these notes is that magnetite and lepidocrocite will result by varying the kinetics of oxidation of Fe^{+2} at the brine-seawater interface. These kinetics would be controlled by the concentration of Fe^{+2} and O_2 as well as pH of the brine and the rate of mixing at the interface.

(C) Geophysicists have long regarded that the magnetic signature in marine sediments is due to high-temperature detrital Fe-Ti oxides. It is well known that iron can be reduced to Fe^{+2} at some depth in some sediments (especially those more organic rich). The upward gradient in Fe (II) concentration produces a flux which in most cases is consumed by oxidation of Fe (II). The discussion presented here suggests that magnetite and maghemite form by the slow oxidation of Fe (II) and thus could form authigenically in marine sediments.

IRON OXIDES: REFERENCES

Adamcik, J.A. (1963) The water vapor content of the Martian atmosphere as a problem of chemical equilibrium. Planet. Space Sci. *11*, 355-359.

Atkinson, R.J., Posner, A.M., and Quirk, J.P. (1968) Crystal nucleation in Fe(III) solutions and hydroxide gels. J. Inorg. Nucl. Chem. *30*, 2371-2381.

Bernal, J.D., Dasgupta, D.A., and MacKay, M.L. (1959) The oxides and hydroxides of iron and their structural interrelationships. Clay Mineral. Bull. *4*,15-30.

Berner, R.A. (1969) Goethite stability and the origin of red beds. Geochim. Cosmochim. Acta *33*, 267-273.

________ (1971) *Principles of Chemical Sedimentology.* McGraw-Hill, New York, 240 pp.

Bertline, K.K. (1974) Origin of Lau Basin Rise sediments. Geochim. Cosmochim. Acta *38*, 629-640.

Biedermann, G. and Chow, T.J. (1966) Studies on the hydrolysis of metal ions. Part 57. The hydrolysis of the iron (II) ion and the solubility product of $Fe(OH)_{2.70}Cl_{0.3}$ in 0.5 M (Na^+) Cl^- medium. Acta Chem. Scand. *20*, 1376-1388.

Bischoff, J.L. (1969a) Red Sea geothermal brine deposits, their mineralogy, chemistry and genesis. *In*, E.T. Degens and D.A. Ross (eds.), *Hot Brines and Recent Heavy Metal Deposits in the Red Sea.* Springer-Verlag, New York, p. 368-401.

________ (1969b) Goethite-hematite stability relations with relevance to sea water and the Red Sea Brine system. *In*, E.T. Degens and D.A. Ross (eds.), *Hot Brines and Recent Heavy Metal Deposits in the Red Sea.* Springer-Verlag, New York, p. 402-406.

Blakemore, R. (1975) Magnetotactic bacteria. Science *190*, 377-379.

Burns, R.G. and Burns, V.M. (1977) Mineralogy of manganese nodules. *In*, G.P. Glasby (ed.), *Marine Manganese Deposits.* Elsevier, New York, p. 185-248.

________ and ________ (1981) Authigenic oxides. *In*, C. Emiliani (Ed.) *The Sea, Vol. 7, The Oceanic Lithosphere.* John Wiley, New York, p. 875-914

Büser, W. and Grütter, A. (1956) Über die Natur der manganknollen. Schweiz. Min. Petrogr. Mitt. *36*, 49-62.

Byrne, R.H. and Kester, D.P. (1976) Solubility of hydrous ferric oxide and iron speciation in seawater. Mar. Chem. *4*, 255-274.

Calvert, S.E. (1978) Geochemistry of oceanic ferromanganese deposits. Phil. Trans. Roy Soc. London *A290*, 43-73.

Carpenter, R., Johnson, H.P., and Twiss, E.S. (1972) Thermomagnetic behavior of manganese nodules. J. Geophys. Res. *77*, 7163-7174.

________ and S. Wakeham (1973) Mössbauer studies of marine and fresh water manganese nodules. Chem. Geol. *11*, 109-116.

Chester, R. and Aston, S.R. (1976) The geochemistry of deep-sea sediments. *In*, J.P. Riley and R. Chester (eds.), *Chemical Oceanography*, 2nd ed., vol. 6. Academic Press, New York, p. 281-390.

Chukhrov, F.V., Gorshkov, A.I., Zirjagin, B.B., and Yernulova, L.P. (1980) Iron oxides as minerals of sedimentary environments and chemogenic eluvium. *In*, I.M. Varentsov and Gy. Grasselly (Eds.) *Geology and Geochemistry of Manganese* (E. Schweizerbart'sche Verlagsbuchhandlung Publ., Stuttgart), Vol. 1, p. 231-258.

________, Zoyagin, B.B., Gorshkov, A.I., Yermilova, L.P., and Balashova, V.V. (1973) O ferrigidrite (ferrihydrite). AN SSSR Izvestiya, ser. geol. 23-33. Also (in English) Internat. Geol. Rev. *16*, 1131-1143.

________, ________, ________, ________, Korovushkin, V.V., Rudnitskaya, Ye.S., and Yakubovskava, N.Yu. (1976) Feroksigit-novaya modifikatsiya FeOOH (feroxyhyte, a new modification of FeOOH). AN SSSR Izvestiya, ser. geol. 5-24. Also (in English) Internat. Geol. Rev. *19*, 873-890.

Coey, J.M.D. and Readman, P.W. (1973) Characterization and magnetic properties of natural ferric gel. Earth Planet. Sci. Lett. *21*, 45-51.

Cronan, D.S. (1976) Manganese nodules and other ferro-manganese oxide deposits. *In*, J.P. Riley and R. Chester (eds.), *Chemical Oceanography*, 2nd ed., vol. 6. Academic Press, New York, p. 217-265.

Dousman, J. and DeBruyn, P.L. (1976) Hydrolysis-precipitation studies of iron solutions. I. Model for hydrolysis and precipitation from Fe(III) nitrate solutions. J. Colloid Int. Sci. *56*, 527-539.

Dymond, J., Corliss, J.B., Heath, G.R., Field, C.W., Dasch, E.J., and Veeh, H.H. (1973) Origin of metalliferous sediments from the Pacific Ocean. Geol. Sci. Am. Bull. *84*, 3355-3372.

Elderfield, H. (1976) Hydrogenous material in marine sediments: excluding manganese nodules. *In*, J.P. Riley and R. Chester (eds.), *Chemical Oceanography*, vol. 5, 2nd ed. Academic Press, New York, p. 137-216.

Ellis, R., R. Giovanoli and W. Stumm (1976) Anion exchange properties of β-FeOOH. Chimia *30*, 194-197.

Enüstün, B.V. and Turkevich, J. (1960) Solubility of fine particles of strontium sulfate. J. Am. Chem. Soc. *82*, 4502-4509.

Ewing, F.J. (1935) The crystal structure of diaspore. J. Chem. Phys. *3*, 203-207.

Fasiska, E.J. (1967) Structural aspects of the oxides and oxyhydroxides of iron. Corrosion Sci. 7, 833-839.

Feitknecht, W. (1959) Über die oxydation von festen hydroxyverbindungen des eisens in wässrigen losungen. Z. Electrochem. *63*, 34-43.

________ and Keller, G. (1950) Über die dunkelgrünen hydroxyverbindungen des eisens. Z. Anorg. Allg. Chem. *262*, 61-68.

________, Giovanoli, R., Michaelis, W., and Müller, M. (1973) Über die hydrolyse von eisen (III) salzlösungen. 1. Die hydrolyse der lösungen von eisen (III) chlorid. Helv. Chim. Acta *56*, 2847-2856.

________ and Michaelis, W. (1962) Über die hydrolyse von eisen (III)-perchlorat-lösungen. Helv. Chim. Acta *45*, 212-224.

Ferrier, A. (1966) Influence de l'état de division de la goethite et de l'oxyde ferrique sur leurs chaleurs de réaction. Rev. Chimie Minérale *3*, 587-615.

Fischer, A.G. (1963) Essay review of descriptive paleoclimatology. Am. J. Sci. *261*, 281-293.

Fischer, W.R. and Schwertmann, U. (1975) The formation of hematite from amorphous iron (II) hydroxide. Clays and Clay Minerals *23*, 33-37.

Fish, F.F., Jr. (1966) The stability of goethite on Mars. J. Geophys. Res. *71*, 3063-3068.

Frankel, R.B., Blakemore, R.P., and Wolfe, R.S. (1978) Magnetite in fresh water magnetotactic bacteria. Science *203*, 1355-1356.

Gallagher, K.J. (1970) The atomic structure of tubular subcrystals of β-iron (III) oxide hydroxide. Nature *226*, 1225-1228.

________ and Phillips, D.N. (1969) Hydrogen exchange studies and proton transfer in β iron (III) oxyhydroxide. Chimia *23*, 465-469.

Garrels, R.M. and Christ, C.L. (1965) *Solutions, Minerals and Equilibria*. Freeman, Cooper and Co., San Francisco, 450 pp.

Gayer, K.H. and Woontner, L. (1956) The solubility of ferrous hydroxide and ferric hydroxide in acidic and basic media at 25°C. Phys. Chem. J. *60*, 1569-1571.

Glasby, G.P. (1972) The nature of the iron oxide phase in marine manganese nodules. N. Z. Sci. *15*, 232-239.

Goncharov, G.N., Kalyamin, A.V., and Lure, B.G. (1973) Iron-manganese concretions from the Pacific Ocean studied by a nuclear γ-resonance method. Dokl. Akad. Nauk, SSSR, *212*, 720-723.

Goodell, H.G., Meylan, M.A. and Brant, G. (1971) Ferromanganese deposits of the South Pacific Ocean, Drake Passage and Scotia Sea. *In*, J.L. Reid (ed.), *Antarctic Oceanology, 1*, 27-92.

Grütter, A. and Buser, W. (1957) Untersuchangen an mangansedimenten. Chimia *11*, 132-133.

Hackett, J.P. and Bischoff, J.L. (1973) New data on the stratigraphy, extent and geologic history of the Red Sea geothermal deposits. Econ. Geol. *68*, 563-564.

Haggerty, S.E. (1970) Magnetic minerals in pelagic sediments. Carnegie Inst. Wash. Year Book *68*, 332-336.

Harrison, C.G.A. and Peterson, M.N.A. (1965) A magnetic mineral from the Indian Ocean. Am. Mineral. *50*, 704-712.

Heath, G.R. and Dymond, J. (1977) Genesis and transformation of metalliferous sediments from the East Pacific Rise, Bauer Deep, and Central Basin, Northwest Nazca Plate. Geol. Soc. Am. Bull. *88*, 723-733.

Hein, J.R., Gutmacher, C., and Miller, J. (1976) DOMES Area C: General statement about mineralogy, diagenesis and sediment classification. *In*, Open File Report 76-548, Deep Ocean Mining Environmental Study, N. E. Pacific Nodule Province, Site C, Geology and Geochemistry.

Hem, J.D. and Cropper, W.H. (1959) Survey of ferrous-ferric chemical equilibria and redox potentials. U.S. Geol. Surv. Water-Supply Paper *1459-A*, 31 pp.

Howe, A.T. and Gallagher, K.J. (1975) Mössbauer studies in the colloid system β-FeOOH-β-Fe_2O_3: structures and dehydration mechanism. J. Chem. Soc. Faraday Trans. I., *71*, 22-34.

Johnson, C.E. and Glasby (1969) Mössbauer effect determination of particle size in microcrystalline iron-manganese nodules. Nature *222*, 376-377.

Johnston, J.H. and Glasby, G.P. (1978) The secondary iron oxide hydroxide mineralogy of some deep-sea and fossil manganese nodules: a Mössbauer and x-ray study. Geochem. J. *12*, 153-164.

Kester, D.R., Byrne, R.H., and Liang, Y.J. (1975) Redox reactions and solution complexes of iron in marine systems. *In*, T.M. Church (ed.), *Marine Chemistry in the Coastal Environment*. ACS Symposium Series 18, Washington, DC, p. 56-79.

King, E.G. and Weller, W.W. (1970) Low-temperature heat capacities and entropies at 298.15°K of goethite and pyrophyllite. U.S. Bur. Mines Rept. Inv. *7369*, 6 pp.

Kiyama, M. (1969) Commentary experiments on the formation of Fe_3O_4 precipitate from aqueous solution. Bull. Inst. Chem. Res. Kyoto Univ. *47*, 607-612.

Kobayashi, K. and Nomura, M. (1974) Ferromagnetic minerals in the sediment cores collected from the Pacific Basin. J. Geophys. *40*, 501-512.

Kulgawczuk, D.S., Obuazko, Z., and Szytula, A. (1968) Susceptibility and magnetization of β- and δ-FeOOH. Phys. Stat. Sol. *26*,

Langmuir, D. (1969) The Gibbs free energies of substances in the system Fe-O_2-H_2O-CO_2 at 25°C. U.S. Geol. Surv. Prof. Paper *650-B*, 180-184.

________ (1971) Particle size effect on the reaction goethite = hematite + water. Am. J. Sci. *271*, 147-156.

Larson, J.W., Cerutti, P., Garber, H.K., and Hepler, L.G. (1968) Electrode potentials and thermodynamic data for aqueous ions, copper, zinc, cadmium, iron, cobalt and nickel. Phys. Chem. J. *72*, 2902-2907.

Latimer, W.M. (1952) *Oxidation Potentials*, 2nd ed. Prentice-Hall, New York.

Lengweiler, H., Buser, W., and Feitknecht, W. (1961) Die ermittlung der Löslichkeit von eisen (III)--hydroxiden mit ^{59}Fe. I. Fällungsund auflösungsversuche. Helv. Chim. Acta *44*, 796-805.

Lindsley, D.H. (1976) The crystal chemistry and structure of oxide minerals as exemplified by the Fe-Ti oxides. *In*, D. Rumble, III (ed.), *Oxide Minerals*, Reviews in Mineralogy 3, Chapter 1.

Logan, N.E., Johnston, J.H., and Childs, C.W. (1976) Mössbauer spectroscopic evidence for akaganéite (β-FeOOH) in New Zealand soils. Aust. J. Soil. Res. *14*, 217-224.

Lowenstam, H.A. (1962) Goethite in radular teeth of recent marine gastropods. Science *137*, 279-280.

________ (1967) Lepidocrocite, an apatite mineral and magnetite in teeth of chitons (Polyplacophora). Science *156*, 1373-1375.

Lyle, M., Dymond, J. and Heath, G.R. (1977) Copper-nickel-enriched ferromanganese nodules and associated crusts from the Bauer Basin, Northwest Nazca Plate. Earth Planet. Sci. Lett. *35*, 55-64.

MacKay, A.L. (1960) β-ferric oxyhydroxide. Mineral. Mag. *32*, 545-557.

Manheim, F.T. (1965) Manganese-iron accumulations in the shallow marine environment. *In*, D.R. Schink and J.T. Corliss (eds.), *Symposium on Marine Chemistry*. Occas. Publ. Univ. Rhode Island, *3*, 217-276.

Matizevic, E. and Scheiner, P. (1978) Ferric hydrous oxide sols. III. Preparation of uniform particles by hydrolysis of Fe(III)-chloride, -nitrate, and -perchlorate solutions. J. Colloid Int. Sci. *63*, 509-524.

Misawa, T., Hashimoto, K. and Shimodaira, S. (1973) Formation of Fe(II)- and Fe(III) intermediate green complex on oxidation of ferrous ion in neutral and slightly alkaline sulphate solutions. J. Inorg. Nucl. Chem. *35*, 4164-4174.

________, ________ and ________ (1974) The mechanisms of formation of iron oxide and oxyhydroxides in aqueous solutions at room temperature. Corrosion Sci. *14*, 131-149.

________, ________, Suetaka, W. and Shimodaira, S. (1973) Formation of Fe(II)-Fe(III) green complex on oxidation of ferrous ion in perchloric acid solution. J. Inorg. Nucl. Chem. *35*, 4159-4166.

Murphy, P.J., Posner, A.M., and Quirk, J.P. (1976a) Characterization of partially neutralized ferric nitrate solutions. J. Colloid Int. Sci. *56*, 270-283.

________, ________ and ________ (1976b) Characterization of partially neutralized ferric chloride solutions. J. Colloid Int. Sci. *56*, 284-297.

________, ________ and ________ (1976c) Characterization of partially neutralized ferric perchlorate solutions. J. Colloid Int. Sci. *56*, 298-311.

________, ________ and ________ (1976d) Characterization of hydrolyzed ferric ion solutions. A comparison of the effects of various anions on the solution. J. Colloid Int. Sci. *56*, 312-319.

Murray, J.W. (1978) β-FeOOH in marine sediments (abstr.). EOS Trans. Am. Geophys. Union *59*, 411-412.

________ and Gill, G. (1978) The geochemistry of iron in Puget Sound. Geochim. Cosmochim. Acta *42*, 9-19.

Oades, J.M. and Townsend, W.N. (1963) The detection of ferromagnetic minerals in soils and clays. J. Soil Sci. *14*, 179-187.

O'Connor, J.T. (1968) Mineral stability at the Martian surface. J. Geophys. Res. *73*, 5301-5311.

Okada, A., Okada, T. and Shinia, M. (1972) Study on the manganese nodules. IV. Some apsects of the chemical form of iron in the manganese nodule. Sci. Papers IPCR *66*, 178-183.

Paterson, E. and Tait, J.M. (1977) Nitrogen adsorption of synthetic akaganéite and its structural implications. Clay Minerals *12*, 345-351.

Pollack, J.B., Wilson, R.N., and Goles, G.G. (1970) A re-examination of the stability of goethite on Mars. J. Geophys. Res. *75*, 7491-7500.

Posjnak, E. and Merwin, H.E. (1922) The system Fe_2O_3-SO_3-H_2O. J. Am. Chem. Soc. *44*, 1965-1994.

Riley, J.P. and Chester, R. (1971) *Introduction to Marine Chemistry.* Academic Press, New York, 465 pp.

Robie, R.A. and Waldbaum, D.R. (1968) Thermodynamic properties of minerals and related substances at 298.15°K (25.0°C) and one atmosphere (1.013 bars) pressure and at higher temperatures. U.S. Geol. Survey Bull. *1259*, 256 pp.

Schwertmann, U. and Fischer, W.R. (1973) Natural "amorphous" ferric hydroxide. Geoderma *10*, 237-347.

________ and Taylor, R.M. (1972a) The transformation of lepidocrocite to goethite. Clays and Clay Minerals *20*, 151-158.

________ and ________ (1972b) The influence of silicate on the transformation of lepidocrocite to goethite. Clays and Clay Minerals *20*, 159-164.

________ and Thalmann, H. (1976) The influence of Fe(II), Si and pH on the formation of lepidocrocite and ferrihydrite during oxidation of aqueous $FeCl_2$ solutions. Clay Minerals *11*, 189-204.

Sillén, L.G. and Martell, A.E. (1964) *Stability Constants of Metal-ion Complexes.* Special Publ. No. 17, The Chemical Society, London.

Singer, P.C. and Stumm, W. (1970) Acidic mine drainage: the rate-determining step. Science *167*, 1121-1123.

Smith, F.G. and Kidd, D.J. (1949) Hematite-goethite relations in neutral and alkaline solutions under pressure. Am. Mineral. *34*, 403-412.

Söderquist, R. and Jansson, S. (1966) On an x-ray and electron microscope study of precipitates formed by the hydrolysis of iron (III) in 0.5 M NaCl ionic medium. Acta Chem. Scand. *20*, 1417-1418.

Stacey, F.D. and Banerjee, S.K. (1974) *The Physical Principles of Rock Magnetism.* Elsevier, New York.

Strangway, D.W. (1970) *History of the Earth's Magnetic Field.* McGraw-Hill, New York, 168 pp.

________, Honea, R.M., McMahon, B.E., and Larson, E.E. (1968) The magnetic properties of naturally occurring goethite. Geophys. J. R. Astr. Soc. *15*, 345-359.

________, McMahon, B.E., and Bischoff, J.L. (1969) Magnetic properties of minerals from the Red Sea thermal brines. *In*, E.T. Degens and D.A. Ross (eds.), *Hot Brines and Recent Heavy Metal Deposits in the Red Sea.* Springer-Verlag, New York, p. 460-473.

Stumm, W. and Lee, G.F. (1960) Oxygenation of ferrous iron. Ind. Eng. Chem. *53*, 143-146.

________ and Morgan, J.J. (1970) *Aquatic Chemistry*. John Wiley and Sons, New York, 583 pp.

Tamura, H., Gato, K., and Nagayama, M. (1976) Effect of anions on the oxygenation of ferrous ion in neutral solutions. J. Inorg. Nucl. Chem. *38*, 113-117.

Taylor, R.M. and Schwertmann, U. (1974a) Maghemite in soils and its origin. I. Properties and observations on soil maghemites. Clay Minerals *10*, 289-298.

________ and ________ (1974b) Maghemite in soils and its origin. II. Maghemite synthesis at ambient temperature and pH 7. Clay Minerals *10*, 299-310.

Towe, K.M. and Bradley, W.F. (1967) Mineralogical constitution of colloidal "hydrous ferric oxides." J. Colloid Interface Sci. *24*, 384-392.

van der Giessen, A.A. (1966) The structure of iron (III) oxide-hydrate gels. J. Inorg. Nucl. Chem. *28*, 2155-2159.

van der Kraan, A.M. and Medema, J. (1969) The nature of fine particles of α-FeOOH. J. Inorg. Nucl. Chem. *31*, 2039-2044.

Wagman, D.D., Evans, W.H. Parker, V.B., Halow, I., Bailey, S.M., and Schumm, R.H. (1969) Selected values of chemical thermodynamic properties. Natl. Bur. Stand. Tech. Note *270-4*, 141 pp.

Watson, J.A. and Angino, E.E. (1969) Iron-rich layers in sediments from the Gulf of Mexico. J. Sed. Petrol. *39*, 1412-1419.

Watson, J.H.L., Cardwell, R.R., and Heller, W. (1962) The internal structure of colloidal crystals of β-FeOOH and remarks on their assemblies in Schiller layers. J. Phys. Chem. *66*, 1757-1763.

Chapter 3

SILICA POLYMORPHS

Miriam Kastner

MINERALOGY

Opal-CT and quartz are the two most common authigenic silica phases in deep-sea sediments. Opal-C, the high-temperature polymorphs of cristobalite and tridymite, and authigenic opal-A are very rare polymorphs in this environment.

Opal-CT

Siliceous rocks and the silica phase that have the characteristic x-ray diffraction pattern shown in Figure 1 were originally referred to in the literature as lussatite (Mallard, 1890; Flörke, 1955), and subsequently just as cristobalite, α-cristobalite, or disordered cristobalite. In 1971, Jones and Segnit introduced the term opal-CT, which since has been widely accepted in the literature. The three peaks near 4.3, 4.1, and 2.5 A were interpreted by Flörke (1955) to represent a unidimensionally disordered three-layered structure of low cristobalite (the peaks near 4.1 and 2.5 A), with stacking disorder in the [111] direction, resulting in a tridymite maximum near 4.3 A. Jones and Segnit (1971) have shown that stacking disorder in opal-CT causes

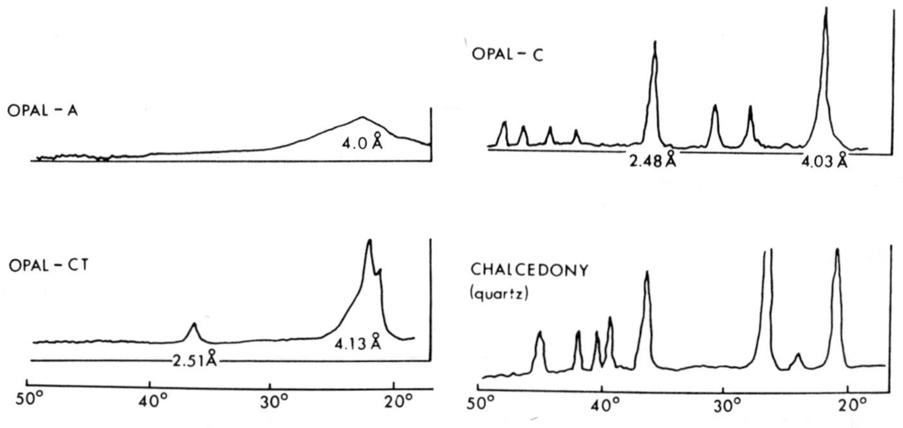

Figure 1. X-ray diffraction patterns of opal-A, opal-CT, opal-C, and quartz, CuK_α radiation, according to the classification of Jones and Segnit (1971) (from Kastner, 1979).

shifts in the *d*(101) cristobalite peak, at about 4.1 A. A progressive sharpening and a shift towards smaller *d*(101) spacings of opal-CT with increasing burial depth has been observed by Murata and Nakata (1974) and Murata and Larson (1975) in the Monterey Formation, California. These observations were interpreted by them to represent solid-state "ordering" of the "disordered" opal-CT. Similar progressive sharpening and a decrease in the *d*(101) opal-CT spacing was observed experimentally by Mizutani (1977) and Kastner and Gieskes (in preparation).

Wilson *et al.* (1974), however, suggested that on the basis of crystal morphology and electron diffraction patterns of opal-CT, it is disordered low tridymite. But the x-ray diffraction pattern of tridymite does not resemble the one shown in Figure 1 for opal-CT, and according to Jones and Segnit (1975), disordered low cristobalite may also be responsible for the electron diffraction patterns observed by Wilson *et al.* (1974).

Opal-CT has also a characteristic infrared absorption spectrum (Jones and Segnit, 1971). The amount of water in opal-CT ranges from a few to about 13 weight percent (Keene, 1976).

Keene (1976) has observed the following forms of opal-CT in deep-sea sediments:

1) Massive opal-CT, the most common form. It occurs as cement and matrix replacement.
2) Opal-CT lepispheres which are spherical aggregates of intersecting platy or bladed euhedral crystals (Wise and Kelts, 1972), shown in Figure 2a. They range in size from about 4 to 30 μm.
3) As casts of radiolarian tests with granular to partly bladed texture.
4) Optically fibrous opal-CT.

The last two forms of opal-CT are less common than the massive and lepispheric forms.

Quartz

Quartz occurs as: (1) chalcedony, a cryptocrystalline variety of quartz which shows a fibrous structure under the microscope and has a somewhat lower index of refraction than micro- or mega-quartz; (2) massive quartz without distinct crystal faces, composed of crypto- to micro-crystals, smaller than 20 μm; and (3) mega-quartz, subhedral to euhedral crystals, greater than 20 μm, occurs as drusy quartz or as vein or crack fillings.

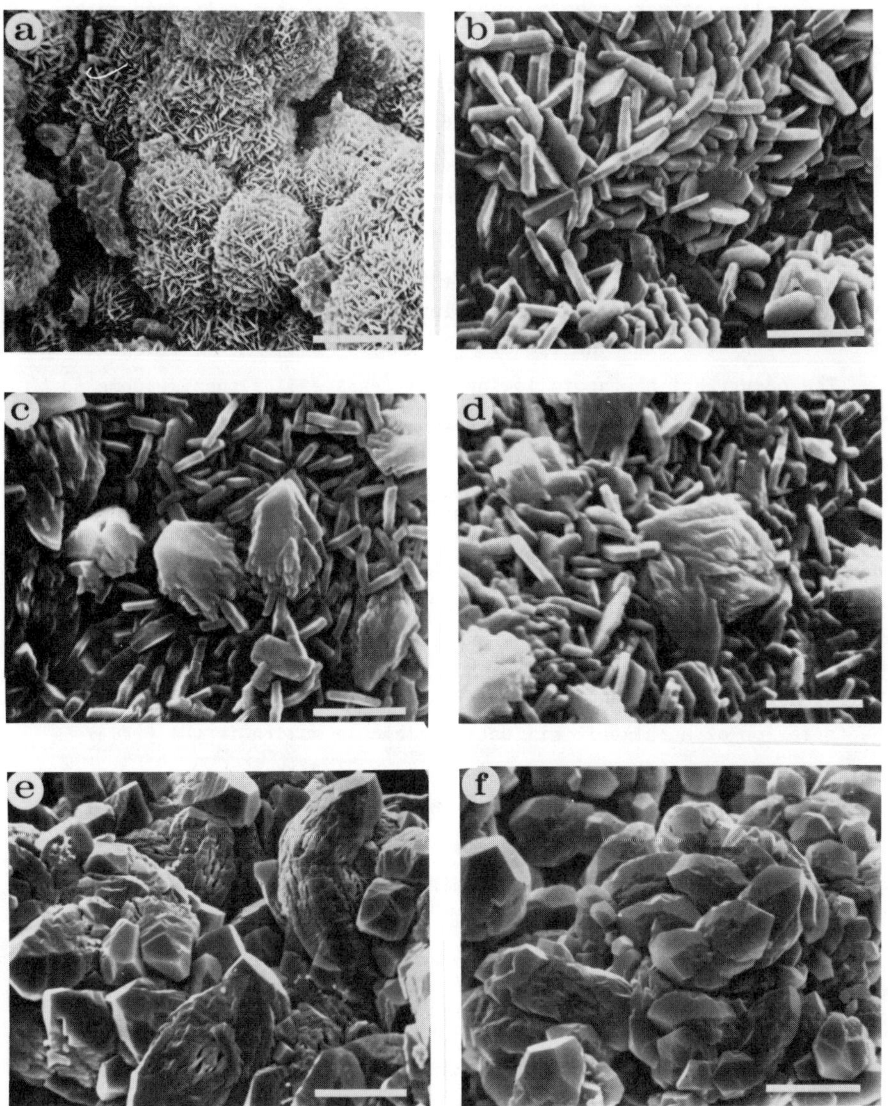

Figure 2. Scanning electron microscope photographs of: (a) opal-CT lepispheres from Eltanin core 47-15, 57°17.3'S, 78°48.5'E, the starting material for hydrothermal experiments of opal-CT to quartz transformation; (b) through (e), various stages of quartz crystallization from opal-CT lepispheres; from platy quartz through coalescing plates and to almost euhedral quartz crystals; (f) euhedral quartz crystals in a pseudo-lepispheric arrangement, crystallized under hydrothermal conditions from the lepispheres shown in Figure 2a. White bars on lower right of (a) and (f) represent 10 μm, of (b), (c), and (d) represent 3 μm, and of (e) represent 24 μm.

The following varieties of chalcedony in deep-sea siliceous rocks were observed by Keene (1975, 1976): length-fast, length-slow, and lutecite, which is a variety of length-slow chalcedony, with the *c*-axis forming an angle of 30° with the axis of the fibers. Folk and Pittman (1971), Pittman and Folk (1971), and Folk (1973) suggested that length-slow chalcedony indicates a shallow-water origin and evaporite conditions. The occurrence of length-slow chalcedony in the deep-sea environment in association with Mg-calcite, dolomite, and/or barite suggests a chemical control on the formation of these various habits of chalcedony, but not necessarily evaporitic conditions.

The occurrence of the following additional silica phases seems to be very rare in deep-sea sediments: opal-C (Jones and Segnit, 1971), which is associated with lava flows, is a well-ordered α-cristobalite with only slight line broadening and minor evidence of tridymite stacking; euhedral crystals of high-temperature cristobalite and tridymite (Klasik, 1975), and authigenic opal-A (Heath, 1969; Hein *et al.*, 1978).

SEQUENCES AND FORMATION RATES OF OPAL-CT AND QUARTZ

Although volcanogenic and hydrothermal silica locally participate in the formation of authigenic silicates, biogenic silica is the primary source of most opal-CT and quartz in deep-sea sediments. The order of decreasing importance of organisms that contribute large quantities of siliceous remains to the sediments is: diatoms, radiolarians, sponge spicules, and silicoflagellates. Their distribution in the sediments reflects surface productivity.

Evidence from deep-sea sediments supports the following most common recrystallization sequence of siliceous oozes: siliceous ooze (opal-A) → porcelanite (mainly opal-CT) → chert (mainly chalcedony and quartz). These transformations proceed by a solution-precipitation mechanism (e.g. Carr and Fyfe, 1958; Kastner *et al.*, 1977; Stein and Kirkpatrick, 1976). Murata and Larson (1975) and Murata *et al.* (1977) suggested a solution-precipitation mechanism for the opal-CT to quartz polymorphic transformation on the basis of oxygen isotope studies of porcelanites and cherts from the Monterey Formation, California.

The assumption that temperature and pressure are the only physical-chemical factors which control the above transformation sequence requires that, stratigraphically, siliceous ooze should occur above and chert beneath porcelanite. However, numerous exceptions to this stratigraphic sequence have been observed; opal-CT is more common in clay-rich sediments and quartz in carbonate sediments of the same geologic age (for example, Lancelot, 1973; Keene, 1975, 1976; Kastner *et al.*, 1977). Kastner *et al.* (1977) observed that in hydrothermal experiments between 50° and 150°C, the rate of opal-A to opal-CT transformation is enhanced in the presence of carbonate by the formation of an unidentified nucleus, a compound which contains Mg^{+2} and OH^- in the ratio 1:2, which serves as a template for opal-CT crystallization. Carbonate dissolution is a continuous source for the necessary alkalinity (OH^-), and sea water is the source of Mg^{+2} for the nucleus. In the presence of clay minerals, in particular smectite, this transformation is retarded. The clay minerals actively compete for the Mg^{+2}, and especially for the OH^-, and there is no readily available source of OH^- to replenish the consumed sea water alkalinity. Clay minerals do not have a direct effect on the transformation rate of opal-CT to quartz (Kastner and Gieskes, in preparation).

Temperature, pressure, and pH of solution strongly affect the transformation rate of opal-CT to quartz (Campbell and Fyfe, 1960; Fyfe and McKay, 1962; Mizutani, 1966; Ernst and Calvert, 1969; Kastner and Gieskes, in preparation). In addition, Kastner and Gieskes observed that ionic strength and the solution chemistry (for example, ratios of various anions) also strongly affect the rate of this polymorphic transformation. A decrease in the $d(101)$ spacing of opal-CT is observed prior to its transformation to quartz. Figures 2b to 2f show the various steps of quartz formation from a precursor of opal-CT lepispheres, the starting material (Fig. 2a).

Direct precipitation of quartz from solution without an opal-CT precursor is possible in geochemical environments in which the dissolved silica concentration is below the inferred solubilities of opal-CT. This has been shown experimentally by Mackenzie and Gees (1971) and by Kastner *et al.* (1977). In the deep-sea environment, this is possible either during the earliest diagenetic stages of particularly permeable carbonate oozes with disseminated biogenic silica, or during the latest diagenetic stages of siliceous sediments in which opal-A has been consumed. The very early direct precipitation of quartz is represented by chalcedony filling of foraminiferal chambers described by Heath and Moberly (1971), and the occurrence of drusy quartz in several deeper sections of deep-sea siliceous sediments represents the later diagenetic stages, during which quartz formation may proceed without an opal-CT precursor.

Reprecipitation of opal-A seems to be an extremely rare phase in deep-sea sediments. The only two reports of reprecipitated opal-A are by Heath (1969) as cement of clay minerals, and implied by Hein *et al.* (1978) in diatomaceous sediments from the Bering Sea.

The transformation of siliceous sediments to porcelanite and chert results in a volume reduction by a factor of 4 to 10 (Keene, 1976) and in a decrease in porosity from 70 to 80% to less than 10% in cherts. Consequently, brecciation may develop.

A compilation of data from 37 DSDP volumes of rates and temperature of formation of opal-CT in deep-sea sediments show the following relationships between time (t in millions of years) and temperature (T,° C) of opal-CT formation (Kastner, 1979):

I. $t = 80 - 2T + 0.01T^2$

II. $t = 83 - 4.15T + 0.1T^2 - 0.001T^3$

The scarcity of reliable data for opal-CT to quartz transformation did not allow the derivation of similar equations for this polymorphic transformation.

SOLUBILITIES AND DISSOLUTION RATES OF SILICA POLYMORPHS

In unsaturated solutions at ordinary temperatures and at pH values < 9, silica is mostly in true solution as monosilicic acid, $Si(OH)_4$ (e.g. Stumm and Morgan, 1970). In addition to pH, the solubility of silica is a function of the crystalline state of the solid phase and of temperature and pressure.

The solubility of quartz at 25°C and 1 bar is ∿6 ppm (100 μM) (Morey *et al.*, 1962). Mackenzie and Gees (1971) determined the solubility of quartz in sea water at 20°C and 1 atmosphere as 4.4 ± 0.3 ppm (73 ± 5 μM), a value which is close to the above value obtained by Morey *et al.* (1962). The somewhat higher solubility of chalcedony most probably results from its small grain size and porous nature. The solubility of amorphous silica in sea water at 2°C and 1 atmosphere is 56 ppm (933 μM) and 70 ppm (1167 μM) at 1000 atmospheres (Jones and Pytkovicz, 1973). At 25°C and 1 atmosphere, the solubility of vitreous silica is 78 ppm (1300 μM) (Jorgensen, 1968) and of silica gel 114 ppm (1900 μM) (Morey *et al.*, 1964). The solubility of acid-cleaned radiolarian tests from surface sediments, central equatorial Pacific, equals amorphous silica solubility. The solubility of radiolarian tests, however, decreases with increasing geologic age (Hurd, 1972, 1973; Hurd and Theyer, 1975).

The solubilities of opal-CT are inferred to lie between the solubilities of chalcedony and amorphous silica.

Most natural waters are undersaturated with respect to amorphous silica, and much of the ocean is also undersaturated with respect to quartz.

The average value of dissolved silica in river water is 13.1 ppm (218.3 μM; Livingstone, 1963). Dissolved silica concentration in the oceans is highly variable. Surface waters are extremely depleted due to rapid biological fixation. Most of the biogenic silica is redissolved and recycled in the water column; thus, dissolved silica concentrations increase with water depth.

North Pacific deep waters are saturated and even supersaturated with respect to quartz. Only about 4% of biogenic silica remains survive in young siliceous sediments. About 2% subsequently redissolve in the interstitial waters. Consequently, a maximum of about 1-2% of the biogenic silica enters the geological record (Heath, 1974; Wollast, 1974).

According to Goto (1958), the dissolution rate of amorphous silica is proportional to the square of the surface area. The rate of amorphous silica dissolution is strongly affected by both pH and ionic strength in NaCl solutions (Wirth and Gieskes, 1979). For example, these authors have measured an increase in the dissolution rate of amorphous silica by a factor of 360, from pH 5 to 11, at sea water ionic strength of 0.7. The observed increased rates of amorphous silica dissolution as a function of pH and ionic strength were interpreted by Wirth and Gieskes (1979) to result from silica surface charge effects.

Organic matter films and multivalent metal ions retard the dissolution rate of amorphous silica (Iler, 1955, 1973; Lewin, 1961; Hurd, 1972, 1973).

SUMMARY

The most common authigenic silica polymorphs in deep-sea sediments are opal-CT and quartz (including chalcedony). Other non-common silica polymorphs are opal-C and high-temperature tridymite and cristobalite.

Most quartz crystallizes from an opal-CT precursor, but during early, and in particular, late diagenesis, quartz may form without an opal-CT precursor.

The rate of diagenesis and of silica polymorphic transformations of siliceous sediments is strongly affected by temperature, pressure, and time. Geochemical factors, however, often reverse the expected relation between burial depth and the nature of the silica polymorph. In the pelagic environment of high dissolved magnesium concentrations and in areas of average oceanic geothermal gradients, the transformation rate of opal-A to opal-CT generally decrease as follows: siliceous ooze + carbonate > pure siliceous ooze > siliceous ooze + clay minerals and/or pyroclastic sediment. Indeed, in most cases the quartz to opal-CT ratio is greater in carbonate versus clay-rich sediments of the same geologic age.

Opal-CT and quartz preserved in deep-sea sediments represent a maximum of 1-2% of the originally-produced siliceous skeletal remains.

Campbell, A.S. and Fyfe, W.S. (1960) Hydroxyl ion catalysis of the hydrothermal crystallization of amorphous silica: a possible high-temperature pH indicator. Am. Mineral. *45*, 464-468.

Carr, R.M. and Fyfe, W.S. (1958) Some observations on the crystallization of amorphous silica. Am. Mineral. *43*, 908-916.

Ernst, W.G. and Calvert, S.E. (1969) An experimental study of the recrystallization of porcelanite and its bearing on the origin of some bedded cherts. Am. J. Sci. *267A*, 114-133.

Flörke, O.W. (1955) Zur Frage des "Hoch"-Cristobalit in Opalen Bentoniten und Gläsern. N. Jahr. Mineral. Monat. *10*, 217-223.

Folk, R.L. (1973) Evidence for peritidal deposition of Devonian Caballos Novaculite, Marathon Basin, Texas. Am. Assoc. Petrol. Geol. Bull. *57*, 702-725.

________ and Pittman, J.S. (1971) Length-slow chalcedony: a new testament for vanished evaporites. J. Sed. Petrol. *41*, 1045-1058.

Fyfe, W.S. and McKay, D.S. (1962) Hydroxyl ion catalysis of the crystallization of amorphous silica at 330 °C and some observations on the hydrolysis of albite solutions. Am. Mineral. *47*, 83-89.

Goto, K. (1958) Estimation of specific surface area on particles in colloidal silica sols from the rate of dissolution. Chem. Soc. Japan Bull. *31*, 900-903.

Heath, G.R. (1969) Mineralogy of Cenozoic deep-sea sediments from the equatorial Pacific Ocean. Geol. Soc. Am. Bull. *80*, 1997-2018.

________ (1974) Dissolved silica and deep-sea sediments. *In*, W.W. Hay (ed.), *Studies in Paleo-oceanography*, Soc. Econ. Paleontol. Mineral. Spec. Pub. *20*, 77-93.

________ and Moberly, R., Jr. (1971) Cherts from the western Pacific, Leg 7, Deep Sea Drilling Project. *In*, E.L. Winterer *et al.* (eds.), *Initial Reports of the Deep Sea Drilling Project*, vol. 7, U.S. Government Printing Office, Washington, DC, p. 991-1007.

Hein, J.R., Scholl, D.W., Barron, J.A., Jones, M.G., and Miller, J. (1978) Diagenesis of late Cenozoic diatomaceous deposits and formation of the bottom simulating reflector in the southern Bering Sea. Sedimentol. *25*, 155-181.

Hurd, D.C. (1972) Factors affecting the solution rate of biogenic opal in sea water. Earth Planet. Sci. Lett. *15*, 411-417.

________ (1973) Interaction of biogenic opal, sediment and seawater in the Central Equatorial Pacific. Geochim. Cosmochim. Acta *37*, 2257-2282.

________ and Theyer, F. (1975) Changes in the physical and chemical properties of biogenic silica from the Central Equatorial Pacific. I. Solubility, specific surface area, and solution rate constants of acid-cleaned samples. Advanc. Chem. Ser. No. *147*, 211-230.

Iler, R.K. (1955) *The Colloid Chemistry of Silica and Silicates.* Cornell Univ. Press, Ithaca, New York, 324 pp.

________ (1973) Effect of adsorbed alumina on the solubility of amorphous silica in water. J. Coll. Interf. Sci. *43*, 399-408.

Jones, M.M. and Pytkowicz, R.M. (1973) The solubility of silica in sea water at high pressures. Soc. Royale. Sci. Liege. Bull. *42*, 118-120.

Jones, J.B. and Segnit, E.R. (1971) The nature of opal. I. Nomenclature and constituent phases. J. Geol. Soc. Australia *18*, 56-68.

Jorgensen, S.S. (1968) Solubility and dissolution kinetics of precipitated silica in 1 m $NaClO_4$ at 25°C. Acta Chem. Scand. *22*, 335-341.

Kastner, M. (1981) Authigenic silicates in deep-sea sediments--formation and diagenesis. *In,* C. Emiliani (ed.), *The Sea, Vol. 7: The Oceanic Lithosphere.* John Wiley, New York.(1981).

________ and Gieskes, J.M. (1979) Diagenesis of siliceous oozes--II. Chemical controls on the rate of opal-CT to quartz transformation--an experimental study (in preparation).

________, Keene, J.B., and Gieskes, J.M. (1977) Diagenesis of siliceous oozes--I. Chemical controls on the rate of opal-A to opal-CT transformation--an experimental study. Geochim. Cosmochim. Acta *41*, 1041-1059.

Keene, J.B. (1975) Cherts and porcelanites from the North Pacific, DSDP Leg 32. *In*, R.L. Larson *et al.* (eds.), *Initial Reports of the Deep Sea Drilling Project*, vol. 32, U.S. Government Printing Office, Washington, DC, p. 429-507.

________ (1976) *The distribution, mineralogy, and petrography of biogenic and authigenic silica from the Pacific Basin.* Ph.D. dissertation, Scripps Inst. Oceanography, 264 pp.

Klasik, J.A. (1975) High cristobalite and high tridymite in Middle Eocene deep-sea cherts. Science *189*, 631-632.

Lancelot, Y. (1973) Chert and silica diagenesis in sediments from the central Pacific. *In*, E.L. Winter *et al.* (eds.), *Initial Reports of the Deep Sea Drilling Project*, vol. 17, U.S. Government Printing Office, Washington, DC, p. 377-405.

Lewin, J.C. (1961) The dissolution of silica from diatom walls. Geochim. Cosmochim. Acta *21*, 182-198.

Livingstone, D.A. (1963) Chemical composition of rivers and lakes. U.S. Geol. Surv. Prof. Paper *440-G*, 64 pp.

Mackenzie, F.T. and Gees, R. (1971) Quartz: synthesis at earth-surface conditions. Science *173*, 533-534.

Mallard, M.E. (1890) Sur la lussatite, nouvelle variétè minerale cristallisée de silice. Soc. Fr. Mineral. Cristallogr. Bull. *13*, 63-66.

Mizutani, S. (1966) Transformation of silica under hydrothermal conditions. J. Earth Sci., Nagoya Univ., *14*, 56-88.

_______ (1977) Progressive ordering of cristobalitic silica in the early stage of diagenesis. Contrib. Mineral. Petrol. *61*, 129-140.

Morey, G.W., Fournier, R.O., and Rowe, J.J. (1962) The solubility of quartz in water in the temperature interval 25 to 300°C. Geochim. Cosmochim. Acta *26*, 1029-1043.

________, ________, and ________ (1964) The solubility of amorphous silica at 25°C. J. Geophys. Res. *69*, 1995-2002.

Murata, K.J., Friedman, I., and Gleason, J.D. (1977) Oxygen isotope relations between diagenetic silica minerals in Monterey Shale, Temblor Range, California. Am. J. Sci. *277*, 259-272.

________ and Larson, R.R. (1975) Diagenesis of Miocene siliceous shales, Temblor Range, California. J. Res. U. S. Geol. Surv. *3*, 553-566.

________ and Nakata, J.K. (1974) Cristobalitic stage in the diagenesis of diatomaceous shale. Science *184*, 567-568.

Pittman, J.J. and Folk, R.L. (1971) Length-slow chalcedony after sulphate evaporite minerals in sedimentary rocks. Nature *230*, 64-65.

Stein, C.L. and Kirkpatrick, R.J. (1976) Experimental porcelanite and recrystallization kinetics: a nucleation and growth model. J. Sed. Petrol. *46*, 430-435.

Stumm, W. and Morgan, J.J. (1970) *Aquatic Chemistry*. Wiley-Interscience, New York, 583 pp.

Wilson, M.J., Russel, J.D., and Tait, J.M. (1974) A new interpretation of the structure of disordered α-cristobalite. Contrib. Mineral. Petrol. *47*, 1-6.

Wirth, G.S. and Gieskes, J.M. (1979) The initial kinetics of the dissolution of vitreous silica in aqueous media. J. Coll. Interf. Sci. *68*, 492-500.

Wise, S.W., Jr. and Kelts, K.R. (1972) Inferred diagenetic history of a weakly silicified deep-sea chalk. Trans. Gulf Coast Ass. Geol. Soc. *22*, 177-203.

Wollast, R. (1974) The silica problem. *In*, Goldberg, E.D. (ed.), *The Sea*, vol. 5, Wiley-Interscience, New York, p. 359-392.

Chapter 4

ZEOLITES

Miriam Kastner

MINERALOGY

Zeolites are hydrated framework aluminosilicates of the alkali and alkaline earth metals. They are known and are important for their reversible dehydration, ion exchange, catalysis, and molecular sieve properties. A general formula for zeolites is $X_uY_vZ_nO_{2n}\cdot mH_2O$, where *X* represents Na and K, *Y* represents Ca and also Sr, Ba, and Mg, and *Z* represents Al and Si. Na + K + (2Σ divalent cations) = Al, and (Si+Al):O = 1:2. The channels in their structures have diameters on the order of 2-12 A (Meier and Olson, 1971, and references therein; Steele *et al.*, 1975).

Clinoptilolite and phillipsite are among the most important diagenetic silicates in deep-sea sediments. An average representative formula of marine clinoptilolite is $K_{2.3}Na_{0.8}Al_{3.1}Si_{14.9}O_{36}\cdot 12H_2O$ (Kastner, 1979), and it constitutes about 2% of deep-sea sediments. An average representative formula of marine phillipsite is $K_{2.8}Na_{1.6}Al_{4.4}Si_{11.6}O_{32}\cdot 10H_2O$ (Kastner, 1979), and it constitutes about 1.5% of deep-sea sediments. Analcite is also a common zeolite in the marine environment, but is much less abundant than phillipsite. The range of its chemical composition has not been determined as yet. Other relatively rare zeolites in deep-sea sediments are: heulandite, mordenite, harmotome, chabazite, erionite, gmelinite, laumontite, thomsonite, and natrolite. The Ca-rich zeolites and natrolite are associated with basaltic material.

Clinoptilolite

Clinoptilolite is a high-silica member of the heulandite group (Mumpton, 1960), and crystallizes in the monoclinic system *Cm* (Merkle and Slaughter, 1968). It forms small euhedral platy crystals (Fig. 1) and occurs mostly as discrete crystals less than 45 μm long. The Si/Al ratio of marine clinoptilolites ranges between 4.2-5.2 (Stonecipher, 1977, 1978; Kastner and Stonecipher, 1978). It is length slow with a mean index of refraction of 1.484 (Boles, 1972). With increasing Si or divalent cation substitution for monovalent ions, the *ac* plane decreases

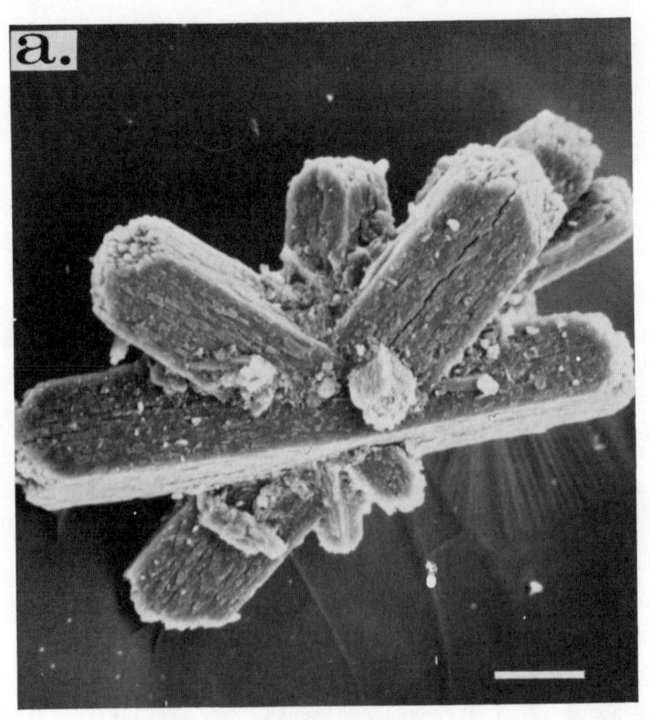
a.

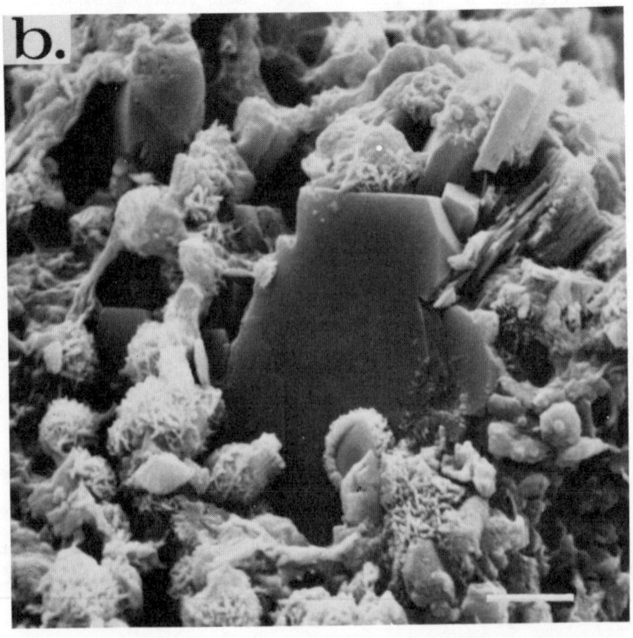
b.

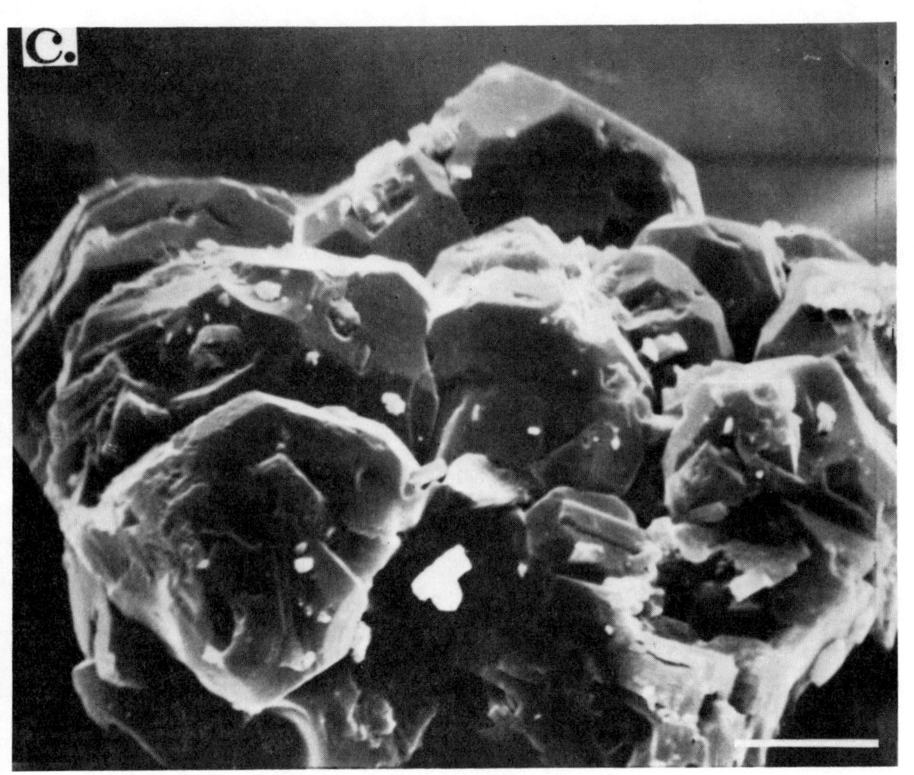

Figure 1 (see facing page and above). Scanning electron photographs of (a) complex sector twinned phillipsite crystals from radiolarian clay, 16°25'N, 164°24'W; (b) clinoptilolite crystals in a silicified limestone, 57°39'N, 15°55'W. Calcite has been dissolved to reveal clinoptilolite and opal-CT lepispheres; and (c) euhedral analcite crystals, DSDP Leg 21, 206-3-1, 101-108 cm. White bar in lower right of (a) represents 25 μm, of (b) represents 5 μm, and of (c) represents 30 μm. (a) and (b) from Kastner and Stonecipher (1978, Fig. 1); (c) was photographed by Dr. A. Desprairies.

(Boles, 1972). Matrix inclusions are rare, which indicates slow growth rates. Clinoptilolite crystals were observed within partially dissolved casts of radiolarian tests (Arrhenius, 1963; Berger and von Rad, 1972; von Rad and Rösch, 1974).

Clinoptilolite coexists with opal-A and opal-CT (Jones and Segnit, 1971), quartz, K-feldspar, palygorskite, sepiolite, smectite, rhyolitic glass, and basaltic glass + opal-A and/or opal-CT (Hathaway and Sachs, 1965; Weaver, 1968; Cook and Zemmels, 1972; Kolla and Biscaye, 1973; Stonecipher, 1977; Couture, 1977; Kastner and Stonecipher, 1978). The association of clinoptilolite with volcanic matter is not always evident.

The importance of clinoptilolite as a diagenetic zeolite in the Atlantic Ocean was recognized by Biscaye (1965), Hathaway and Sachs (1965), Turekian (1965), and Peterson *et al.* (1970). Its importance in Pacific and Indian Ocean sediments was recently summarized by Kolla and Biscaye (1973), Cronan (1974), Stonecipher (1977), and Kastner and Stonecipher (1978).

Heulandite

Heulandite, the low-silica member of the heulandite group (Mumpton, 1960), was reported by Boles and Wise (1978) to occur in the Indian Ocean in organic-rich sediments. It is not certain, however, that heulandite is indeed extremely rare in deep-sea sediments. Using x-ray diffraction patterns, the distinction between heulandite and clinoptilolite requires heat treatment of the sample (Mumpton, 1960; Alietti, 1972), although this has not been routinely performed.

Phillipsite

Phillipsite crystallizes in the monoclinic system ($P2_1$ or $P2_1/m$) with no evidence of order/disorder in the Al/Si distribution (Rinaldi *et al.*, 1974). It forms euhedral to subhedral elongated prismatic crystals up to about 250 μm long. The crystals are either colorless or yellowish from numerous inclusions of Fe-oxyhydroxides and clay minerals, mainly smectite. The great abundance of inclusions indicates rapid growth rates, and common etched crystal faces, even at shallow burial depths, indicate rapid dissolution. Phillipsite occurs as discrete crystals, as complex sector-twinned crystals (Fig. 1a), cement, cavity, and fracture fillings, and replaces plagioclase (e.g. Morgenstein, 1967; Rex, 1967, Bass *et al.*, 1973).

It has also been observed to occur and to form within manganese nodules (Murray and Renard, 1891; Bonatti, 1963; Burns and Burns, 1978) (see Ch. 1, Fig. 12, this volume). The Si/Al ratio of marine phillipsites ranges between 2.3-2.8, intermediate between phillipsites in mafic igneous rocks with a ratio between 1.3-2.4 and those formed from silicic glass in saline alkaline lakes with a ratio of 2.6-3.4 (Sheppard *et al.*, 1970; Stonecipher, 1977, 1978). The mean index of refraction varies between 1.477-1.486 (Sheppard *et al.*, 1970). Sheppard *et al.* did not observe any correlation between the SiO_2 content and the index of refraction of phillipsite, but according to Hay (1966), an inverse relationship between these two parameters exists. Barium, a common trace element in phillipsite, causes an increase in its mean index of refraction, but chemical variations in phillipsite do not appreciably affect its unit cell dimensions (Sheppard *et al.*, 1970).

Phillipsite coexists with smectite, Fe-oxyhydroxides and Mn-oxides, and almost always with basaltic glass (Murray and Renard, 1891; Bonatti, 1963; Stonecipher, 1977; Kastner and Stonecipher, 1978).

Harmotome

Harmotome is isostructural with phillipsite, and it contains Ba as a major cation (e.g. Goldberg and Arrhenius, 1958; Arrhenius, 1963; Bonatti, 1963; Rinaldi *et al.*, 1974). Harmotome seems to be an early alteration product of palagonite. Subsequent growth of phillipsite is quantitatively the important product of basaltic glass alteration in the deep-sea environment and not harmotome (Bonatti and Arrhenius, 1965).

Phillipsite and harmotome were recognized as diagenetic minerals in deep-sea sediments associated with volcanic matter in the Pacific and Indian Oceans by Murray and Renard (1891). Young (1939) described the occurrence of phillipsite in Atlantic Ocean sediments.

Analcite

Analcite crystallizes in the isometric system (*Ia3d*) (Taylor, 1930) and has the ideal formula of $NaAlSi_2O_6 \cdot H_2O$. It occurs as discrete euhedral crystals, shown in Figure 1c. The Si/Al ratio of sedimentary zeolites ranges between 2.0-2.9 (Hay, 1966; Coombs and Whetten, 1967; Sheppard and Gude, 1969); the range of Si/Al ratio for marine analcite has not been determined as yet. Analcite is most commonly associated with basaltic matter and coexists with smectite, chlorite, celadonite, phillipsite, clinoptilolite, and K-feldspar. Its occurrence in deep-sea sediments has already been observed by Murray and Renard (1891).

OCCURRENCE, FORMATION, AND STABILITY OF PHILLIPSITE, CLINOPTILOLITE, AND ANALCITE

Phillipsite is most common in Oligocene and younger sediments. Except for a decrease in the abundance of clinoptilolite in Late Jurassic sediments, its frequency of occurrence increases with increasing geologic age. The frequency of analcite occurrence also increases with increasing geologic age (Fig. 2). Phillipsite is prevalent at shallow burial depths of <100 meters, and clinoptilolite occurs mainly in sediments deeper than 100 meters from the sediment/water interface. No correlation between analcite occurrence and burial depth has been observed (e.g. Kastner and Stonecipher, 1978). Phillipsite and analcite are more common in Pacific and Indian Ocean sediments than in Atlantic Ocean sediments. Clinoptilolite, however, is more common in Atlantic Ocean sediments. Both phillipsite and clinoptilolite occur in most types of deep-sea sediments. For phillipsite, the lithologic order of decreasing frequency of occurrence is: clay-rich > clay-volcanic > calcareous > siliceous; and for clinoptilolite: calcareous > clay-rich > calcareous-siliceous (opal-CT) > volcanic > siliceous (opal-A). Thus, phillipsite forms generally in regimes of lower sedimentation rate than clinoptilolite. Hence, time (kinetics) and lithology (chemistry) are very important variables that determine the distribution of phillipsite and clinoptilolite in deep-sea sediments (Stonecipher, 1977; Kastner and Stonecipher, 1978). Analcite occurs primarily in volcaniclastic and ash-bearing calcareous sediments.

Phillipsite grows rapidly. It often nucleates and starts to grow at the sediment/water interface and continues to grow within the sediment column. Larger and sector-twinned crystals are more common below the sediment/water interface (Bernat and Goldberg, 1969; Bernat *et al.*, 1970; Czyscinski, 1973). Phillipsite also dissolves rapidly; etched crystal faces are observed, and the amount of phillipsite decreases with burial depth until it disappears. It is rarely present below 500 meters within

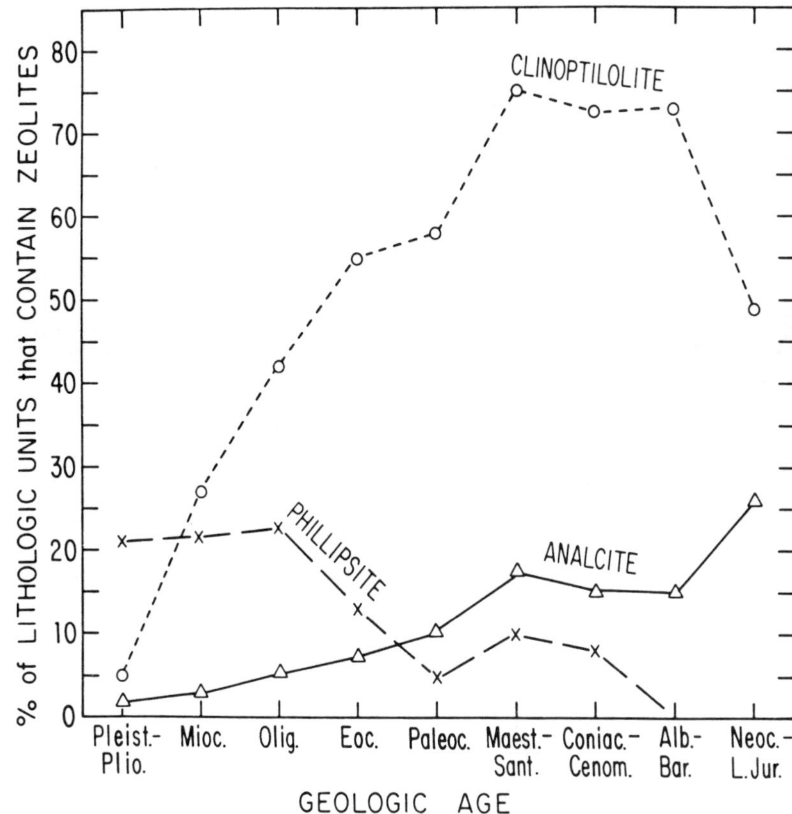

Figure 2. Changes in frequencies of phillipsite, clinoptilolite, and analcite occurrences with age of deep-sea sediments. From Kastner (1979, Fig. 7).

the sediment column. In a few cores in the Pacific Ocean, phillipsite has already disappeared at four meters depth below the sediment/water interface (Stonecipher, 1977; Boles and Wise, 1978; Kastner and Stonecipher, 1978). According to Czyscinski (1973) and Bernat *et al.* (1970), phillipsite formation in deep-sea sediments requires 10^5 to 10^6 years, respectively. The rapid growth and subsequent dissolution of phillipsite strongly suggest that its formation is kinetically controlled, and that phillipsite is not a thermodynamically stable phase in the deep-sea environment (Kastner, 1979).

As mentioned already, clinoptilolite occurs more frequently in sediments deeper than 100 meters below the sediment/water interface and persists even at burial depths of greater than one kilometer. No etched crystal faces similar to those of phillipsite were observed for clinoptilolite. These observations suggest that clinoptilolite may be a thermodynamically stable phase in the deep-sea environment. The sharp decrease in the frequency of clinoptilolite in Late Jurassic (Fig. 2) may be interpreted, however, to indicate that this zeolite also disappears with geologic age and that it, too, is not a thermodynamically stable phase. Indeed, in sediments exposed on land, the oldest known clinoptilolite is of Triassic age (Boles and Coombs, 1975). Thermochemical data for phillipsite and clinoptilolite are not available. In the experiments of Lerman *et al.* (1975) on the dissolution rates of silicates, the maximum dissolved silica value for phillipsite was $\sim$9 ppm ($\sim$150 μM) and for clinoptilolite $\sim$50 ppm ($\sim$830 μM). Assuming that these dissolved silica values approximate the solubilities of the two phases, phillipsite could form at the sediment/water interface in the North Pacific. In other deep-sea regions, an additional silica source is required. Dissolution of biogenic silica (opal-A) causes rapid increases in dissolved silica concentrations of interstitial waters in the uppermost few tens of centimeters of deep-sea sediments. Indeed, Bernat *et al.* (1970) and Czyscinski (1973) observed a positive correlation between the extent of biogenic silica dissolution and the amount of phillipsite formation. The high experimental dissolved silica value for clinoptilolite indicates that its formation is not favored at the sediment/water interface, but instead within the sediment column, and in particular in siliceous sediments.

Basaltic material, in particular basaltic glass and/or palagonite, is the main precursor of phillipsite (Murray and Renard, 1891; Bonatti, 1963; Rex, 1967; Kolla and Biscaye, 1973; Stonecipher, 1977; Kastner and Stonecipher, 1978). Smectite which formed from altered basaltic glass has also been suggested as a precursor of phillipsite (Bonatti and Arrhenius, 1965; Hay, 1966; Kastner, 1979). The reaction basaltic glass + dissolved silica (primarily biogenic) $\rightarrow$ phillipsite has been suggested by Bernat *et al.* (1970) and Czyscinski (1973).

The immediate precursor(s) of clinoptilolite is (are) often difficult to determine. The following are observed and suggested reactions for clinoptilolite formation in deep-sea sediments (Kastner, 1979):

Rhyolitic or andesitic glass + H_4SiO_4 → clinoptilolite

Phillipsite + H_4SiO_4 → clinoptilolite

Opal-A + $Al(OH)_4^-$ + K^+ → opal-CT + clinoptilolite

Opal-A + $Al(OH)_4^-$ + K^+ → clinoptilolite

Phillipsite + smectite + H_4SiO_4 → clinoptilolite + palygorskite

and for analcite formation:

Clinoptilolite + Na^+ → analcite + K^+ + quartz

Phillipsite + Na^+ → analcite + K^+(?)

Volcanogenic material → analcite

CONCLUDING REMARKS

Non-marine zeolites have been studied extensively. In addition to the quantitative importance of marine zeolites, research on the formation of these zeolites will allow conclusions about a range of physical-chemical conditions of sedimentary zeolite formation which was not covered by the extensive studies of non-marine zeolites. The main differences between most of the non-marine and marine zeolites are: Much of the non-marine zeolites formed in saline alkaline lakes with pH values of nine and above and with salinities which exceed sea water salinity. Their rate of formation is thus much faster than that of zeolites in the deep-sea environment of pH around eight and salinity of 34°/oo to 35°/oo. In addition, the saline alkaline lakes are generally ephemeral; diagenetic reactions over a time span of about 150 million years can be studied in the deep-sea sedimentary record.

ZEOLITES: REFERENCES

Alietti, A. (1972) Polymorphism and crystal-chemistry of heulandites and clinoptilolites. Am. Mineral. *57*, 1448-1462.

Arrhenius, G. (1963) Pelagic sediments. *In*, N.M. Hill (ed.), *The Sea*, Wiley-Interscience, New York, p. 655-727.

Bass, M.N., Moberly, R., Rhodes, J.M., Shih Chi-yu, and Church, S.E. (1973) Volcanic rocks cored in the Central Pacific, Leg 17, Deep Sea Drilling Project. *In*, E.L. Winterer *et al.* (eds.), *Initial Reports of the Deep Sea Drilling Project*, vol. 17, U. S. Government Printing Office, Washington, DC, p. 429-503.

Berger, W.H. and von Rad, U. (1972) Cretaceous and Cenozoic sediments from the Atlantic Ocean. *In*, D.E. Hayes *et al.* (eds.), *Initial Reports of the Deep Sea Drilling Project*, vol. 14, U. S. Government Printing Office, Washington, DC, p. 787-954.

Bernat, M., Bieri, R.H., Koide, M., Griffin, J.J., and Goldberg, E.D. (1970) Uranium, thorium, potassium and argon in marine phillipsites. Geochim. Cosmochim. Acta *34*, 1053-1071.

________ and Goldberg, E.D. (1969) Thorium isotopes in the marine environment. Earth Planet. Sci. Lett. *5*, 308-312.

Biscaye, P.E. (1965) Mineralogy and sedimentation of recent deep-sea clay in the Atlantic Ocean and adjacent seas and oceans. Geol. Soc. Am. Bull. *76*, 803-832.

Boles, J.R. (1972) Composition, optical properties, cell dimension, and stability of some heulandite group zeolites. Am. Mineral. *57*, 1463-1493.

________ and Coombs, D.S. (1975) Mineral reactions in zeolite Triassic tuff, Hokonui Hills, New Zealand. Geol. Soc. Am. Bull. *86*, 163-173.

________ and Wise, W.S. (1978) Nature and origin of deep-sea clinoptilolite. *In*, L.B. Sand and F.A. Mumpton (eds.), *Natural Zeolites, Occurrence, Properties, Use*, Pergamon Press, p. 235-243.

Bonatti, E. (1963) Zeolites in Pacific pelagic sediments. Trans. N. Y. Acad. Sci. *25*, 938-948.

________ and Arrhenius, G. (1965) Eolian sedimentation in the Pacific off Northern Mexico. Mar. Geol. *3*, 337-348.

Burns, V.M. and Burns, R.G. (1978) Post-depositional metal enrichment processes inside manganese nodules from the North Equatorial Pacific. Earth Planet. Sci. Lett. *39*, 341-348.

Cook, H.E. and Zemmels, I. (1972) X-ray mineralogy studies, DSDP Leg 9. *In*, Hays, J.D. *et al.* (eds.), *Initial Reports of the Deep Sea Drilling Project*, vol. 9, U. S. Government Printing Office, Washington, DC, p. 707-778.

Coombs, D.S. and Whetten, J.T. (1967) Composition of analcime from sedimentary and burial metamorphic rocks. Geol. Soc. Am. Bull. *78*, 269-282.

Couture, R.A. (1977) Composition and origin of palygorskite-rich and montmorillonite-rich zeolite-containing sediments from the Pacific Ocean. Chem. Geol. *19*, 113-130.

Cronan, D.S. (1974) Authigenic minerals in deep-sea sediments. *In*, E.D. Goldberg (ed.), *The Sea*, vol. 5, Wiley Interscience, p. 491-525.

Czyscinski, K. (1973) Authigenic phillipsite formation rates in the central Indian Ocean and the equatorial Pacific Ocean. Deep-Sea Res. *20*, 555-559.

Goldberg, E.D. and Arrhenius, G. (1958) Chemistry of Pacific pelagic sediments. Geochim. Cosmochim. Acta *13*, 153-212.

Hathaway, J.C. and Sachs, P.L. (1965) Sepiolite and clinoptilolite from the Mid-Atlantic Ridge. Am. Mineral. *50*, 852-867.

Hay, R.L. (1966) Zeolites and zeolitic reactions in sedimentary rocks. Geol. Soc. Am. Spec. Pap. *85*, 130 pp.

Jones, J.B. and Segnit, E.R. (1971) The nature of opal. I. Nomenclature and constituent phases. J. Geol. Soc. Australia *18*, 56-68.

Kastner, M. (1981) Authigenic silicates in deep-sea sediments--formation and diagenesis. *In*, C. Emiliani (ed.), *The Sea, Vol. 7: The Oceanic Lithosphere*. John Wiley, New York. (1981).

________ and Stonecipher, S.A. (1978) Zeolites in pelagic sediments of the Atlantic, Pacific, and Indian Oceans. *In*, L.B. Sand and F.A. Mumpton (eds.), *Natural Zeolites, Occurrence, Properties, Use*, Pergamon Press, p. 199-220.

Kolla, V. and Biscaye, P.E. (1973) Deep-sea zeolites; variations in space and time in the sediments of the Indian Ocean. Mar. Geol. *15*, 11-17.

Lerman, A., Mackenzie, F. T., and Bricker, O.P. (1975) Rates of dissolution of aluminosilicates in seawater. Earth Planet. Sci. Lett. *25*, 82-88.

Meier, W.M. and Olson, D.H. (1971) Zeolite frameworks. Advances in Chemistry Ser. 101, *Molecular Sieve Zeolites I*, p. 155-170.

Merkle, A.B. and Slaughter, M. (1968) Determination and refinement of the structure of heulandite. Am. Mineral. *53*, 1120-1138.

Morgenstein, M. (1967) Authigenic cementation of scoriaceous deep-sea sediments west of the Society Ridge, South Pacific. Sedimentol. *9*, 105-118.

Mumpton, F.A. (1960) Clinoptilolite redefined. Am. Mineral. *45*, 351-369.

Murray, J. and Renard, A.F. (1891) *Report on the Scientific Results of the Voyage of H.M.S.* Challenger *during the years 1873-76: deep-sea deposits*. Johnson Reprint Co., London, 525 pp.

Peterson, M.N.A., Edgar, N.T., von der Borch, C.C., and Rex, R.W. (1970) Cruise leg summary and discussion. *In*, M.N.A. Peterson *et al.* (eds.), *Initial Reports of the Deep Sea Drilling Project*, vol. 2, U. S. Government Printing Office, Washington, DC, p. 413-427.

Rex, R.W. (1967) Authigenic silicates formed from basaltic glass by more than 60 million years' contact with sea water, Sylvania Guyot, Marshall Islands. Clays and Clay Minerals, Proc. Natl. Conf. *15*, 195-203.

Rinaldi, R., Pluth, J.J., and Smith, J.V. (1974) Refinement of crystal structure of phillipsite and harmotome. Acta Crystallogr. *B30*, 2426-2433.

Sheppard, R.A. and Gude, A.J., 3rd (1969) Diagenesis of tuffs in the Barstow Formation, Mud Hills, San Bernardino County, California. U. S. Geol. Surv. Prof. Paper *634*, 35 pp.

________, ________, and Griffin, J.J. (1970) Chemical composition and physical properties of phillipsite from the Pacific and Indian Oceans. Am. Mineral. *55*, 1053-2062.

Steele, I.M., Smith, J.V., Pluth, J.J., and Solberg, T.N. (1975) Quantitative analysis of zeolites using an energy dispersive system. Proc. 10th Ann. Conf. Microbeam Anal. Soc., 37A-D.

Stonecipher, S.A. (1977) *Origin, distribution, and diagenesis of deep-sea clinoptilolite and phillipsite*. Ph.D. dissertation, Scripps Inst. Oceanography, 223 pp.

________ (1978) Chemistry of deep-sea phillipsite, clinoptilolite, and host sediment. *In*, L.B. Sand and F.A. Mumpton (eds.), *Natural Zeolites, Occurrence, Properties, Use*, Pergamon Press, p. 221-234.

Taylor, W.H. (1930) The structure of analcite ($NaAlSi_2O_6 \cdot H_2O$). Z. Kristallogr. *74*, 1-19.

Turekian, K.K. (1965) Some apsects of the geochemistry of marine sediments. *In*, Riley, J.P. and Skirrow, G. (eds.), *Chemical Oceanography*, vol. 2, Academic Press, London, p. 81-126.

von Rad, U. and Rösch, H. (1974) Petrography and diagenesis of deep-sea cherts from the central Atlantic. *In*, K.J. Hsü and H.C. Jenkyns (eds.), Int. Ass. Sedimentol., Spec. Pub. *1*, 327-347.

Weaver, C.E. (1968) Mineral facies in the Tertiary of the continental shelf and Blake Plateau. Southeastern Geol. *9*, 57-63.

Young, J.A. (1939) Minerals from deep-sea core surface deposits of Bermudian calcareous sediments. Am. J. Sci. *237*, 798-810.

Chapter 5

CLAY MINERALS

John C. Hathaway

INTRODUCTION

Clay minerals are major constituents of materials deposited in the marine environment. Of some 10,000 sediment samples analyzed by the Deep Sea Drilling Project (DSDP), more than 99% contain materials classified as clay or clay minerals. However, the term clay carries two connotations: one of particle size and one of particle composition. This discussion will be concerned mainly with clay as a mineral composition, although consideration of particle size and association with "non-clay" minerals of comparable size is inescapable.

Clay minerals are considered by most workers to be the phyllosilicates (layer silicates) that usually occur in particle sizes of 2 μm or less. Many specimens of these minerals are also of larger particle size (mica and chlorite, for example), and many "non-clay" minerals (e.g., quartz, feldspar, amphiboles, and various carbonates) may occur in particles considerably less than 2 μm in diameter. Since the phyllosilicates of fine particle size most strongly influence the characteristic physical property of "clay," plasticity, these constitute the basic groups that we call "clay minerals." These materials can be detrital, authigenic, or diagenetic in origin.

CLASSIFICATION AND NOMENCLATURE

Basic Structures

The two basic building blocks of the phyllosilicates are octahedra layers of gibbsite- and brucite-like structure and sheets of corner-sharing $[SiO_4]$ tetrahedra. These have been illustrated very clearly by Grim (1968), and Figures 1 through 8 are from his book, which I recommend for a much more thorough discussion of the subject of clay mineralogy than is possible within the scope of the present paper.

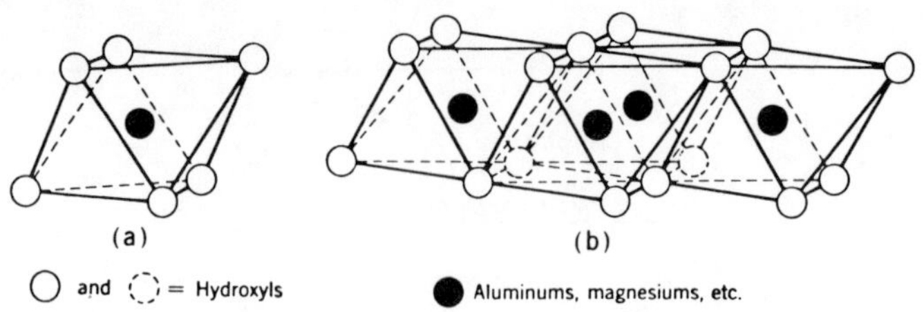

Figure 1. Diagram of (a) a single octahedral unit and (b) the sheet structure of the octahedral units. From *Clay Mineralogy* by R. E. Grim, copyright 1968 by McGraw-Hill, Inc. Used with permission of McGraw-Hill Book Company.

The octahedral portions of the structures are shown in Figure 1. Other cations, notably iron, may substitute for aluminum or magnesium in these octahedra, where trivalent cations (e.g., Al^{3+} or Fe^{3+}) occupy the octahedra, only two out of every three octahedra may contain a cation and still produce a balanced charge for the sheet.[1] Structures composed of such octahedral sheets are called "dioctahedral." Conversely, where divalent ions (e.g., Mg^{2+} or Fe^{2+}) are present, all three octahedra can be filled; the resulting sheets are termed "trioctahedral." The aluminum and magnesium hydroxide octahedral sheets form minerals in their own right, gibbsite and brucite, respectively. These minerals are allied to clay minerals but are not usually considered to be clays. The "traditional" clay minerals incorporate both the octahedral sheet and one or more tetrahedral sheets, like that shown in Figure 2. This figure illustrates an $[SiO_4]$ tetrahedron and the sheet structure formed by a hexagonal network of corner-shared tetrahedra. Aluminum and, rarely, other cations may substitute for silicon in these tetrahedra.

----

[1] In these discussions, I have followed the recommendations of the Nomenclature Committee of the Clay Minerals Society (Brindley *et al.*, 1968; Bailey *et al.*, 1971a,b; Bailey *et al.*, 1979) with regard to classification and terminology. "Recommended usage is as a single *plane* of atoms, a tetrahedral or octahedral *sheet*, and a 1:1 or 2:1 *layer*. Thus plane, sheet, and layer refer to increasingly thicker arrangements. A sheet is a combination of planes and a layer is a combination of sheets. In addition, layers may be separated from one another by various *interlayer* materials, including cations, hydrated cations, organic molecules, and hydroxode octahedral groups and sheets" (Bailey *et al.*, 1971).

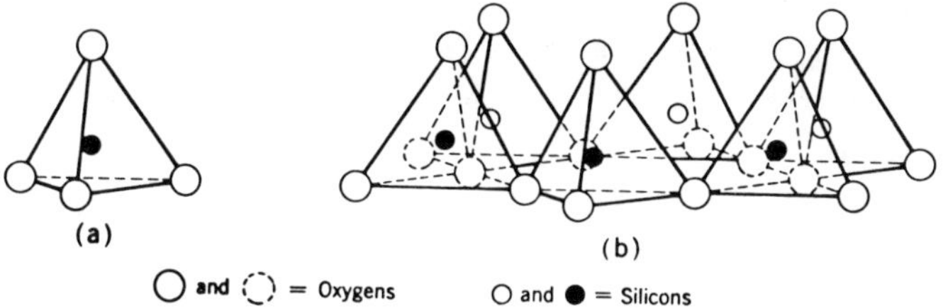

Figure 2. Diagram of (a) a single $[SiO_4]$ tetrahedron and (b) the sheet structure of $[SiO_4]$ tetrahedra arranged in a hexagonal network. From *Clay Mineralogy* by R. E. Grim, copyright 1968 by McGraw-Hill, Inc. Used with permission of McGraw-Hill Book Company.

The 1:1 Layer Group

Layers composed of one sheet of $[SiO_4]$ tetrahedra combined with one sheet of Al-octahedra form the kaolinite group of clay minerals (Fig. 3). The fit between the hexagonal network of the tetrahedral sheet and the octahedral sheet is not exact, and various distortions of these sheets give rise to different mineral species and polytypes within the group, and may impose a limitation on the size to which the individual crystallites may grow. Members of the kaolinite (or 1:1 layer) group are:

Kaolinite
Halloysite (tubular morphology)
Dickite } polytypes of kaolinite
Nacrite

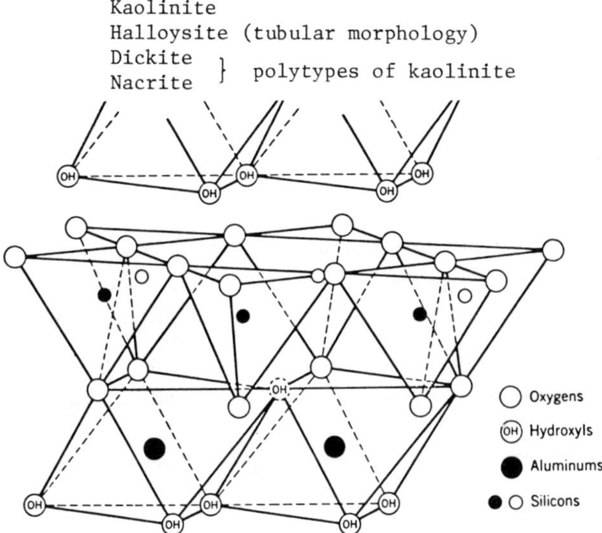

Figure 3. Diagram of the kaolinite structure. From *Clay Mineralogy* by R. E. Grim, copyright 1968 by McGraw-Hill, Inc. Used with permission of McGraw-Hill Book Company.

Of these, kaolinite is the only species generally reported in marine occurrences, although in some places halloysite may be present but unrecognized because of the similarities of its x-ray diffraction pattern to that of kaolinite.

The ideal chemical composition of the kaolinite group minerals and other minerals discussed in this paper are given in Table 1; halloysite has two forms, one with the same composition as kaolinite, and a second called halloysite (hydrated) having two additional water molecules which occupy the interlayer positions. I am not aware of any report of halloysite (hydrated) in marine samples. Inasmuch as drying causes irreversible dehydration of halloysite (hydrated) to halloysite, the hydrated form would not be detectable unless the sample were kept moist before and *during* analysis. As this is seldom, if ever, done in the analyses of marine samples, it is possible that some halloysite (hydrated) may exist in the marine environment but has gone undetected.

Dickite and nacrite are rare polytypes of kaolinite; that is, they are identical in composition but vary in stacking sequence of the 1:1 layers to give unit cells that differ in size and symmetry from the unit cell of kaolinite. To my knowledge, they have never been reported in marine samples.

The 2:1 Layer Groups

Structures composed of two sheets of $[SiO_4]$ tetrahedra with one octahedral sheet (gibbsite- or brucite-like) sandwiched between them comprise the 2:1 layer groups. These include pyrophyllite and talc, the illite or mica groups, which include glauconite, the smectite (montmorillonite) group, vermiculite, and palygorskite and sepiolite. The latter two minerals have, in the past, been grouped with the chain silicates rather than the phyllosilicates, but substantial reasons exist (Bailey *et al.*, 1971b) for classifying them as 2:1 layer silicates. Pyrophyllite, although not reported in marine samples, represents the ideal (no substitutions and therefore no net layer charge) dioctahedral (Al^{3+}) 2:1 structure and talc, the ideal trioctahedral (Mg^{2+}) structure. No substitutions of cations occur in either of these minerals in either the octahedral or tetrahedral layers, and no ions or molecules occupy the interlayer positions. Figure 4 illustrates the basic structure for the mica groups wherein a cation, usually potassium, occupies the interlayer

Table 1. Classification and ideal chemical formulas of clay minerals and other closely associated phyllosilicates.

Structural Groups	Ideal Formula[1]
<u>1:1 layers</u>	
Kaolinite group: *trioctahedral*	
Kaolinite (dickite and nacrite)	$Al_2Si_2O_5(OH)_4$
Halloysite	$Al_2Si_2O_5(OH)_4$
Halloysite (hydrated)	$Al_2Si_2O_5(OH)_4 \cdot 2H_2O$
Serpentine group: *trioctahedral*[2]	
<u>2:1 layers</u>	
Pyrophyllite	$Al_2Si_4O_{10}(OH)_2$
Talc	$Mg_3Si_4O_{10}(OH)_2$
Mica group: *dioctahedral*	
Muscovite	$KAl_2(Si_3Al)O_{10}(OH)_2$
Glauconite	$K(Fe^3_{1.33}Mg_{0.67})(Si_{3.67}Al_{0.33})O_{10}(OH)$
Celadonite	$K(MgFe^{3+})Si_4O_{10}(OH)_2$
Mica group: *trioctahedral*	
Biotite (phlogopite)	$KMg_3(Si_3Al)O_{10}(OH)_2$
Illite[3] (mostly dioctahedral but can include trioctahedral forms)	$K_{\leqslant 1}Al_{<2}(Fe^{3+}Mg^{2+})_{>0}Si_{4-x}Al_x)O_{10}(OH)_2$
Smectite: *dioctahedral*	
Montmorillonite	$Na_{0.33}(Al_{1.67}Mg_{0.33})Si_4O_{10}(OH)_2$
Beidellite	$Na_{0.33}Al_{2.22}(Si_3Al)O_{10}(OH)_2$
Nontronite	$Na_{0.33}Fe^{3+}_2(Si_{3.67}Si_{0.33})O_{10}(OH)_2$
Smectite: *trioctahedral*	
Saponite	$Na_{0.33}Mg_3(Si_{3.67}Al_{0.33})O_{10}(OH)_2$
Stevensite	$Na_{0.33}Mg_{2.83}Si_4O_{10}(OH)_2$
Vermiculite:	
trioctahedral	$(Mg,Ca)_{x/2}Mg_3(Si_{4-x}Al_x)O_{10}(OH)_2$
dioctahedral	$(Al)_{x/3}Al_2(Si_{4-x}Al_x)O_{10}(OH)_2$
Palygorskite	$(OH_2)_4Mg_5Si_8O_{20}(OH)_2 \cdot 4H_2O$
Sepiolite	$(OH_2)_4Mg_8Si_{12}O_{30}(OH)_2 \cdot 4H_2O$

...**CONTINUED, next page**...

Table 1 (continued).

Structural Groups	Ideal Formula
2:1:1 layers	
Chlorite group: *trioctahedral*[4]	
Clinochlore	$(Mg_5Al)(Si_3Al)O_{10}(OH)_8$
Chamosite	$(Fe^{2+}_5Al)(Si_3Al)O_{10}(OH)_8$

[1]All formulas have been reduced to the smallest unit formula for the sake of comparison; they do not reflect unit cell composition. The formulas are highly idealized; many variations occur, particularly in the interlayer cation positions of the smectite and vermiculite groups.

[2]Minerals of the serpentine group are not usually considered clay minerals.

[3]See text.

[4]Other minerals such as nimite (Ni-rich), pennantite (Mn-rich), and donbassite (dioctahedral) exist, but I consider them outside the scope of this paper.

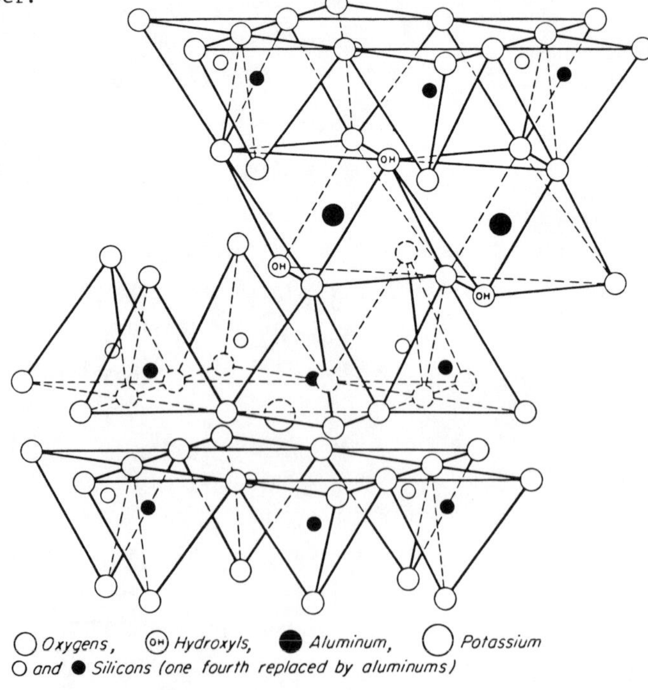

Figure 4. Diagram of the muscovite structure. From *Clay Mineralogy* by R. E. Grim, copyright 1968 by McGraw-Hill, Inc. Used with permission of McGraw-Hill Book Company.

space between adjacent 2:1 "sandwiches." This cation balances a positive charge deficiency caused by the substitution of an aluminum in one of every four "silica" tetrahedra. As the aluminum has a charge of +3 *vs* a charge of +4 for the silicon atom that it replaces, the layer is left with a net negative charge, balanced in muscovite, biotite and in most material classified as illite, by potassium. "Illite" differs from muscovite, a dioctahedral mineral, or biotite, the trioctahedral analog, in two ways: (1) "Illite" has been used to denote material which has turned out to be a variety of minerals having a considerable range of composition; whereas, muscovite and biotite are mineral species having compositions that vary within narrowly prescribed limits, and having structures that can be identified as specific crystallographic polytypes. Muscovite and biotite also have potassium contents that closely approximate the ideal of one atom per formula unit (not to be confused with unit cell) as exemplified in Table 1. Many "illites" yield chemical analyses that show a deficiency in potassium from this ideal, and many samples fail to exhibit enough crystallographic information to determine the polytypes involved. (2) Muscovite and biotite usually occur in crystallites of relatively large size--large enough to display the megascopic characteristics commonly associated with "mica" (flaky habit, highly reflective cleavage planes). Most material described as illite, on the other hand, occurs in very fine particle sizes and, in massive form, displays the characteristics associated with "clay," mainly that of plasticity. However, the differences are gradational and, in any gradational sequence, definition of a precise boundary is not always feasible. The term "mica" has been used to cover both categories of material, particularly where chemical information is lacking, the sample includes a variety of particle sizes; and x-ray powder diffraction analyses are the only means of identification; but "illite" has been used by many workers in the same way. Although the terms are not technically synonymous, one must interpret many reports as though they were.

Glauconite is defined by the Nomenclature Committee of the Clay Minerals Society (Bailey *et al.*, 1979) as an Fe-rich dioctahedral mica with tetrahedral Al (or Fe^{3+}) usually greater than 0.2 atoms per formula unit, and with trivalent octahedral cations correspondingly greater than 1.2 atoms. Similar micas with tetrahedral substitutions of less than

1.2 atoms. Similar micas with tetrahedral substitutions of less than 0.2 atoms are defined as celadonite. The committee further states: "The species glauconite is single-phase and ideally is nonstratified.[2] Mixtures containing an iron-rich mica as a major component can be called *glauconitic*. Specimens with expandable layers can be described as randomly interstratified glauconite-smectite. Mode of origin is not a criterion, and a green fecal pellet in a marine sediment that meets the definition of celadonite should be called celadonite."

I emphasize these nomenclatural considerations because glauconitic material is an important constituent of many marine samples and the term glauconite has been widely applied to any green fine-grained pelletal material, even though many samples of this material are actually mixtures of mineral species. Also, many analyses that report "illite" or "mica" may contain glauconite which, because it has similar crystallographic characteristics, may have gone unrecognized.

Smectite (Montmorillonite-saponite) Group

The term smectite has gained general acceptance among most, although not all, workers as a name for the group of 2:1 layer expandable minerals which in the past have often been called montmorillonite. Montmorillonite, however, is also a species name, and to avoid confusion, the Nomenclature Committee of the Clay Minerals Society, the Association International Pour L'étude Des Argiles, and the International Mineralogical Association (Bailey *et al*., 1971a) approved smectite as an alternate name for the group, expecting that a preferred term would emerge by popular usage. In keeping with the major usage thus far, the term smectite will be used in this paper. Table 1 lists the various species of smectite, and Figure 5 illustrates the most common view of their structure. Substitution of various cations in both the octahedral and tetrahedral sheets results not only in differentiation into the species of the group but, by producing net negative charges on the layers, such substitutions are responsible for the peculiar characteristic of smectites to incorporate various exchangeable cations and water molecules in the interlayer positions. These are manifest in the expandable or swelling capabilities of the smectites and their capacity to adsorb organic compounds in the inter-

---

[2]See discussion of mixed-layered minerals below.

layer space, characteristics which give rise to a multitude of commerical uses ranging from drilling mud to "Kitty Litter."

In marine sediments, it is not usually possible to determine the species of smectite present, especially in mixtures of other clay minerals or in mixed-layered assemblages, and identification is made by the group name only. Most marine smectites probably have compositions in the montmorillonite-beidellite range, although iron-rich smectite and saponite have been observed in the hot-brine deposits of the Red Sea (Bischoff, 1969; J.C. Hathaway, unpub. data).

Vermiculites are also 2:1 layer minerals and differ from smectites mainly in their lack of ability to swell when treated with certain

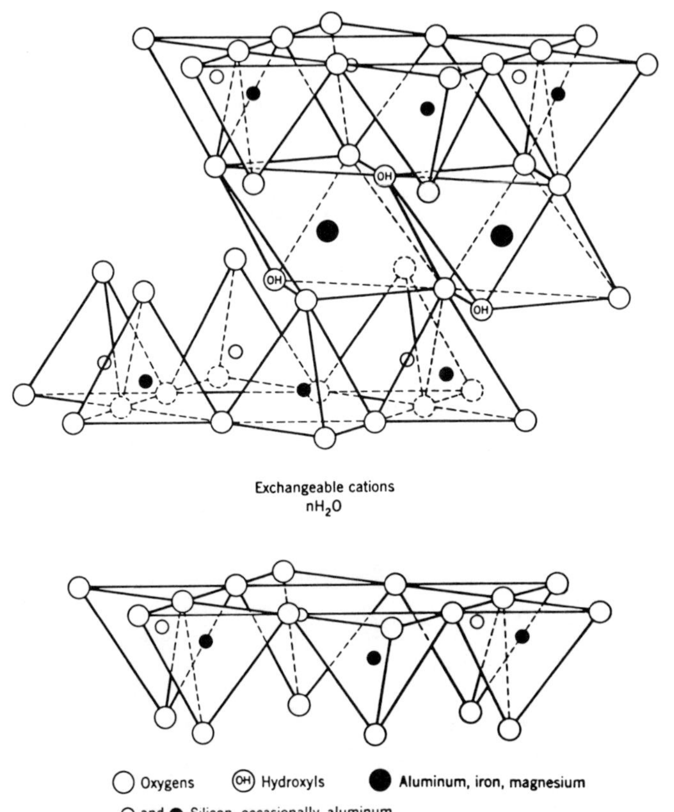

Figure 5. Diagram of the smectite structure. From *Clay Mineralogy* by R. E. Grim, copyright 1968 by McGraw-Hill, Inc. Used with permission of McGraw-Hill Book Company.

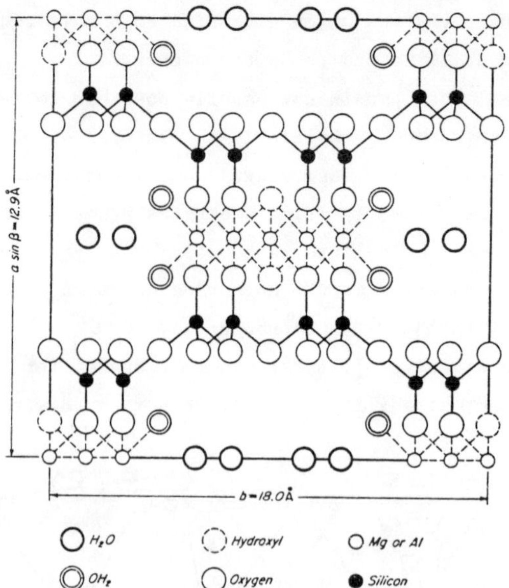

Figure 6. Diagram of the palygorskite structure. From *Clay Mineralogy* by R.E. Grim, copyright 1968 by McGraw-Hill, Inc. Used with permission of McGraw-Hill Book Company.

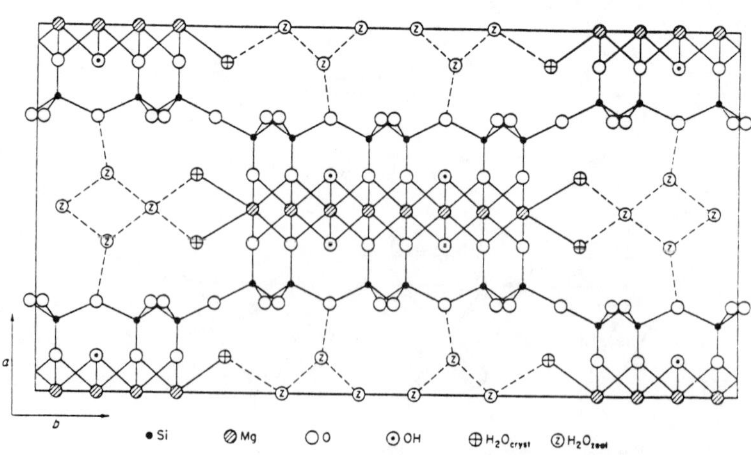

Figure 7. Diagram of the sepiolite structure. From *Clay Mineralogy* by R.E. Grim, copyright 1968 by McGraw-Hill, Inc. Used with permission of McGraw-Hill Book Company.

organic compounds. However, some trioctahedral vermiculites of very small crystallite size show characteristics that grade into those of saponite. Vermiculites are not reported as abundant in marine sediments, and may have gone unrecognized in many analyses.

Palygorskite and sepiolite are also 2:1 layer structures but, instead of forming continuous layers, the apices of the tetrahedra periodically reverse in direction to give alternating bands of 2:1 layers (Figs. 6 and 7). The spaces or channels between these bands are occupied by "structural" water molecules attached to the octahedral layers and by "zeolitic" or loosely held water molecules. The unit structures of the two minerals differ in size. Palygorskite usually contains aluminum which may substitute for magnesium; whereas, sepiolite containing virtually no Al has been reported (Hathaway and Sachs, 1965, and references therein). The presence of aluminum may be a factor in determining which of the two structures forms. Both minerals have been reported in marine sediments in recent years (Hathaway *et al.*, 1970; Deep Sea Drilling Project, 1970-1979; Hathaway *et al.*, in press) with palygorskite observed much more commonly and in greater abundance than sepiolite (Deep Sea Drilling Project, unpub. data file).

The 2:1:1 Layer Group

Minerals composed of the 2:1 layer arrangement but with an octahedral sheet incorporated in the interlayer position comprise the 2:1:1 layer, or chlorite, group. The interlayer sheet may be either magnesium or iron hydroxide (Fig. 8).

Although dioctahedral chlorites are known, most chlorites observed in marine sediments are probably trioctahedral. As in the case of the smectites, it is usually not possible to determine the species in most samples owing to the relatively small amounts present and their occurrence as very fine particles intimately mixed with other clay minerals. The group name chlorite is generally used for these minerals, most of which are detrital in origin and common in the marine sediments of high latitudes. However, chamosite has been reported as an authigenic or diagenetic mineral in some samples (Porrenga, 1967; Leclaire, 1968; Velde, 1977).

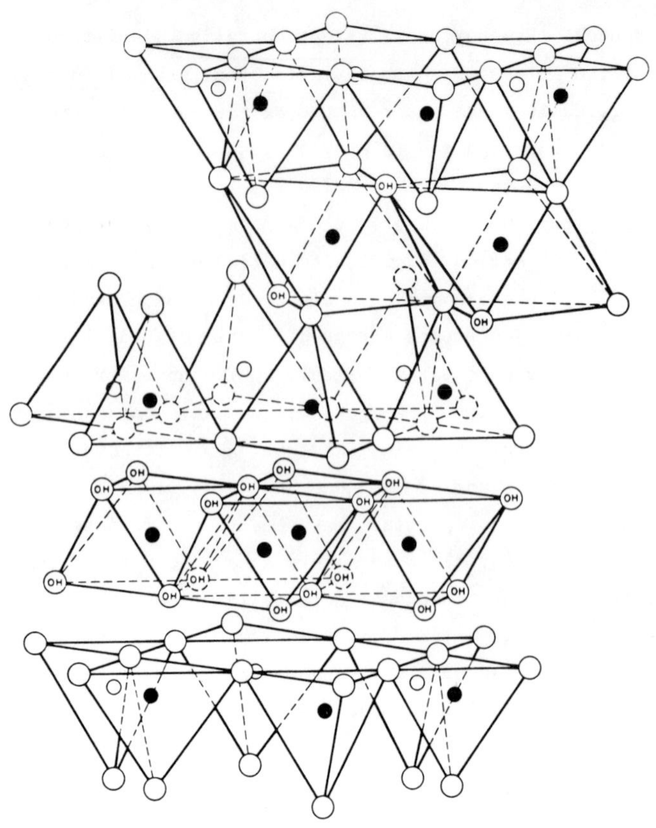

Figure 8. Diagram of the chlorite structure. From *Clay Mineralogy* by R. E. Grim, copyright 1968 by McGraw-Hill, Inc. Used with permission of McGraw-Hill Book Company.

Mixed-layered Minerals

Commonly, the 2:1 layer minerals occur interstratified with one another. Such interstratifications occur at the unit cell level and can be either regular or random. Regularly interstratified minerals are rare in marine sediments (except where burial diagenesis has occurred) but random mixed-layering is common. A usual combination is smectite-illite in which most layers are smectite (expandable layers) and fewer are illite. This produces x-ray diffraction characteristics that are much like those of "pure" smectite, and many analyses fail to distinguish between the two. Other combinations include chlorite or vermiculite layers. Only rarely, if ever, does kaolinite interstratify with the

2:1 layer or other minerals and, most probably, such assemblages are absent from marine sediments.

Amorphous Material

Amorphous silica is common in the marine environment (Lisitsin, 1974), and clay-size aluminosilicate material which also displays no detectable crystal structure probably occurs in many marine sediments. Heath and Pisias (1979) estimate that as much as 60% of the material in some samples from the north Pacific is amorphous to x-ray diffraction (the most common method of clay mineral analysis), and they point out that the effects of such material have too often been neglected in analyses of marine samples. They suggest an internal standard method to avoid the analytical errors resulting from failure to recognize this material. However, the method does not distinguish amorphous silica from amorphous silicates of clay mineral composition. Several methods have been proposed for the analysis of amorphous silica (Goldberg, 1958; Leinen, 1977; Eggiman *et al.*, in press, and references therein).

Palagonite, the product of the alteration in sea water of volcanic glass, may contribute considerable amorphous material although smectite is also one of the observed products.

Non-clay Minerals

The clay-size fraction of marine samples may contain fine-grained particles of minerals that are usually not considered clay minerals. Table 2 lists some that have been observed in marine samples and which may complicate a clay mineral analysis.

IDENTIFICATION AND THE PROBLEM OF QUANTIFICATION

Table 3 lists the principal methods used in the analysis of clay minerals. Of these, optical microscopy and x-ray powder diffraction are the principal tools routinely used for marine samples. The other methods are usually employed only in special or unusual cases. Optical microscopy is usually performed using what is called a smear slide, prepared by puddling a small amount of sample in water and spreading the suspension as a thin smear on a glass slide, drying it, and by immersing the grains in a medium such as heated, artificial Canada-balsam and covering with a

Table 2. Some non-clay minerals observed in marine sediments.[1]

Mineral or Group	Frequency of occurrence[2] (percent)	Mineral or Group	Frequency of occurrence[2] (percent)
Quartz	86		
Plagioclase feldspar	65	Rhodochrosite	<1
Potassium feldspar	37	Ilmenite	<1
Calcite	24	Chabazite	<1
Clinoptilolite	20	Cuprite	<1
Pyrite	16	Celestite	<1
Amphibole	11	Psilomelane	<1
Phillipsite	9	Bassanite[5]	<1
Cristobalite (opal CT; α)	9	Sphalerite	<1
Augite	5	Erionite	<1
Dolomite	5	Lepidocrocite	<1[4]
Tridymite	3	Akaganeite	<1[4]
Magnetite	3	Chalcopyrite	<1[4]
Goethite	3	Pyrrhotite	<1[4]
Gypsum	3	Melanterite[5]	<1
Halite	3	Ankerite	<1[4]
Analcite	3	Manganosiderite	<1[4]
Hematite	2	Manganite	<1[4]
Anatase	2	Todorokite	<1[4]
Aragonite	1[3]	Groutite	<1[4]
Gibbsite	<1	Heulandite	<1[4]
Siderite	<1	Serpentine	<1[4]
Apatite	<1	Rutile	<1[4]
Mg-calcite	<1[3]	Ilvaite	<1[4]
Ca-dolomite	<1	Various unknowns	5-7[4]

[1] Data from Bischoff (1969), J. C. Hathaway (unpub. data), Hathaway *et al.* (1970), Hathaway (1971), Peter Stoffers (pers. comm., 1973), Deep Sea Drilling Project (unpub. data file). J. C. Hathaway and L. J. Poppe (unpub. data).

[2] Number of times the mineral is reported in all fractions analyzed by [DSDP.

[3] DSDP samples only, probably higher for all marine samples.

[4] Estimated from data other than DSDP.

[5] May be artifacts of oxidation on drying.

Table 3. Methods of analysis of clay minerals.

Optical microscopy
X-ray powder diffraction (XRD)
Electron microscopy: transmission (TEM), scanning (SEM), electron diffraction
Chemical analyses: wet chemistry, x-ray fluorescence (XRF), emission spectrography, atomic absorption (AA), neutron activation, electron microprobe - scanning electron microscope (SEM), including wavelength and energy dispersive x-ray analysis.
Differential thermal analysis (DTA)
Infrared absorbtion spectrometry (IR)
Mössbauer spectrometry
Nuclear magnetic resonance (NMR)

cover glass. From such a slide can be determined or estimated: particle size, index of refraction, birefringence, color, presence of organic remains, and shape and state of aggregation of the particles. Glauconitic material can readily be detected. However, x-ray powder diffraction remains the principal tool in the analysis of clay minerals, and a thorough discussion of the subject is given in Brown (1961). Inasmuch as crystal structure differences are distinctive characteristics of clay minerals and fine crystallite size and mixing of phases do not adversely affect the resolution of the method, powder diffraction provides the most definitive information on the makeup of a natural mixture. The very small particle sizes and intimate mixing of phases render the results of most of the other analytical methods ambiguous. Differentiation of species may require mineral separation (not often possible with clay minerals) and chemical analysis, but identification of the clay mineral groups present is relatively easy by x-ray diffraction.

Typical x-ray diffractograms for a marine sample include the pattern of a powder of randomly oriented particles, which gives the full diffraction pattern for the sample, and the pattern of an oriented aggregate in which the flake shapes or basal cleavage planes of the layer silicates cause them to lie flat on the substrate, thus enhancing the diffraction peaks caused by interplanar spacings parallel to the flake surfaces. Since these peaks, called the basal reflections, behave in distinctive ways under various treatments, they are the most useful identifying characteristics of each clay mineral group. Table 4 lists some of these criteria.

Quantitative evaluation is not as easy, and Brindley (1961) presents a detailed discussion of the problem. One would guess intuitively that the intensities of the diffraction peaks would be a function of amount, and might lead one to a quantitative analysis of the sample. However, many factors in addition to the quantity of a mineral affect x-ray diffraction intensity. Prime among these is chemical composition. Isomorphous variation of elements within the structure represented by a mineral group can strongly affect both the relative and absolute intensities of the peaks produced. Mass absorbtion coefficient is heavily dependent on chemical composition, and it strongly affects the overall intensity of the entire pattern. Microabsorbtion and extinction effects play a part. Preferred orientation of particles has dramatic effect on the relative peak intensities. Crystallite size and perfection have strong effects. There is no assurance that any chosen standard will have the same set of all these characteristics as a given unknown; therefore, the best that one can hope for is an estimate, not a determination, of the amounts of the minerals detected. If a chemical analysis is available, several of the factors can be controlled and definite limits placed on possible composition. Internal standards can help control still other factors, but one is still faced with the fact that materials called by the same mineral name may be quite different from one sample to the next, simply because the name usually covers a variety of compositions. The chlorite of one sample is not necessarily the chlorite of another, and a quantitative comparison may be virtually meaningless. However, a ray of hope exists. Marine deposits usually vary only slightly over long distances, and ocean mixing processes probably insure that radical differences in minerals are smoothed out. The relative amounts (estimated) from sample to sample may reflect real and significant trends even though the absolute amounts estimated may be considerably in error. Thus, it is worthwhile to make estimates as carefully as possible in spite of the uncertainties but with full acknowledgment of them.

Table 4. Criteria for identification by x-ray powder diffraction of clay minerals commonly found in marine sediments.

Observed *d* spacings (in Angstroms) under various treatments				
Air dried	Treated ethylene glycol	Heated to 400°C	Heated to 550°C	Mineral or Group
>12 and usually <15	17	10	10	Smectite
	shift to larger spacing <17 and >14	shift to smaller spacing >10	possible intensity increase	Mixed-layered smectite-chlorite
		10	10	Mixed-layered smectite-illite
	14	14	14 with possible intensity increase	Chlorite
		10-14	10	Vermiculite
	12 or 12+	12 or may disappear	-	Sepiolite
>10 <12	shift to larger spacing	10	10	Mixed-layered illite-smectite
	10.5	10.5 or may disappear	-	Palygorskite
10	10	10	10	Illite
9.3	9.3	9.3	9.3	Talc
9	9	9	9	Clinoptilolite*
7.1	7.1	7.1	disappears	Kaolinite and chlorite
		disappears	-	Phillipsite*
5	5	5	5	Illite
4.7	4.7	4.7	disappears	Chlorite
			4.7	Talc
3.58**	3.58	3.58	disappears	Kaolinite
3.54**	3.54	3.54	disappears	Chlorite

*Although clinoptilolite and phillipsite are zeolites, not clay minerals, they are common constituents of marine samples and their peaks might be mistaken for those of clay minerals.

**These two peaks may not be resolved from one another.

Methods such as those proposed in recent papers by Pearson (1978) (in which the combined use of full chemical and x-ray analyses is detailed) and by Heath and Pisias (1979), using talc as an internal standard to allow determination of the effect of amorphous material, give promise of improving the accuracy of clay mineral quantification.

DISTRIBUTION AND SOURCES

The most comprehensive synthesis of data on the distribution of clay minerals in the surficial sediments of marine realm is that of Griffin *et al.* (1968). They incorporate their own data and that of Biscaye (1964, 1965) who had first presented such information for the Atlantic Ocean. The Deep Sea Drilling Project has since recovered and analyzed much material from greater depths of penetration than the surficial sediments previously sampled by grabs or piston cores. The mineralogic data from this program are contained in the more than 40 volumes of *The Initial Reports of the Deep Sea Drilling Project*. A data-file tape also contains the mineralogic information, and Table 5 shows the frequencies of occurrence and abundance of the clay minerals in samples represented by this file. As of this writing, a computer analysis of these data is not yet complete, so stratigraphic distributions are not presented here.

The general distribution shows mica to be the most frequently encountered and, in bulk samples, the most abundant clay mineral group. Within the clay (<2 μm) fraction, however, smectite is considerably more abundant. This reflects the difference in particle sizes between these materials; smectite occurs almost exclusively in particles less than 2 μm in diameter; whereas, silt- and sand-size mica occur in many samples. Chlorite is found more frequently than kaolinite but the total quantity of chlorite is smaller.

In this compilation, palygorskite surprisingly shows more abundance than either kaolinite or chlorite. However, a sampling bias may be present; an unrepresentative number of special samples may have been analyzed because of unusual appearance, or attempts to better delineate palygorskite-bearing zones. Sepiolite is much less common than palygorskite despite their structural and chemical similarity.

Table 5. Clay mineral distribution in marine samples.[a]

Mineral or Group	Frequency of occurrence[b] (percent)	Mean abundance in bulk samples (percent)[f]	Mean abundance in clay (<2 μm) fraction (percent)[f]
Mica group[c]	70	10.8	16.3
Smectite[d]	60	8.8	40.5
Chlorite	43	1.4	2.5
Kaolinite	32	1.7	4.9
Palygorskite	13	2.0	5.0
Sepiolite	<1	0.01	0.03
Talc	<1	*g*	*g*
Vermiculite[e]	<1[d]	*h*	*h*
Total clay mineral abundance[f]		24.7	69.2

[a] Calculated from unpublished data-file tape, Deep Sea Drilling Project, Legs 10-37.
[b] As these values are freuqencies, not mean quantities, the sum exceeds 100 percent. These frequencies reflect all size fractions.
[c] Includes illite and glauconite.
[d] Includes mixed-layered minerals.
[e] Estimated from Hathaway (1971).
[f] Not adjusted for amorphous (non-diffracting material).
[g] No abundances estimated.
[h] No data suitable for comparable calculation available.

Talc is exceedingly rare in spite of numerous reports in traces of this mineral in suspended material in seawater. These reports may reflect the modern introduction of this mineral into the atmosphere and hence the ocean through industrial and agricultural use of talc as a vehicle for pesticides.

Vermiculite is also rare and is probably restricted primarily to estuarine and continental shelf deposits (Hathaway, 1972).

Figures 9-12 show the geographic distribution of the four major clay minerals in surficial samples given in Griffin *et al.* (1968). Although later works (Goldberg and Griffin, 1970; Kolla *et al.*, 1976; Lisitsin, 1978; Heath and Pisias, 1979) illustrate modifications in areas such as the Indian Ocean and the North Pacific, these figures present the general outlines of worldwide distributions. They emphasis the strong influence of detrital processes.

The distribution of mica (illite) is illustrated in Figure 9. The influence of terrigenous source is illustrated especially well in the North Pacific by the large zone of high concentration of illite stretching from the Asian mainland eastward across the Pacific Ocean, a reflection of wind transport of terrigenous detritus. The North Atlantic sediments are also high in illite, which Griffin *et al.* (1968) also attribute primarily to wind transport. They point out that K-Ar ages of marine illite reflect not only a detrital rather than authigenic source but also, because of argon retention, indicate that little diagenesis in the form of crystallo-chemical alteration has taken place. River-borne illite concentrations also occur and are best delineated along the eastern coast of South America.

Smectite (montmorillonite) concentrations (Fig. 10) are higher in the southern regions than in the northern oceans. Griffin *et al.* (1968) suggest that this reflects a probable higher input of volcanic material in the southern oceans. They point out that the highest concentration area traverses the volcanic provinces of the South Pacific, and that the smectite is in close association in this region with phillipsite and volcanic shards. They also note that the low sedimentation rates reflect a lesser input of solid phases from the continents compared with contributions to the northern Pacific and other oceans.

Chlorite (Fig. 11) occurs in greatest concentrations generally in the high latitude parts of the ocean. This distribution reflects the

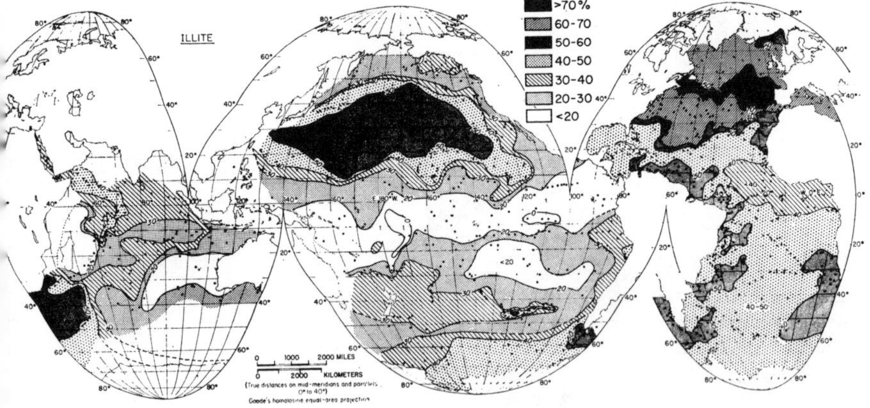

Figure 9. Illite (mica) concentrations in the <2 μm size fraction of surficial marine sediments. From Griffin, Windom and Goldberg (1968). Copyright 1968, Pergamon Press. Used with permission of Pergamon Press.

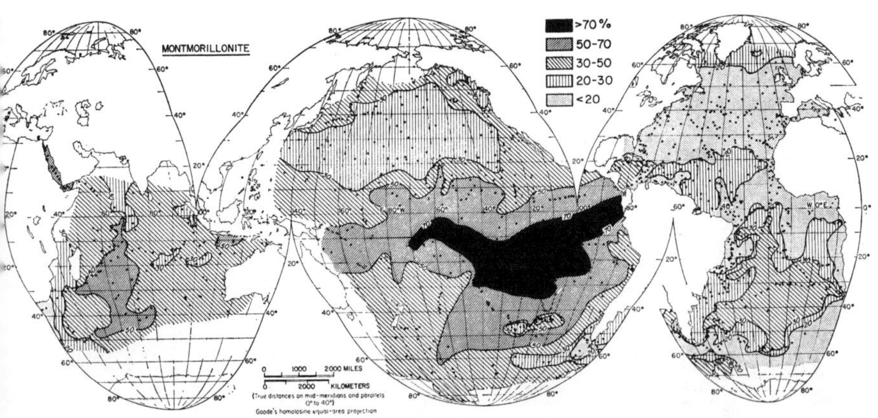

Figure 10. Montmorillonite (smectite) concentrations in the <2 μm size fraction of surficial marine sediments. From Griffin, Windom and Goldberg (1968). Copyright 1968, Pergamon Press. Used with permission of Pergamon Press.

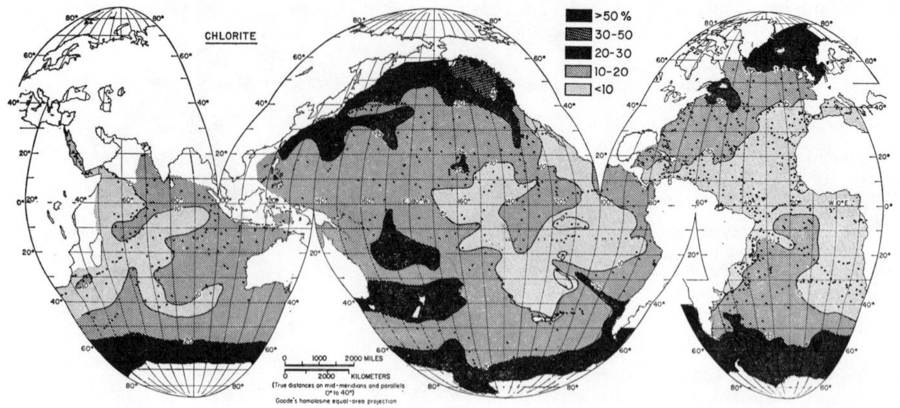

Figure 11. Chlorite concentrations in the <2 µm size fraction of surficial marine sediments. From Griffin, Windom and Goldberg (1968). Copyright 1968, Pergamon Press. Used with permission of Pergamon Press.

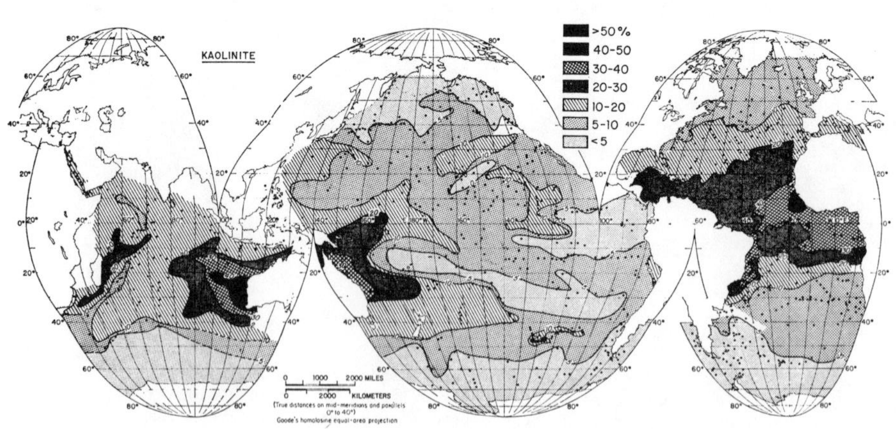

Figure 12. Kaolinite concentrations in the <2 µm size fraction of surficial marine sediments. From Griffin, Windom and Goldberg (1968). Copyright 1968, Pergamon Press. Used with permission of Pergamon Press.

abundance of chlorite in mechanically weathered rocks of the colder parts of the continents where soil-forming processes do not result in the chemical alteration of chlorite into other minerals. These chlorites have reached the oceans by wind, river, or glacial transport. However, the area of chlorite concentration northeast of Australia may be derived from volcanic material inasmuch as the chlorite is associated with volcanic glass, pyroxenes, and opaque minerals, and is not accompanied by terrigenous detrital material. A similar origin is proposed by Griffin *et al.* (1968) for the chlorite around the Hawaiian Islands although they note that Swindale and Fan (1967) reported chlorite formed around detrital gibbsite grains in sediments off Hawaii, suggesting that chlorite can form authigenically. However, in other areas where gibbsite occurs in marine sediments such as those off the east coast of Brazil (Biscaye, 1965), no concentrations of chlorite appear. Apart from this, the concentrations, x-ray diffraction characteristics, and particle size distribution of marine chlorites led Griffin *et al.* (1968) to conclude that most deep-sea deposits of chlorite are "products of continental erosion of chlorite-bearing metamorphic and sedimentary rocks and not the products of marine authigenic or of soil-forming processes."

Kaolinite distribution (Fig. 12) is almost the complement of that of chlorite. Highest concentrations occur in the lower latitudes, especially in the Atlantic. Kaolinite is well known as a product of chemical weathering in equatorial to sub-tropical climates so its occurrence in the seas adjacent to these regions is not surprising. River-borne material accounts for much of the distribution, except for the highest concentration off equatorial Africa which is probably the result of eolian transport. The large concentration west of Australia also probably results from the fall-out of wind-borne material derived from the western Australian desert. Griffin *et al.* (1968) explain the relative dearth of kaolinite off the western coasts of the Americas as the result of the mountain ranges extending the length of the continents which do not provide the proper conditions for the formation of kaolinite-rich soils. The meager runoff is characterized by high elevation weathering products (cold climate) which are normally very low in kaolinite. These mountain ranges also block further westward transport of dust from the North African deserts.

No map of palygorskite distribution is given here. Although it has been reported as occurring in surficial sediments from the western Indian Ocean region (e.g., Heezen *et al.*, 1965), most occurrences in marine sediments are reported from deeper and older (chiefly Mesozoic and Tertiary) material obtained by various drilling programs (Hathaway *et al.*, 1970; von Rad and Rösch, 1972; Weaver and Beck, 1977; Deep Sea Drilling Project, unpub. data file; Hathaway *et al.*, in press). Whether the palygorskite in marine sediments can be detrital, authigenic, or diagenetic in origin and where it forms is still the subject of study and argument (Millot, 1970; von Rad and Rösch, 1972; Weaver and Beck, 1977). Millot (1970, p. 326) favors authigenesis ("neoformation") in sedimentary basins in alkaline environments in either lakes or seas, although he recognizes (p. 329) that palygorskite ("attapulgite") as well as sepiolite can also result from later diagenesis. From their study of deep-sea drill cores from sites off northwestern Africa, von Rad and Rösch (1972) favor a diagenetic origin from degraded bentonitic (smectite-rich) clays by precipitation from magnesium-rich solutions with excess silica derived from the devitrification of silicic ash and/or the dissolution of opaline organisms. Weaver and Beck (1977) maintain strongly that most, if not all, palygorskite forms by chemical precipitation in coastal, brackish, lagoonal environments and that deep-sea occurrences reported from various Deep Sea Drilling Project sites, such as those discussed by von Rad and Rösch (1972), are detrital and composed of material transported from nearshore brackish lagoons. Sepiolite commonly accompanies palygorskite, but it is usually minor in amount (see Table 5). Only rarely does it occur alone or as a major mineral in deep-sea materials (Hathaway and Sachs, 1965; Zemmels *et al.*, 1972). Arguments concerning its origin are similar to those for palygorskite.

The Problem of Diagenesis

Twenty-five or more years ago, a prevalent concept held that clay minerals were quite sensitive to their environment and were susceptible to considerable change upon entering the marine environment from terrigenous sources. This idea stemmed largely from studies of estuarine or nearshore samples that showed differences in composition with distance toward or into the sea (Grim *et al.*, 1949; Grim and Johns, 1954; Powers,

1954, 1957) or from theoretical considerations which held that loss of ions, such as potassium, from the clay structures during weathering in soils would create "degraded" structures capable of reincorporating the lost ions upon entering the marine environment which was known to contain a supply of the lost ions. An assumption was made in these early studies and in interpretation of them (e.g., Millot, 1970) that transportation into the marine environment particularly via estuaries is a one-way street, from land to sea. That this was not the case was demonstrated later by Meade (1969) and supported by clay mineral studies of continental margin sediments (Hathaway, 1972); but other workers (Riviere and Vissé, 1954; Weaver, 1958) had, meanwhile, challenged the concept that alteration by diagenesis was responsible for the clay mineral assemblages observed in marine sediments. The studies by Biscaye (1965), Griffin *et al.* (1968), and various Russian workers (Rateev, 1971) on the worldwide distribution of clay minerals in the oceans made it clear that source is a far more important factor in determining the clay mineral composition of marine sediments than is diagenesis. Distinct changes in clays do occur upon deep burial (Hower *et al.*, 1976) and during incipient metamorphism (Dunoyer de Segonzac, 1970) largely as the result of temperature effects; but the concept prevalent in the 1950's, that diagenesis upon contact with sea water accounts for major distributions of marine clay minerals, has not survived the scrutiny of later studies.

CLAY MINERALS: REFERENCES

Bailey, S.W., Brindley, G.W., Johns, W.D., Martin, R.T., and Ross, M. (1971a) Summary of national and international recommendations on clay mineral nomenclature. Clays and Clay Minerals *19*, 129-132.

________, ________, ________, ________, and ________ (1971b) Clay Minerals Society. Report of nomenclature committee 1969-1970. Clays and Clay Minerals *19*, 132-133.

________, ________, Kodama, H., and Martin, R.T. (1979) Report of the Clay Minerals Society Nomenclature Committee for 1977 and 1978. Clays and Clay Minerals *27*, 238-239.

Biscaye, P.E. (1964) *Mineralogy and sedimentation of the deep-sea sediment fine fraction in the Atlantic Ocean and adjacent seas and oceans.* Ph.D. dissertation, Yale University.

________ (1965) Mineralogy and sedimentation of recent deep-sea clay in the Atlantic Ocean and adjacent seas and oceans. Bull. Geol. Soc. Am. *76*, 803-832.

Bischoff, J.L. (1969) Red Sea geothermal brine deposits: their mineralogy, chemistry, and genesis. *In*, E.T. Degens and D.A. Ross (eds.), *Hot Brines and Recent Heavy Metal Deposits in the Red Sea.* Springer-Verlag, New York, p. 368-406.

Brindley, G.W. (1961) Quantitative analysis of clay mixtures. *In*, G. Brown (ed.), *The X-ray Identification and Crystal Structure of Clay Minerals.* The Mineralogical Society, London, p. 489-516.

________, Bailey, S.W., Faust, G.T., Forman, S.A., and Rich, C.S. (1968) Report of the Nomenclature Committee (1966-1967) of the Clay Minerals Society. Clays and Clay Minerals *16*, 322-324.

Brown, G. (ed.) (1961) *The X-ray Identification and Crystal Structures of Clay Minerals.* The Mineralogical Society, London, 544 p.

Dunoyer de Segonzac, G. (1970) The transformation of clay minerals during diagenesis and low-grade metamorphism, a review. Sedimentology *15*, 281-346.

Eggiman, D.W., Manheim, F.T., and Betzer, P.R. (in press) Dissolution and analysis of amorphous silica in marine sediments. *J. Sed. Petrol.*

Goldberg, E.D. (1958) Determination of opal in marine sediments. Marine Res. *17*, 178-182.

________ and Griffin, J.J. (1970) The sediments of the northern Indian Ocean. Deep Sea Research *17*, 53-537.

Griffin, J.J., Windom, H., and Goldberg, E.D. (1968) The distribution of clay minerals in the world ocean. Deep Sea Research *15*, 433-459.

Grim, R.E. (1968) *Clay Mineralogy.* McGraw-Hill, New York, 596 p.

________, Dietz, R.S., and Bradley, W.F. (1949) Clay mineral composition of some sediments from the Pacific Ocean off the California coast and the Gulf of California. Bull. Geol. Soc. Am. *60*, 1785-1808.

Grimm, R.E. and Johns, W.D. (1954) Clay mineral investigations of sediments in the northern Gulf of Mexico. *Clays and Clay Minerals*, 2nd Natl. Conf. on Clays and Clay Minerals, Columbia, MO, 1953, Natl. Res. Council Pub. *327*, 81-103.

Hathaway, J.C. (1971) Data File, Continental Margin Program, Atlantic Const. of the United States *2*, Sample collection and analytical data. Woods Hole Oceanographic Inst., Ref. no. 71-15, 496 p.

________ (1972) Regional clay mineral facies in estuaries and continental margin of the United States east coast. Geol. Soc. Am. Mem. *133*, 293-316.

________, McFarlin, P.F., and Ross, D.A. (1970) Mineralogy and origin of sediments from drill holes on the continental margin off Florida. U. S. Geol. Surv. Prof. Paper 581-E, E1-E26.

________, Poag, C.W., Valentine, P.C., Miller, R.E., Shultz, D.M., Manheim, F.T., Kohout, F.A., Bothner, M.H., and Sangrey, D.A. (in press) U. S. Geol. Survey Core Drilling on the Atlantic Shelf. Science.

________, and Sachs, P.L. (1965) Sepiolite from the mid-Atlantic Ridge. Am. Mineral. *50*, 852-857.

Heath, G.R. and Psisias, N.G. (1979) A method for the quantitative estimation of clay minerals in North Pacific deep sea sediments. Clay and Clay Minerals *27*, 174-184.

Heezen, D.C., Nesteroff, W.D., Oberlin, A., and Sabatier, G. (1965) Decouverte d'attapulgite dans les sediments profonds du Golfe d'Aden et de la Mer Rouge. Compt. Rend. Acad. Sci. Paris *260*, 5819.

Hower, J., Eslinger, E.V., Hower, M.E., and Perry, E.A. (1976) Mechanism of burial metamorphism of argillaceous sediment: 1. Mineralogical and chemical evidence. Geol. Soc. Am. Bull. *87*, 725-737.

Kolla, V., Henderson, L., and Biscaye, P.E. (1976) Clay mineralogy and sedimentation in the western Indian Ocean. Deep Sea Res. *23*, 949-961.

Leclaire, L. (1968) Détermination du degrée d'oxidation d'un sédiment (boues ou vases actuelle et récents, marines, etc.) par l'etude de l'etat du fer dans ses formes minerals authigenes. Comt. Rend. Acad. Sci. Paris *266*, 452-454.

Leinen, M. (1977) A normative calculation technique for determining opal in deep sea sediments. Geochim. Cosmochim. Acta *41*, 671-675.

Lisitsin, A.L. (1974) *Osadkoobrazovanie v Okeanakh*. Izdat. "Nauka," Moscow, 438 p.

________ (1978) *Protsessy Okeanskoi Sedimentatsii*. Izdat. "Nauka," Moscow, 392 p.

Meade, R.H. (1969) Landward transport of bottom sediments in estuaries of the Atlantic Coastal Plain. J. Sed. Petrol. *39*, 222-234.

Millot, G. (1970) *Geology of Clays*. Springer-Verlag, New York, 429 p.

Pearson, J.M. (1978) Quantitative clay mineral analyses from the bulk chemistry of sedimentary rocks. Clay and Clay Minerals *26*, 423-433.

Porrenga, D.H. (1967) Glauconite and chamosite as depth indicators in the marine environment. Marine Geology *5*, 495-501.

Powers, M.C. (1954) Clay diagenesis in the Chesapeake Bay area. *Clays and Clay Minerals*, 2nd Natl. Conf. on Clays and Clay Minerals, Columbia, MO, 1953, Natl. Res. Council Pub. *327*, 68-80.

________ (1957) Adjustment of land-derived clays to the marine environment. J. Sed. Petrol. *27*, 355-372.

von Rad, U. and Rösch, H. (1972) Mineralogy and origin of clay minerals, silica and authigenic silicates in Leg 14 sediments. *In*, D.E. Hayes, A.C. Pimm *et al.* (eds.), *Initial Reports of the Deep Sea Drilling Project, vol. XIV*, Washington (U. S. Gov't. Printing Office), p. 727-751.

Rateev, M.A. (1971) Recent views on the distribution of clay minerals in the sediments of the world ocean [in Russian]. *Istoriya Miravogo Okeana*, Izdat. "Nauka," Moscow, p. 220-236.

Riviere, A. and Vissé, L. (1954) L'origine des mineraux des sediments marins. Bull. Soc. Geol. France 6e. serie *4*, 467-473.

Swindale, L.D. and Pow-Foong Fan (1967) Transformation of gibbsite to chlorite in ocean bottom sediments. Science *157*, 799-800.

Velde, B. (1977) *Clays and Clay Minerals in Natural and Synthetic Systems*. Elsevier, Amsterdam, 218 p.

Weaver, C.E. (1958) Geologic interpretation of argillaceous sediments, Part 1: Origin and significance of clay minerals in sedimentary rocks. Am. Assoc. Petrol. Geol. Bull. *42*, 254-271.

________ and Beck, R.C. (1977) Miocene of the SE United States, a model for chemical sedimentation in a peri-marine environment. *Developments in Sedimentology 22*, Elsevier, Amsterdam, 234 p.

Zemmels, I., Cook, H.E., Hathaway, J.C. (1972) X-ray mineralogy studies Leg 11. *In*, C.D. Hollister, J.I. Ewing *et al.* (eds.), *Initial Reports of the Deep Sea Drilling Project, vol. XI*, Washington (U. S. Gov't Printing Office), p. 729-789.

Chapter 6

MARINE PHOSPHORITES

F.T. Manheim and R.A. Gulbrandsen

INTRODUCTION

Phosphorite is a sedimentary rock composed largely of calcium phosphate minerals, dominated by carbonate fluorapatite.

The great bulk of the world's resources of phosphorus on land, estimated at 67 x 10^9 tonnes (Stowasser, 1979) occurs in the form of bedded sedimentary phosphorites of marine origin. These are found in strata from Precambrian age onward. Seventy-five percent of phosphorus resources are in the great Cenozoic deposits of North Africa (Tunisia, Morocco, and Algeria),[1] and U.S. southeast Atlantic, as well as any smaller but historically important coralline islands (e.g., Christmas Island in the Indian Ocean, Nauru, Makatea and Ocean Islands in the Pacific).

Aside from ocean island guano deposits, which were utilized by the Incas for fertilizer (Hutchinson, 1950), counterparts to the land phosphorites were first found in the ocean by the pioneering *Challenger* Expedition of 1873-76 (Murray and Renard, 1891), and again by Murray (1885) on the Blake Plateau off Florida-South Carolina. Both of these deposits had a harder nodular and abraded character that suggested foxxil (rather than currently forming) origin.

In contrast to the richly elaborated studies of phosphorites on land (Cayeux, 1939, 1941, 1950; McKelvey *et al.*, 1967; Bushinskii, 1966; and Gimmelfarb, 1965; and numerous references cited by these authors), sea-floor phosphorites were long poorly known except for the Agulhas Plateau deposits found by Murray, and phosphorites found on submarine banks off southern California (Dietz *et al.*, 1942). The work of Kazakov (1937), followed by McKelvey *et al.* (1953), McKelvey (1963), and Brongersma-Sanders (1957) pointed out the potential importance of upwelling, nutrient-rich zones in the world oceans to formation of sea floor phosphorites. As may be seen in Figure 1, there is

[1] Include Cretaceous phosphorites as well.

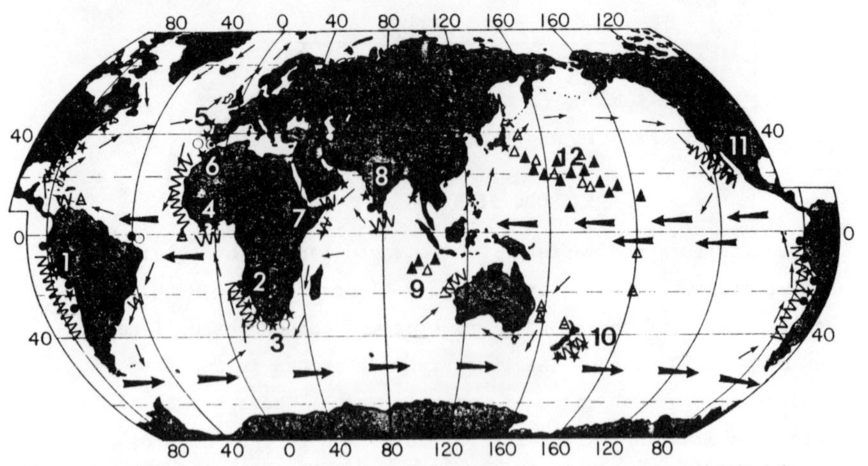

Figure 1. Distribution of upwelling zones and phosphorites in the ocean. Modified from Brongersma-Sanders (1957) and Baturin (1968). Zigzag lines show major areas of upwelling, corresponding in many cases to cold water currents along coast (small arrows). Fat arrows show major trade and other currents. Solid circles refer to Holocene (contemporary) phosphorites along continental margins. Stars indicate Middle Tertiary (Oligo-Miocene) phosphorites along continental margins. Hollow circles indicate Cretaceous phosphorite. Triangles refer to island phosphorites (horizontal bar, Middle Tertiary; solid, Paleogene; hollow, Cretaceous). Numbers designate following areas:

1. Peru-Chile slope and guano islands
2. Southwest Africa
3. Agulhas Plateau
4. Central Africa (Togo-Dahomey-Nigeria)
5. Spain (Cape Ortega)
6. North Africa (Morocco)
7. East Africa (Socotra)
8. Southwest India (also Arabian Sea)
9. Indian Ocean island deposits (e.g. Ocean and Christmas Island)
10. Chatham Rise
11. Southern California offshore banks
12. Pacific Ocean island phosphates
13. Southeast (U.S.) Atlantic margin

[Erratum added in proof: Fig. 1. Delete solid circle at equator and 40°W longitude.]

remarkable coincidence between some of the most prominent upwelling zones related to cold surface currents and submarine occurrences of phosphorite. This is especially true off the west coasts of North and South America, and Africa.

The distribution of high nutrient, and especially phosphate concentrations is clearly not the only factor involved in phosphorite deposition, as may be seen in Figure 2. Very high concentrations of nutrients and large bird populations occur in waters around Antarctica and in the Gulf of Alaska, where phosphorites are absent in bottom sediments. Moreover, phosphorite deposits occur off the U.S. Atlantic continental margin (Blake Plateau, Georgia-South Carolina, and North Carolina), off northern

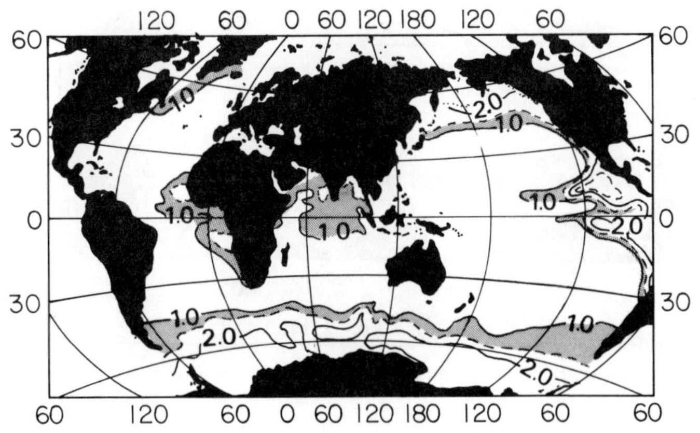

Figure 2. Distribution of dissolved (orthophosphate) phosphate in world oceans at 100 meters depth. Shaded area indicates 1.0 μgram/atom P per liter contour.

Spain (Lucas *et al.*, 1978), on the Chatham Rise off New Zealand, and elsewhere where there is no present upwelling.

Dramatic turns in knowledge of the formation of phosphorites have come since 1969. Kolodny (1969) showed by uranium-daughter dating of known phosphorite specimens that most sea floor phosphorites are not currently forming, but are relics. Thus the riddle remained: if upwelling was a major factor in phosphorite formation phenomena in various settings occurring in the world's oceans, why could we not find major areas of ongoing phosphorite formation? The answer was soon provided by concurrent studies of a Russian geochemist, G.N. Baturin at the Institute of Oceanology, Moscow, and his coworkers (Baturin, 1969, 1971; Baturin *et al.*, 1972). Baturin showed that, in fact, two classic sites of massive upwelling: off southwest Africa and the Peru-Chile margin, had both relict phosphorite in the form of hard concretionary and nodular phosphorites, as well as softer, phosphate-enriched diatomaceous and coprolitic sediments in which primary phosphorite was apparently being accumulated. The Holocene age was confirmed by Baturin and coworkers and by subsequent investigators through $^{234}U/^{238}U$ dating, maximum age determination using $^{230}Th/^{238}U$ activity ratios (Fig. 3; Veeh *et al.*, 1977; Burnett, 1974), gradational phosphatization of Holocene organisms (Manheim *et al.*, 1975) and other

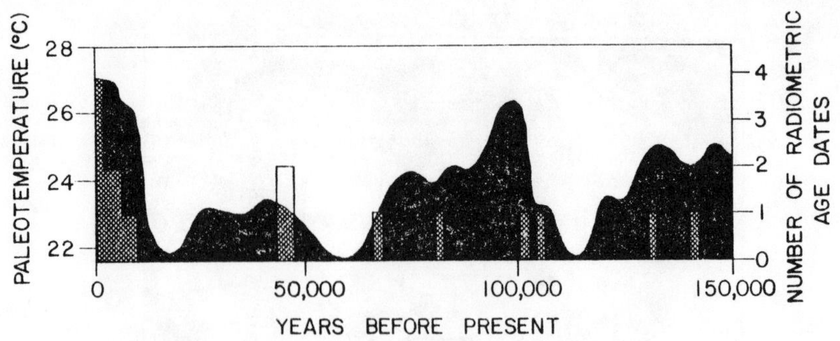

Figure 3. Radioactive age dates on Peru-Chile phosphorites and isotopic temperature measurements of the oceans in Late Pleistocene time (from Burnett, 1974).

geological analysis (i.e., observation of coprolitic phosphatic remains that must have been formed *in situ* because they were too fragile to permit transport). The Southwest Africa and Peru-Chile continental margins have become areas of intensive investigation (e.g., see Price and Calvert, 1978; Baturin, 1978).

MINERALOGY

Guano and biologically-derived phosphates.

Guano, the excreta of birds (or occasionally other mammals) is an indeterminate, organic-rich compound substance dominated by uric acid, calcium, phosphate, and potassium, with lesser sodium, magnesium and trace elements. Upon reacting with calcium carbonates in typical island substrates, the phosphate minerals listed below are commonly observed (see especially Gulbrandsen, 1975);

brushite	$CaHPO_4 \cdot 2H_2O$ (isostructural with gypsum)	
monetite	$CaHPO_4$	
whitlockite	$\beta Ca_3(PO_4)$	
dahllite	$Ca_{10}(PO_4)_6(OH)_2$	[with CO_3 substituted for OH] [carbonate hydroxyapatite]

These phases are often accompanied by other salts, such as struvite ($NH_4MgPO_4 \cdot 6H_2O$), gypsum, etc., and the apatites are inherently low in fluorine. On exposure to sea water by immersion, spray or wash, alteration and recrystallization take place. Altschuler (1973) suggests that most of the fluorine (F) of insular apatite is acquired by replacement of OH, the amount increasing with the length of time hydroxyapatite is exposed to sea water. Thus, a conditional test for guano origin of phosphorite is whether it has a F/P_2O_5 ratio less than an empirical value of 0.089. Marine phosphorites almost invariably exceed this ratio. A full complement of F corresponding to fluorapatite has a molar ratio F/(P + C) of 0.333. The main deposits of such islands as Nauru, Ocean and Christmas, are older and higher in F than younger guano phosphorites on Enderbury Island, Pacific Ocean (Gulbrandsen, 1975).

McConnell (1973) has pointed out that the mineral of bone and teeth has been incorrectly called hydroxyapatite ($Ca_{10}(PO_4)_6(OH)_2$) by bioscientists. Rather, it is a dahllite, or low-fluorine carbonate hydroxyapatite with CO_2 > 1%, and F < 1%, set in an organic matrix. A summary by Kudrin *et al.* (1964) indicated that freshwater fish bone has about 0.02% F, and marine fishes and mammals contain on the order of 0.4% F, corresponding to a F/P_2O_5 ratio of 0.022. Other phosphatic hard parts include conodonts, shells of brachiopods, invertebrate radula (teeth) and parts of crustacean and bryozoan exoskeletons, urinary calculi and renal (kidney) stones. Doyle *et al.* (1978) recently reported concretions in estuarine mollusk kidneys that contained 6% Zn and 0.4% Cu. On death of the organisms in a marine environment, most of these phases ultimately alter to full carbonate fluorapatite or dissolve.

Marine apatite

The main mineral in marine phosphorites is carbonate fluorapatite, whose composition and properties are still imperfectly known. This is due in large part to the mineral's ubiquitous occurrence as impure aggregates of less than micron-size crystals, to its complexity in composition, and also to failure so far to synthesize it in sea water solutions. Because of difficulty in obtaining analyses free from inhomogeneities and mechanical admixtures, we will treat theoretical compositions as a means of illustrating the constraints and probable mineralogical relationships.

Carbonate fluorapatite is a variety of fluorapatite that exhibits a small range of compositional variations. The structure of the mineral fluorapatite is hexagonal, space group $P6_3/m$. The unit cell contains

$Ca_{10}(PO_4)_6F_2$. Thus 20 positive charges of the 10 calcium atoms are balanced by 20 negative charges of the 6 phosphate anion groups plus the 2 fluorines. These therefore, are the basic features of the mineral that define the structural and charge constraints required in considering the substitution of other cations or anions for the main components. Table 1 shows the theoretical compositions and formulas of fluorapatite and some carbonate fluorapatites.

The principal component that characterizes marine apatite is the carbonate (CO_3) that substitutes for phosphate (PO_4), hence the descriptive compositional name of carbonate fluorapatite. The substitution is an unusual one; an essentially planar carbonate anion (a carbon atom in the center of a triangle of 3 oxygen atoms) takes the place of a tetrahedral phosphate anion (a phosphorus atom in the center of a tetrahedral array of oxygens). Legeros *et al.* (1967) have shown that CO_3 has the effect

Table 1. Theoretical chemical compositions and structural formulas of fluroapatite (no. 1) and some carbonate fluorapatites (nos. 2, 3, and 4).

	1	2	3	4
	Weight percent			
CaO	55.60	56.50	54.66	53.78
MgO	—	—	—	.82
Na_2O	—	—	1.59	1.27
P_2O_5	42.22	35.75	36.41	36.92
SO_3	—	—	—	.82
CO_2	—	4.43	4.52	3.59
F	3.77	5.74	4.87	4.85
Total	101.59	102.42	102.05	102.05
O=F	1.59	2.42	2.05	2.04
Total	100.00	100.00	100.00	100.01

STRUCTURAL FORMULAS

	10 cations	6 anion groups	2 F anions
1.	Ca_{10}	$(PO_4)_6$	F_2
		Total cation charges = +20 Total anion charges = -20	
2.	Ca_{10}	$(PO_4)_5\,(CO_3F)$	F_2
		Total cation charges = +20 Total anion charges = -20	
3.	$Ca_{9.5}Na_{.5}$	$(PO_4)_5\,(CO_3)_{.5}\,(CO_3F)_{.5}$	F_2
		Total cation charges = +19.5 Total anion charges = -19.5	
4.	$Ca_{9.4}\,Mg_{.2}\,Na_{.4}$	$(PO_4)_{5.1}\,(SO_4)_{.1}\,(CO_3)_{.3}\,(CO_3F)_{.5}$	F_2
		Total cation charges = +19.6 Total anion charges = -19.6	

of reducing crystallite size to submicroscopic dimensions. This gives rise to an aggregate polarization effect which causes both ancient and modern nodular apatite to appear pseudoisotropic under the polarizing microscope (Fig. 8).* Not only is the structural accommodation of carbonate substitution difficult to perceive, but it also creates an imbalance of charges; a negative charge is lost. If, however, an atom of fluorine should accompany the carbonate and form what would probably be a distorted tetrahedral anion $(CO_3F)^{-3}$, the effect of carbonate substitution upon the structure would be greatly reduced and the charges would be balanced. This solution to the problem requires one mole of fluorine, in addition to the two moles that are essential in the fluorapatite unit cell, for each mole of carbonate that substitutes for phosphate. An apatite with a substitution of one mole each of carbonate and fluorine, for example, has a structural formula of $Ca_{10}(PO_4)_5(CO_3F)F_2$, as shown in Table 1. The carbonate as CO_2 comprises 4.43% by weight of the mineral, an amount probably somewhat higher than that of most marine apatite; and the total of fluorine is 5.74% by weight, an amount higher than any that is known. Since carbonate substitution ranges up to around 8% as CO_2, the appealing concept of $(CO_3F)^{-3}$ anions is not required, at least for all carbonate replacement. The $(OH)^-$ anion, which is similar in size to F^-, may play the same role and make up for the deficiency of fluorine. The presence of $(OH)^-$, however, has not been confirmed in these cases. Another means of compensating for the charge imbalance due to the substitution of $(CO_3)^{-2}$, but not the disparity between the forms of the carbonate and phosphate anions, is found in the nearly ubiquitous occurrence of significant amounts of sodium in the apatite.

The amount of sodium in marine apatite, as shown by the analyses of McClellan and Lehr (1969), ranges up to about 1.6% Na_2O but is generally less than 1%. The sodium substitutes for calcium, being closely similar in ionic radius, but has a positive valence of one rather than two of calcium. Its substitution for calcium acts, therefore, to decrease the positive charges of the structure and to help balance the decrease in negative charges that is brought about by the substitution of carbonate for phosphate, as discussed above. In this case, where sodium and carbonate are the only substitutions of significance, the sum of the moles of sodium and the moles of excess fluorine (in excess of the two moles of essential F) should equal the moles of carbonate, as shown by formula 3

---

*See inside front and back covers for Figure 8.

of Table 1. This equality will balance the charges of the structure.

The substitution of sulfate $(SO_4)^{-2}$ (McArthur, 1978) for phosphate complicates the relationship discussed above. The sulfate anion is similar to the phosphate tetrahedral configuration in size and does not cause strain in the structure like the carbonate anion does. Like carbonate, it is divalent and requires a compensating substitution in order to maintain overall charge balance. Sodium substitution is the most obvious way of satisfying this requirement. The charges are balanced, therefore, when the sum of the moles of sodium and excess fluorine equals the sum of the moles of carbonate and sulfate (formula 4, Table 1). A plausible substitution that could help to balance the decrease in charges due to carbonate and sulfate substitution is the substitution of silicate $(SiO_4)^{-4}$ for $(PO_4)^{-3}$. This would increase the negative charges. The silicate substitution in small amount is a difficult one to prove and so far has not been substantiated.

Magnesium appears to substitute in small amount for calcium in most marine apatite, but its presence in other minerals, particularly dolomite, is not easily precluded. Because magnesium is much smaller in size than calcium, its substitution in large amount is not favored. Magnesium is divalent like calcium and therefore imposes no change in the balance of charges; it is included as a component of apatite in formula 4 of Table 1.

Chemical analyses and structural formulas of real marine apatite are presented in Table 2. The method used for calculation of the structural formulas is illustrated in Table 3. There are several ways to make this calculation, but none is clearly superior to the others. Only the apatite constituents are reported in the analyses. The Florida, Morocco, and western U.S. (Phosphoria) analyses are averages of 30, 8, and 18 samples, respectively, of apatite concentrates (free of calcite and dolomite) from those deposits. They are data of McClellan and Lehr (1969). The other analyses are averages of 3 samples each of the nodular deposits on the continental shelves off Southwest Africa and Peru (Baturin, 1971). The CO_2 and MgO contents are assigned to apatite; no calcite or dolomite is reported and the <u>a</u> cell dimensions determined by Baturin and others (1970) on other samples from the Southwest African shelf are consistent with the CO_2 contents reported here.

Table 2. Chemical analyses of impure marine carbonate fluorapatite concentrates and structural formulas of the apatites. (Only the constituents of apatite are reported.)

	1 Florida[1]	2 Morocco[1]	3 WUSA[1]	4 S-W Africa[2]	5 Peru[2]	6
			weight percent			
CaO	43.7	51.8	44.5	41.5	29.5	
MgO	.32	.37	.26	1.6	1.5	
Na_2O	.51	.73	.54	—	—	
P_2O_5	29.9	33.9	31.1	28.1	20.7	
CO_2	3.1	4.05	1.4	5.7	2.8	
F	3.5	4.2	3.4	2.7	2.1	

Structural formulas

	10 cations	6 anion groups	F anions
1.	$Ca_{9.7}\,Mg_{.1}\,Na_{.2}$	$(PO_4)_{5.1}\,(CO_3)_{.7}\,(CO_3F)_{.2}$ Total cation charges =+19.8 Total anion charges =-19.3	F_2
2.	$Ca_{9.7}\,Mg_{.1}\,Na_{.2}$	$(PO_4)_{5.0}\,(CO_3)_{.7}\,(CO_3F)_{.3}$ Total cation charges = +19.8 Total anion charges = -19.3	F_2
3.	$Ca_{9.7}\,Mg_{.1}\,Na_{.2}$	$(PO_4)_{5.6}\,(CO_3)_{.1}\,(CO_3F)_{.3}$ Total cation charges = +19.8 Total anion charges = -19.9	F_2
4.	$Ca_{9.5}\,Mg_{.5}$	$(PO_4)_{4.5}\,(CO_3)_{1.5}$ Total cation charges = +20 Total anion charges = -18.1	$F_{1.6}$
5.	$Ca_{9.3}\,Mg_{.7}$	$(PO_4)_{4.9}\,(CO_3)_{1.1}$ Total cation charges = +20 Total anion charges = -18.8	$F_{1.9}$

[1]Averages of 30 Florida samples, 8 Morocco samples and 15 western U.S. (Phosphoria) samples. Data from McClellan and Lehr, 1969.

[2]Averages of 3 samples from the South-West Africa shelf and 3 samples from the Peru shelf. Data from Baturin, 1971.

The composition of the apatites of Florida, Morocco, and western U.S. are reasonably representative of Tertiary and older marine apatite. McClellan and Lehr (1969) found, out of 94 analyses of marine apatite, that substituent elements ranged individually only up to 0.34 moles of magnesium, 0.53 moles of sodium, and 1.35 moles of carbonate. They determined sulfate in 17 samples and found a range of 0.02 to 0.10 moles and considered the amounts to be too low generally to be of significance. Significantly greater amounts of sulfate occur in some marine apatite; Bliskovskii *et al.* (1977), for example, report values up to 3.8 percent SO_3.

Table 3. Method of calculation of structural formula, using data of Florida carbonate fluorapatite in Table 2.

	Weight %	Monatomic MW[1]	Moles[2]	Recalc. Moles[3]	Cations	Anions	Charges[4]
CaO	43.7	56.08	.77924	9.696	Ca^{+2}		+19.392
MgO	.32	40.31	.00794	.099	Mg^{+2}		+.198
Na_2O	.51	30.99	.01646	.205	Na^{+1}		+.205
		total	.80364	10.000		total	+19.795
P_2O_5	29.9	70.97	.42131	5.141		PO_4^{-3}	-15.423
CO_2	3.1	44.01	.07044	.859		CO_3^{-2}	-1.718
		total	.49175	6.000			
F	3.5	19.	.18421	2.248		F^{-1}	-2.248
						Total	-19.389

STRUCTURAL FORMULA

$$Ca_{9.7}\ Mg_{.1}\ Na_{.2}\ /\ (PO_4)_{5.1}\ (CO_3)_{.7}(CO_3F)_{.2}\ /\ F_2$$

[1]Molecular weight of 1 atom plus equivalent oxygen, except for fluorine.

[2]Weight percent/monatomic molecular weight.

[3]Moles of Ca, Mg, and Na recalculated to 10 cations of unit cell.
Recalculated moles = (10/.80364) moles.
Moles of P and C recalculated to 6 anion groups of unit cell (P moles = PO_4 moles; C moles = CO_3 moles). Recalculated moles = (6/.49175) moles
Moles of F recalculated with the same factor that is derived for P and C, arbitrarily.

[4]Valence of respective cations and anions multiplied by corresponding number of recalculated moles.

The balance of compensating substitutions (Na moles + excess F moles = (CO_3) moles) in these apatites can be determined from their structural formulas; for example, the Florida apatite shows 0.2 moles of Na, 0.2 moles of excess F and 0.9 moles of (CO_3), an imbalance of 0.5 moles. Similarly, the Morocco apatite has the same imbalance of 0.5 moles, whereas the western U.S. imbalance is only 0.1 moles. Most of the individual analyses of McClellan and Lehr (1969), including other deposits than the three used here, show a deficiency of Na and excess of F.

The compositions of the Holocene and Pleistocene apatites from the Southwest Africa and Peru shelves differ from the older ones in having greater amounts of magnesium and having either a minimum amount of fluorine (2 moles) or a deficiency. Baturin (1978, and references cited) related degree of lithification to compositional changes: CaO, CO_2, F, and Sr increase with lithification; Mg and SO_3 decrease.

Reviewing a broad range of apatites, McClellan and Lehr (1969) found that unit cell a dimensions ranged from 9.322 to 9.392A, and c ranged from 6.877 to 6.900A. Carbonate substitution caused the main changes in the a dimension. Gulbrandsen (1970) developed an X-ray peak pair method of estimating CO_2 weight percent: $y = 23.6341 - 14.7361x$, where y is CO_2 in weight percent and $x = \Delta 2\theta[(004) - (410)]$, and the (410) and (004) peaks occur around 51.6° and 53.1°2θ for Cu*K*α X-radiation. Data from McClelland and Lehr (1969) show, in turn, that a cell dimension is approximated by $a = 9.3730 - CO_2/108.253$, where CO_2 is in weight percent and a in Ångström units. This method is particularly useful for quick estimates, and/or where admixture of calcite or dolomite complicate direct chemical determination of CO_2.

Relationships of chemical parameters to key physical properties are shown briefly in Table 4. Because of small crystallite size, only mean refractive index can normally be determined. This may be partly obscured by organic matter, or modified by radiation damage due to uranium fission. However, the Table points up the controlling influence of CO_2 on both specific gravity, index of refraction and unit cell size, and demonstrates the use of refractive index as an independent method of obtaining compositional information on marine apatite.

Table 4. Selected physical and chemical properties of typical phosphorites and apatites. Values in weight percent except where noted.

Location	F	CO_2	P_2O_5	F/P_2O_5	Mol ratio F/(P + C)	Density	Refr. Index[2]	a(A)[4]	Ref.
Blake Plateau	2.5	11.4[1]	22.2	0.11	0.293	2.90-.95	1.600	9.30	1
Polpinskoe	3.13	6.86	31.77	0.098	0.274	2.91-.95	1.600	9.305	2
Aksai (Karatau)	2.07	4.93	34.62	0.068	0.182	3.05-3.10	1.613	9.325	2
Luga (Baltic)	2.23	3.45	36.13	0.062	0.199	2.90-3.10	1.618	9.32	2
R. Podkamennoi (Tunguska)	3.04	1.63	36.59	0.083	0.290	3.0-3.10	1.619	9.34	2
Chulaktau (Karatau)	2.88	0.50	32.11	0.090	0.328	-	1.627	9.35	2
Fluorapatite:[3] Cerro de Mercado, Mexico	0.07	0.05	40.78	0.0012	0.318	3.216	1.636	9.391	3

1/ Contains physically admixed free $CaCO_3$

2/ ω index.

3/ Also contains 0.41% Cl and 1.43% rare earths.

4/ Mean of 21 crystal measurements.

References:

1. Manheim et al., 1979.
2. Bushinskii, 1966.
3. Young et al., 1969.

Solubility of calcium phosphates

The solubility of calcium phosphates is highly dependent on pH, the mineral and dissolved chemical species involved, and surface effects (Kester and Pytkowicz, 1967; Leckie and Stumm, 1970). The distribution of orthophosphate species is related to pH, as depicted in Fig. 4.

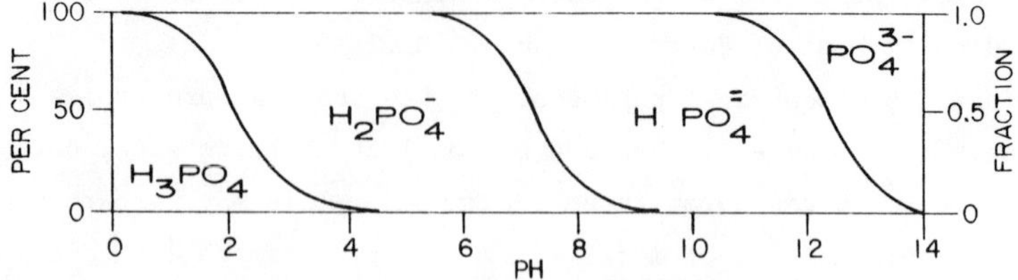

Figure 4. Distribution of main orthophosphate species with pH (Sillen, 1959).

Table 5. Solubility of phosphate compounds at standard temperature and pressure (approximately 1 atmosphere, 25°C).

Species	Name	Solubility product (K_{sp})*	Reference
$Ca(H_2PO_4)_2$		9.3×10^{-3}	Ulrich (1961)
$CaHPO_4$	monetite	1.8×10^{-6}	Ulrich (1961)
$CaHPO_4 \cdot 2H_2O$	brushite	2.2×10^{-7}	Farr (1950)
$Ca_3(PO_4)_2$		1×10^{-25}	D'Ans-Lax (1967)
$Ca_4H_2OH(PO_4)_3$	"defect apatite"	1.3×10^{-47}	Moreno *et al.* (1960)
$Ca_5(OH)(PO_4)_3$	hydroxyapatite	5×10^{-28} 3×10^{-58}	Various authors in Koritnig (1976)
$Ca_5F(PO_4)_3$	fluorapatite	4.0×10^{-61}	Massard (1973)

*e.g. for $CaHPO_4 = Ca^{2+} + HPO_4^{2-}$ $K_{sp} = \dfrac{(Ca^{2+}(HPO_4{}^{2-})}{(CaHPO_4)} = 1.8 \times 10^{-6}$

where $CaHPO_4$ is a solid whose activity = 1.

Replacement of carbonate by phosphate is recognized both in the field, and demonstrated by laboratory experiments on calcite by Ames (1959), using phosphate solutions as dilute as 1ppm. Equilibration of natural phosphate minerals, which may include extraneous minerals, with sea water at temperatures from 2° to 25°C does not show predictable changes in composition (Table 6).

<u>Organic and trace constituents of phosphorites</u>

Many evidences speak for the origin of non-guano marine phosphorites in an organic-rich, reducing environment. The actual concentration of organic matter remaining with apatite, however, may be variable. Analyses of contemporaneous phosphorites are shown in Table 7.

A characteristic element in marine phosphorites is uranium, which ranges in value from a few parts per million to about 500 ppm, with a mean on the order of 80 ppm. Maximum values are observed in the Chatham Rise, off New Zealand (Cullen, 1978). Phosphate nodules exposed to oxidizing conditions and currents at the sea floor may have the more insoluble U(IV) component oxidized to U(VI) and leached away, depleting the sample in uranium (Altschuler *et al.*, 1958; Burnett and Gomberg, 1977). Buried phosphorites generally retain original uranium proportions, and maintain a sufficiently constant U/P_2O_5 ratio such that prospecting for phosphorites by gamma ray logging is a highly effective and sensitive technique. Still

Table 6. Composition of 3 natural paptites and 1 marine apatite (10 g) exposed to 100 g sea water at 2° and 25°C for 192 hours (from MacKnight, 1976).

	Mexico		Ontario		Quebec		Florida	
	2°C	25°C	2°C	25°C	2°C	25°C	2°C	25°C
$\Sigma PO_4 \times 10^{-5}$M	10.1	9.5	10.5	4.5	3.2	6.1	2.3	9.1
F $\times 10^{-5}$M	3.9	3.5	8.9	4.6	6.7	5.0	6.3	6.8
OH $\times 10^{-7}$M	2.2	6.1	4.5	8.0	10.7	14.7	3.7	7.0
Ca $\times 10^{-3}$M	6.8	8.1	8.6	10.6	9.3	10.2	8.0	7.1

Table 7. Selected biogenic constituents in lithifying phosphorites from Southwest Africa (from Romankevich and Baturin, 1972).

Species	Associated sediment	Phosphatic concretions: Soft	Partly lithified	Dense
P_2O_5	1.15	23.8	29.6	32.7
SiO_2 (amorphous)	53.8	7.8	4.05	.90
CO_2	1.5	5.52	5.34	6.33
C_{org}	5.2	1.8	1.03	.92
N_{org}	0.60	0.18	0.097	0.081

higher uranium concentrations have been reported from land phosphorites (e.g., Force *et al.*, 1978; South Carolina) and some of the enrichment is attributable to remobilization or reprecipitation of uranium in the ground water zone. Among minor constituents, some ranges and mean values are shown in Table 8.

Table 8. Ranges and mean values for minor and trace constituents in marine phosphorites (principally from Baturin (1978) and Kolodny (1979). Values in ppm, except where noted. Mean or approximate mode in parenthesis.

SiO_2(%)	.16 - 45 (13)	V	10 - 160 (30)
Al_2O_3(%)	.02 - 6.0 (1.3)	Co	5 - 50 (8)
Fe_2O_3(%)	.07 - 9.8 (1.7)	Mo	2 - 30 (6)
MnO(%)	.1 - 1.5 (.4)	Pb	1 - 100 (10)
TiO_2(%)	.003 - .7 (.1)	Zn	2 - 200 (35)
K_2O(%)	.1 - 2.0 (.4)	Cu	1 - 60 (20)
MgO(%)	.40 - 3.5 (1.5)	Zr	10 - 100 (30)
SrO(%)	.08 - .5 (.15)	Rb	6 - 30 (15)
Cr	3 - 200 (30)	Ba	10 - 900 (100)
As	3 - 29 (11)	I	5 - 20 (8)
Nb	(6)	Th	1 - 42 (1.4)
Ni	2 - 200 (15)	U	5 - 600 (80-100)
	Total rare earths (%)		(.03-.1) (.045)

Trace elements specifically enriched in or associated with phosphorite include the rare earths, with a mean value of about 0.05% ΣRE. These RE elements substitute for calcium in the apatite structure. Baturin (1978) concludes that shelf phosphorites derive their cerium-enriched RE concentrations from associated sediments. The Mid-Pacific guyot deposits, on the other hand display a RE pattern resembling ocean water and are depleted in cerium. Tooth-derived apatite takes an intermediate position.

In special environments, associated phases may raise trace element concentrations. Thus, off Namibia (Southwest Africa) in sulfide-rich environments, molybdenum may reach top-of-the-range values; whereas on the Blake Plateau, replacement and accretion of ferromanganese oxides add well-known metals associated with these phases (Ni, Cu, Co, Mn, Fe, Mo). High silica reflects the usual diatom productivity of upwelling zones, and hence the diatomaceous sediment associated with phosphate formation.

PETROGRAPHY AND ORIGIN

Thanks to the studies of both recent and ancient or paleo-phosphorites, genesis is now much clearer. All workers are now agreed that for the major bedded phosphorites, upwelling of nutrient-rich waters provides the primary source of phosphorus. However, whereas earlier workers conceived that precipitation in the water column might form phosphate, dominant opinion now points to the interstitial water micro-environment as the generating site for primary apatite. Here, phosphate up to 50 μg -atom/ℓ, nearly 20-fold the maximum phosphate found in the principal upwelling waters (Fig. 2), is present in the pore fluids of the sediments (Baturin, 1978 and references cited). Carbonate and calcium as well as a nucleation site may be provided by calcium carbonate detritus. Under such conditions supersaturation with apatite should be present, and phosphate concentrations readily permit replacement of calcium carbonate where present, by apatite. Fluoride (1 mg/ℓ) and uranium (3 μg/ℓ) are likewise supplied by normal sea water. Thus, the requirement for phosphate generation may be not so much phosphate rich bottom water, as organic production, whose debris accumulates in bottom sediment and sustains high phosphate concentrations in pore fluid. An additional requirement for significant phosphorite enrichment is the absence of high inputs of diluting terrigenous (mineral) matter. Limited--but finite--

inputs of carbonate minerals have been found to be an important supporting factor in the Peru-Chile area. These requirements tend to exclude the nutrient-rich polar seas as phosphorite generators. In contrast, cold upwellings off desert environments such as Southwest Africa and Peru-Chile fit the optimum phosphorite-generating pattern.

Figure 5 shows a transect off southwest Africa at 23° S, with a schematic sketch of distribution of recent and fossil phosphorites. Here one can conclude that the conditions suitable for forming apatite have been maintained over much of Neogene-Holocene time. Radioactive dating of phosphorites from the Peru-Chile area shows a similar pattern of episodic phosphorite formation in Pleistocene-Holocene time, as well as during earlier epochs (Fig. 3).

Many of the ocean phosphates, however, are products of paleoenvironments different from those prevailing at present. The nodules and plates that occur on the Blake Plateau, off Florida-South Carolina in the southeast Atlantic coast of the U.S. (Figs. 6 and 7) are now swept and polished by the highly oxygenated Gulf Stream at depths of from 500 to 900 meters. Examination of the nodules in hand specimens, under the microscope, and by their stratigraphic occurrence in drill cores, reveals quite a different primary origin. They were probably formed in organic-rich waters and

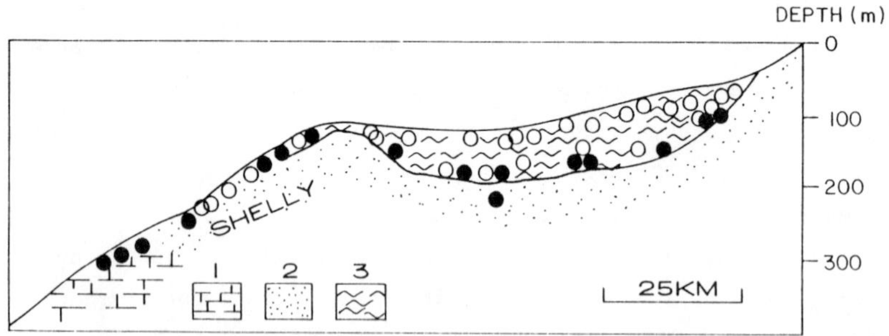

Figure 5. Schematic distribution of recent and fossil phosphorite along a transect at 23°S, Southwest Africa (from Baturin, 1978). Key to symbols: (1) foraminiferal sediment: (2) terrigenous sand and silt, shelly in part; (3) diatomaceous mud, shelly in part. Hollow circles represent Holocene (contemporaneous and subrecent) phosphorite concretions; solid circles represent older phosphorites (to Mid-Tertiary). Vertical scale for sediments is approximate (to within a few meters).

Figure 6. Photograph of phosphorite cobbles on ocean bottom on the Blake Plateau. Location approximately $31^{\circ}36'N$, $79^{\circ}04'W$, 500 m depth. (R.M. Pratt, photo.)

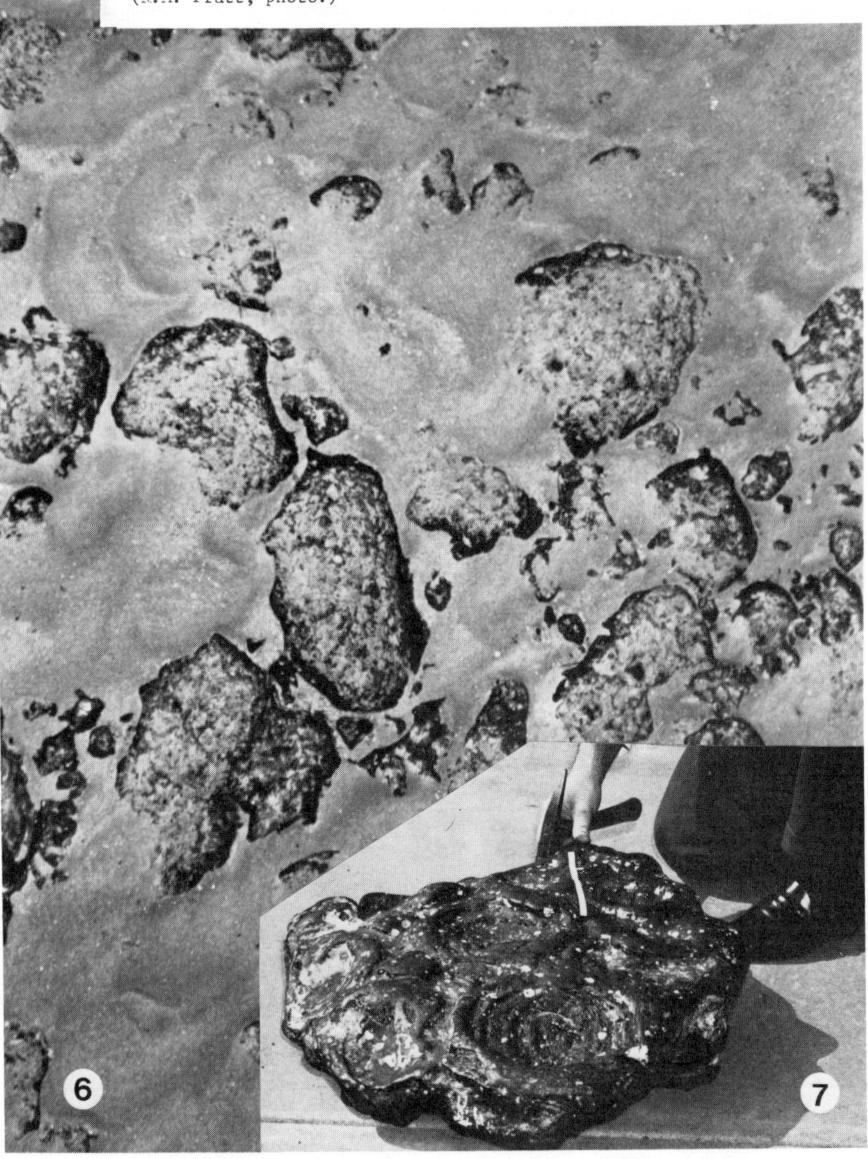

Figure 7. Massive plate of phosphorite recovered from Blake Plateau by Reynolds Aluminum Co. submarine, *Aluminaut*. Note remnants of coral holdfasts (white patches).

anoxic sediments at depths of a few tens of meters to perhaps 200 meters depth, during late Oligocene to early Miocene time. The original source of phosphate was probably replacement of carbonate particles, most readily the fine debris filling carbonate tests (e.g., Fig. 8a and b)*; but ultimately, also shell walls. In middle-late Miocene time, the Gulf Stream established itself, winnowing away the clayey-dolomitic matrix of the phosphatic marls and leaving a lag residue of more solution-resistant and partly recrystallized phosphorite (Fig. 8)*. From this time through at least Pliocene time, recrystallization and erosion in multiple stages may be observed in thin sections (e.g., Fig. 8d, e and f)*. Here, an eroded phosphatized surface is overlain by lithified but unphosphatized *globigerina* ooze containing the same type of organisms. Phosphatized bone of dugong or manatee (Fig. 8g)* can readily be distinguished from replacement phosphate by the characteristic Haversian canal structure. The phosphorites on the Blake Plateau are indistinguishable by X-ray diffraction from the famous Florida pebble phosphorite (Fig. 9).

———

*Figures 8a-d appear inside front cover, 8e-f inside back cover.

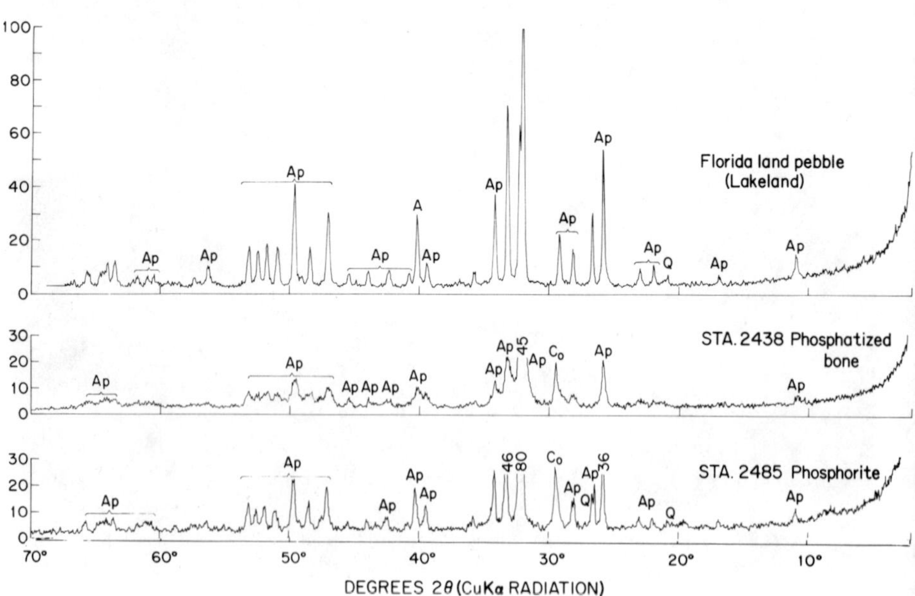

Figure 9. X-ray diffractograms of Blake Plateau and Florida phosphorites. C, Ap and Q refer to calcite, apatite, and quartz, respectively.

X-ray studies summarized by Baturin (1978) show that first-formed apatites have broadened lines signifying an amorphous phase that becomes more crystalline upon aging. Following upon this state, winnowing, erosion and polishing of aggregates of phosphatized organisms and debris may create the typical pebble conglomerates.

In places, the hard substrate of the phosphorites presents a surface for ferromanganese oxide replacement and accretion (Fig. 8h). Such developments occur in oxidizing environments very different from the primary phosphatization conditions. Erosion and recrystallization phenomena in many ocean phosphorites produce structures similar to those shown here. However, phosphorite grains have not always been recognized. They may be mistaken for glauconite, dolomite, and siderite-ankerite, or even carbonate particles.

MARINE PHOSPHORITES: REFERENCES

Altschuler, Z.S. (1973) The weathering of phosphate deposits-geochemical and environmental aspects. In Griffith, E.J., Beeton, A., Spencer, J.M. and Mitchell, D.T. (Eds.), *Environmental Phosphorus Handbook*, John Wiley & Sons, New York, p. 33-96.

_______, Clarke, R.S., and Young, E.J. (1958) Geochemistry of uranium in apatite and phosphorite. U.S. Geol. Survey Prof. Paper, 314-D, 45-90.

Ames, L.J., Jr. (1959) The genesis of carbonate apatite. Econ. Geol., 54, 829-841.

Baturin, G.N. (1968) Authigenic phosphate concretions in recent sediments of the southwest African shelf. Dokl. Akad. Nauk, Earth Sci. Sed. English transl., 189, 227-230.

_______. (1971) Stages of phosphorite formation on the ocean floor. Nature Phys. Sci., 232, 61-62.

_______. (1978) *Fosfority na dne ikeanov.* [Phosphorites on the ocean floor] Izdat. Nauka, Moscow, 231 p.

_______, Merkulova, K.I., and Chalov, P.I. (1972) Radiometric evidence for recent formation of phosphatic nodules in marine shelf sediments. Marine Geol., 13, 37-41.

Bliskovskii, V.Z., Grinenko, V.A., Mgdisov, A.A., and Savina, L.I. (1977) Sulfur isotope composition of the minerals of phosphorites. Geochem. Internat., 14, 148-155.

Brongersma-Sanders, M. (1957) Mass mortality in the sea. In Hedgpeth, J.W. (Ed.), *Treatise on marine ecology and paleoecology, Vol. I.* Geol. Soc. Amer. Mem., 67, 941-1010.

Burnett, W.C. (1974) Phosphorite deposits from the sea floor off Peru and Chile: Radiochemical and geochemical investigations concerning their origin. Hawaii Inst. Geophys., Rept. 74-3, 163 p.

_______. (1977) Geochemistry and origin of phosphorite deposits from off Peru and Chile. Geol. Soc. Am. Bull., 88, 813-823.

_______, and Gomberg, D.N. (1977) Uranium oxidation and probable subaerial weathering of phosphatized limestone from the Pourtales Terrace. Sedimentology, 21, 291-302.

Bushinskii, G.I. (1966) *Drevnie fosfority Azii i ith genezis* [Ancient phosphorites of Asia, and their origin]. Moscow, Nauka, 192 p.

Cayeux, L. (1939)(1941)(1950) Les Phosphates de chaux sedimentaries de France (France metropolitaine et d'outremer), Etude des gites mineraux de la France, Service Carte geol. France, Paris, Imprimerie Nationale, V. I, 350 p., V. II, 310 p., V. III, 360 p.

Cullen, D.J. (1978) The uranium content of submarine phosphorite and glauconite deposits on Chatham Rise, East of New Zealand. Marine Geology, 28, 67-76.

D'Ans-Lax (1967) Reported in Koritnig, S. (1976) [see below].

Dietz, R.S., Emery, K.O., and Shepard, F.P. (1942) Phosphorite deposits on the sea floor off southern California. Geol. Soc. Am. Bull., 53, 815-848.

Doyle, L.J., Blake, N.J., Woo, C.C. and Yevich, P. (1978) Recent biogenic phosphorite: concretious in mollusk kidneys. Science, 199, 1431-1433.

Farr, T.D. (1950) cited in Ulrich, B. (1961) [see below].

Force, E., Gregory, R., Gohn, S., Lucey, M., and Higgins, B. (1978) Uranium and phosphate resources in the Cooper Formation of the Chatham region, South Carolina. U.S. Geol. Survey Open-File Rept. 78-586, 22 p.

Gimmelfarb, B.M. (1965) *Zakonomernosti razmeshcheniya mestorozhdenii fofforitov SSSR i ikh geneticheskaya klassifikatsiya* [Distribution of phosphorites in the USSR and their genetic classification]. Mowcow, Nodra.

Gulbrandsen, R.A. (1970) Relation of carbon dioxide content of apatite of the Phosphoria Formation to regional facies. U.S. Geol. Survey Prof. Paper, 700-B, 9-13.

_______. (1975) Whitlockite and apatite of surficial phosphate occurrences on Enderbury Island, Phoenix Islands, Pacific Ocean. U.S. Geol. Survey Jour. Res., 3, 409-414.

Hutchinson, G.E. (1950) *The biogeochemistry of vertebrate excretion.* Am. Museum Nat. Hist. Bull., 96, 554 p.

Kazakov, A.V. (1973) The phosphorite facies and the genesis of phosphorites [in Russian]. In *Trudy NIUIF* (Nauchno-issledovatel'skii Institut po udobreniyam insektofungicidam), Leningrad, USSR, 142, 93-113.

Kester, D.R., and Pytkowicz, R.M. (1967) Determination of the apparent dissociation constants of phosphoric acid in seawater. Limnol. Oceanogr., 12, 243-252.

Kolodny, Y. (1969) Are marine phosphorites forming today? Nature, 224, 1017-1019.

_______. (1979) Phosphorites, Chapter II.3. In Emiliani, C. (Ed.), *The Sea*, V. 7, Academic Press (in press).

Koritnig, S. (1976) Phosphorus. In Wedepohl, K.H. (Ed.), *Handbook of Geochemistry*, Springer-Verlag, Berlin, 113 p.

Kudrin, L.N., Sivkova, A.S., and Martynova, S.S. (1962) Fluorine, phosphorus and minor elements in the bones of fossil fishes and dolphins. Dokl. Akad. Nauk, Earth Science Sections, in Engl. Transl., 142, 160-161.

Leckie, J. and Stumm, W., 1970. Phosphate precipitation. In Gloyna, E., and Eckenfelder, E.W., (Eds.), *Advances in water quality improvement--Physical and chemical processes.* Univ. of Texas Press, Houston, p. 237-249.

Legeros, R.Z., Trantz, O.R., Legeros, J.P., and Klein (1967) Apatite crystallites: Effect of carbonate on morphology. Science, 155, 1409-1411.

Lucas, J., Prevot, L., and Lamboy, M. (1978) Les phosphorites de la marge nord de l'Espagne. Chimie, mineralogie, genese. Oceanologica Acta, 1, 55-72.

McArthur, J.M. (1978) Systematic variations in the contents of Na, Sr, CO_3 and SO_4 in marine carbonate-fluorapatite and their relation to weathering. Chem. Geol., v. 21, 89-112.

McClellan, G.H., and Lehr, J.R. (1969) Crystal-chemical investigation of natural apatites. Amer. Mineral., 54, 1374-1391.

McConnell, D. (1973) *Apatite*. Springer-Verlag, New York, 111 p.

McKelvey, V.E. (1963) Successful new techniques in prospecting for phosphate deposits. In U.S. Dept. State, Natural Resources--Minerals and mining, mapping and geodetic control. Geneva, United Nations Conference. Applications of Science and Technology for the Benefit of Less Developed Areas, U.S. paper, V. 2, 169-172.

_______. (1967) Phosphate deposits. U.S. Geol. Surv. Bull., 1252-D, 1-21.

_______, Swanson, R.W., and Sheldon, R.P. (1953) The Permian phosphorite deposits of western United States. Cong. Geol. Intern. Compt. Rend. 19th, Algiers 1952, XI (II), 45-64.

MacKnight, S.D. (1976) An investigation into the solubility behavior of phosphorite in sea water. Ph.D. Dissertation, Dalhousie Univ., Appendix Table B-1.

Manheim. F.T., Pratt, R.M., and McFarlin, P.F. (1979) Composition and origin or phosphorite deposits of the Blake Plateau. SEPM Spec. Paper (in press).

_______, Rowe, G.T., and Jipa, D. (1975) Marine phosphorite formation off Peru. J. Sed. Petrol., 45 243-251.

Massard (1973) reported in Koritnig (1976) [see above].

Murray, J. (1885) Report on the specimens of botton deposits. Harvard Mus. Comp. Zoology Bull., 12, 37-61.

_______ and Renard, A.F. (1891) Report on deep-sea deposits based on the specimens collected during the voyage of HMS *Challenger* in the years 1872-1876. In *Report on Scientific Research, III, Deep Sea, Deposits*, 525 p.

Moreno *et al.* (1960) reported in Koritnig (1976) [see above].

Price, N.B., and Calvert, S.E. (1978) The geochemistry of phosphorites from the Manibian shelf: Chem. Geol., 23, 151-170.

Romankevich, E.A., and Baturin, G.N. (1972) Composition of the organic matter in phosphorites from the continental shelf of southwest Africa (Russ.): Geokhimiya, 6, 719-726.

Sillen, L.G. (1959) Graphic presentation of equilibrium data. In Lee, T.S., and Sillen, L.G., (Eds.), *Chemical Equilibrium in Analytical Chemistry, Section B.* Interscience Publishers, New York, 277-317.

Stowasser, W.F. (1979) Phosphate: Mineral Commodity Profiles. U.S. Bur. Mines, Pittsburgh, PA, 19 p.

Ulrich, B. (1961) Die Wechselbeziehungen von Boden und Pflanze in Physikalisch-Chemischer Betrachtuny. Stuttgart, Ferdinand Enke Verlag, reported in Koritnig (1976) [see above].

Veeh, H.H., Burnett, W.C., and Sontar, A. (1973) Contemporary phosphorites on the continental margin of Peru. Science, 181, 844-845.

Young, E.J., Myers, A.T., Munson, E.L., and Conklin, N.M. (1969) Mineralogy and geochemistry of fluorapatite from Cerro de Mercado, Durango, Mexico. U.S. Geol. Surv. Prof. Paper, 659-D, 84-93.

Chapter 7

MARINE BARITE

Thomas M. Church

OCCURRENCE AND ORIGIN

Previous Studies

The mineral barite has often been documented on land as a low temperature geochemical precipitate that is an accessory phase to low grade metamorphic and hydrothermal events (see Hanor, 1966 for review). Barite was also discovered in the deep sea as borderland concretions by the *Challenger* Expedition (Murray and Renard, 1898). The Russians (Vinogradov, 1953) later found barite crystals and nodules off the borderlands of southern India associated with benthic protozoans of the order Xenophyphorida. Marine barite has been detected as rather ubiquitously dispersed microcrystals in deep-sea sediments largely underlying productive areas of the eastern equatorial Pacific and occasionally the Atlantic (Goldberg and Arrhenius, 1958; Griffin and Goldberg, 1964). In fact, barite occurrence in the marine environment was proposed to limit the concentration of soluble barium in the oceans (Arrhenius, 1959). A preliminary calculation of the solubility of barite suggested that in some deeper ocean waters supersaturation may occur (Chow and Goldberg, 1960), and that the generally observed increased concentrations of barium with depth in the oceans may reflect the dissolution of biologically formed barite.

Morphologically marine barite crystals were found commonly small (2-4μm) often localized in yellow aggregates presumed to be fecalized biological debris (Goldberg and Arrhenius, 1958). Occasional large euhedral crystals (30 x 50μm) were reported for these same sediments (Arrhenius, 1963).

Concentrations of barite up to 10% by weight on a carbonate-free basis were reported for sediments of the East Pacific Rise (Arrhenius and Bonatti, 1965). To explain this geographical distribution, these workers proposed hydrothermal transport of barium to the sediment surface associated with deeper magmatic activity typical of oceanic

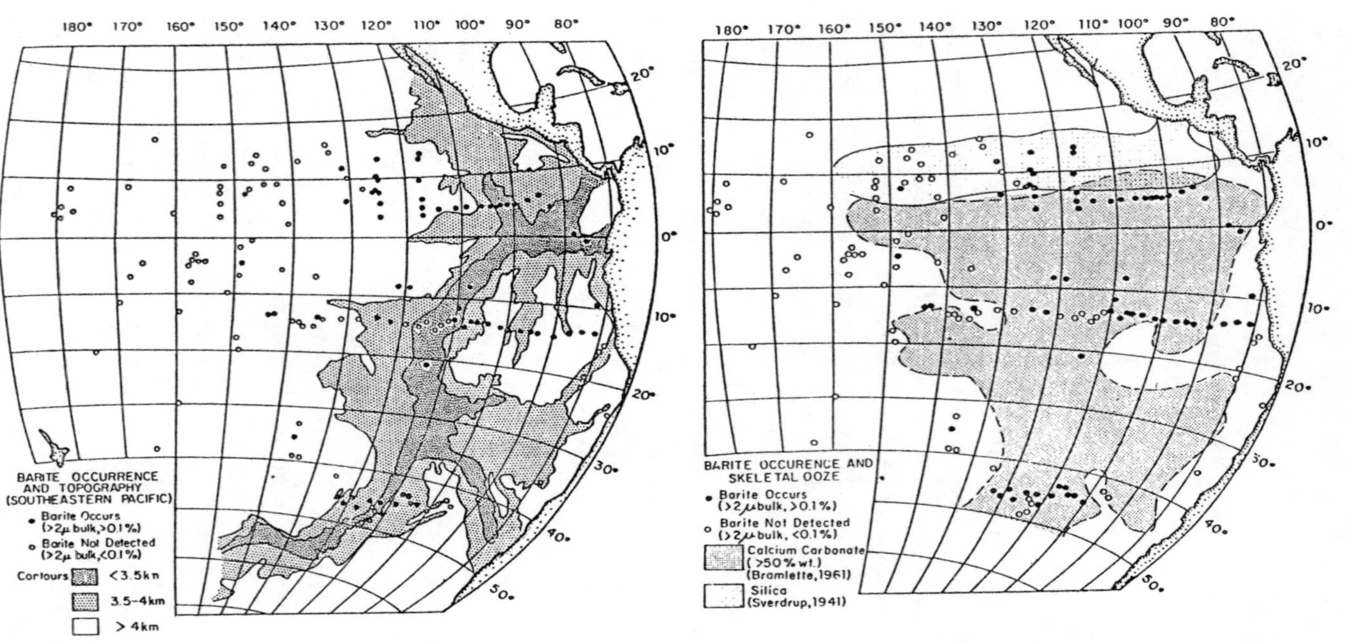

Figure 1. Barite and topography of the East Pacific Rise.

Figure 2. Barite and skeletal oozes in surface sediments of the eastern Pacific.

rises. The released barium enriches these rise sediments by being distributed between pore solutions, smectite phases, barium zeolites (such as harmotone), and barite. Alternatively, oceanic equatorial regions in the vicinity of ridge systems may produce a rain of marine organisms which can extract barium from surface waters or which might be metabolically excreted as granules, such as those associated with Xenophyphorida (Arrhenius, 1963). Thus, marine barites might collect at the sediment surface along with other biological debris.

Recent Studies

In the past decade there has been continuing research and perhaps more accurate information on the occurrence of a marine barite and means for its formation. Church (1970) completed a dissertation on the subject, largely as a sedimentary phase; and in another recent dissertation, Dehairs (1979) investigated barite as a particulate phase in the water column.

Church (1970) used both heavy liquids and acid leaches to isolate marine barite from numerous deep-sea sediments and other phases. Marine barite was detected down to the 500 ppm level in many pelagic sediments largely from the eastern equatorial Pacific (Fig. 1). Abundances averaged about 1% and occasionally up to 3%, most often in the 2-4μm range and occasionally over 5μm. Whereas much barite is found on the East Pacific Rise, any firm correlation with water depth is not evident. Along the equatorial region, marine barite occurrence is broad, but south of 12°S the probability of finding barite becomes less with increasing distance from shore. In the temperate southern latitudes, the occurrence of barite is restricted to the East Pacific Rise, in general agreement with the results of Arrhenius and Bonnati (1965), although it apparently is not the consistently abundant (10%) and uniform phenomenon they suggest. Locally, there are some conspicuous absences of barite, such as at the very crest of the East Pacific Rise, where hydrothermal barium sources are proposed to exist. Between adjacent locations and even within a single core there are variations of a factor two or more. Whereas there is no uniform trend of barite abundance within the sediment column in the equatorial region, there seems to be a decrease with depth for East Pacific Rise sediments.

Sedimentary Mineralogical Associations

According to Church (1970), the occurrence of barite is characterized by three basically different environments: (1) an abundance of biological debris such as calcareous and siliceous tests; (2) abundant manganese and iron phases both in concretionary and dispersed forms; and (3) altered volcanic debris such as predominant montmorillonite, palygorskite, and clinoptilolite facies. This is also in order of frequency of occurrence by about 10:2:1, and each facies is now discussed separately.

A distinct occurrence of marine barite in predominant (>50%) carbonate and siliceous oozes of the eastern Pacific is shown in Figure 2. A correlation is found with 72% of the samples containing barite within the carbonate area. Often such predominant carbonate sediments are well below the lysocline and show distinct corrosion of thin-walled tests. There is likewise a good correlation between barium and silica in ocean waters, sediments, and the occurrence of barite in carbonate-free silica oozes of the eastern Pacific. In opaline sediments, barite occurrence is often concentrated in the darker iron-rich material by a factor of three. Thus the incorporation and dissolution of barium-enriched silica tests (Martin and Knauer, 1973) could provide an equally important means for the fixation of sedimentary marine barite. Most likely the overall factor in marine barite formation is the efficient deposition and benthic degradation of organic carbon rather than the type of individual test, and Figure 3 shows a broad correlation between marine barite and surface organic carbon.

Manganese phases are well known to concentrate barium relative to other alkaline earths (Arrhenius, 1963; Puchelt, 1967; Cronnan and Tooms, 1969; see also Chapter 1, this volume). Since barium concentrations for these minerals are tenths of percent, they probably provide a barium-rich milieu and potential sources of barium for sulfate precipitation, and several occurrences of large crystalline barite in manganiferous material have been observed (Church, 1970; Burns and Burns, 1978).

Altered volcanic debris can provide surface reactive minerals to concentrate and redistribute barium, perhaps of local volcanic origin, from basalt weathering. These phases usually include palygorskite/ atapulgite and occasionally the zeolite clinoptilolite (see Chapter 4,

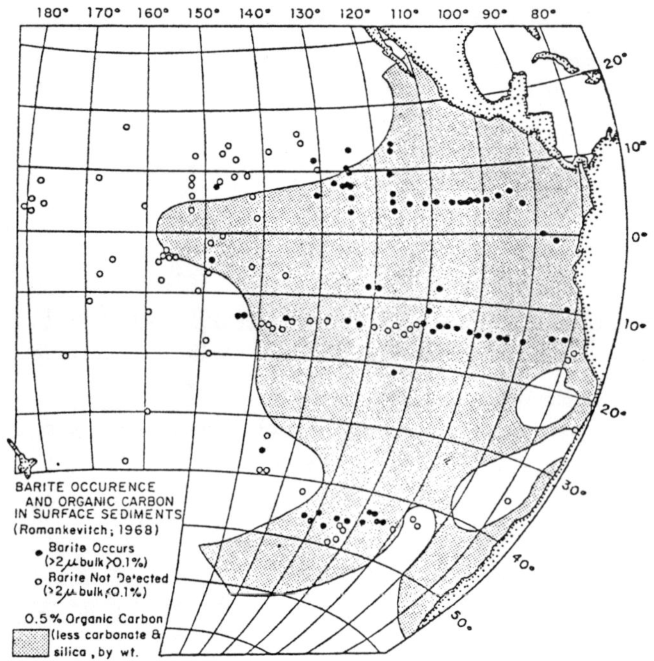

Figure 3. Barite and organic carbon in surface sediments of the eastern Pacific.

this volume). However, in one attempt to correlate the occurrence of barite with carbonate versus phillipsite facies (Church, 1970), there was a factor of ten greater occurrence of barite in carbonate, but phillipsite-free sediments than in phillipsite or carbonate-free sediments alone. Because of its exchange properties, phillipsite could compete for Ba^{2+} ions and limit the growth or perhaps even dissolve barite (one barite dissolution technique uses cation exchange resin). The barium zeolite, harmotome, has been one such potential phase identified on the sea floor (Arrhenius, 1963). Thus there seems to be a rather direct and probable biological link between the occurrence of marine barite and biogenic (calcareous) sediments.

Biological Associations

Barite may arise in the marine environment as a direct metabolic product. In fact, barite has been reported in benthic protozoan animals classed as the order Xenophyphorida (Vinogradov and Kovelisky, 1962), which had been suggested (Arrhenius, 1963) to contribute some, if not all, of the barite to deep-sea sediments. Alternatively, Xenophyphoridae could incorporate pre-existing sedimentary barite along with the tests, spicules, and mineral grains from which they form an arenaceous skeleton (Leoblich, 1966; Tendal, 1972). In one set of specimens netted in the western Atlantic waters, Church (1970) could find no barite. Also, since the geographic range of these organisms is restricted to equatorial regions, while marine barite occurs far outside these bounds, it appears unlikely that Xenophyphoridae are significant biological contributors. However, the group of marine organisms called Acentharia (Botazzi and Schreiber, 1971) have tests of celestite ($SrSO_4$) with several mole percent barium (Vinogradov, 1953). Although celestite should be unstable in sea water, Acantharia or these tests could be important barium suppliers or precursors for marine barite. This hypothesis has gained strong support from the recent discovery of suspended marine barite in most oceanic water columns (Chesselet *et al.*, 1976; Dehairs, 1979) which is discussed next.

Suspended Marine Barite

Barite particles have been discovered as a universal component of suspended water in the Atlantic and Pacific Oceans (Chesselet *et al.*, 1976, Dehairs *et al.*, 1979, Dehairs, 1979). This discovery was made after electron microscope/microprobe analysis of discrete particles collected during the GEOSECS program, and these barite particles of about a micron in size represent by far the largest proportion of total suspended particulate barium. Particulate barite concentrations in the water column usually decreased from 40 (surface) to 10 (bottom), averaging 20 ng/kg sea water.

The morphology of these suspended barite particles include irregularly shaped ellipsoidal or spherically rounded particles more than the distinct crystalline habits typical of sediments, suggesting some water column dissolution. Such dissolution could contribute significantly to the increased concentration and evolution of barium to deep-sea waters related to the apparent undersaturation of barite in all ocean water columns (discussed in the next section). Even in the water column, biological debris appears to be essential for providing a barium rich, sulfate excess micro-environment for the non-equilibrium precipitation of barite. Thus, the basic mechanisms for marine barite precipitation could be the same in sediments and the water column; the organic degradation of barium enriched biological matter. Such a concept is also supported by the particulate barium concentration in the ocean water column which correlates with biological productivity, as indicated by dissolved phosphate contents of corresponding surface waters (Dehairs *et al.*, 1979). In such cases, the highest contents of suspended barite are found just below the euphotic zone. Alternative to indirect marine barite precipitation after death and decay, one could also consider direct metabolic secretion and precipitation. The celestite spines of Acantharia can contain up to 0.5% barium, and upon death could re-crystallize to the more barium-rich forms.

Marine Barite Fluxes

It was proposed by Church (1970) that calcareous fluxes containing up to 200 ppm of barium could provide sufficient barium to produce all the barite observed in such calcareous sediments (Table 1). A similar

Table 1. Barium in carbonate sediments - A mass balance.

Fraction	Accumulation Rate (g/cm^2-10^3 yrs.)	Ba ppm	Ba Accumulation ($\mu g/cm^2-10^3$ yrs.)	(%)
Carbonate (90%)	2	200	400	77
Clay	0.2	600	120	23
"Marine Organics"	0.04	10	< 1	< 1
TOTAL	2.24	-	520	100
Bulk Sediment	90% Carbonate	232		
	Carbonate Free	2320		
Comparisons				
Rispac 68 (Hanor & Brass, 1968)		2200		
Rispac 58 (this work)		3800		

mass balance for siliceous sediments might also be possible were there more conclusive values for barium enrichment. From the failure of several laboratories to grow barium rich diatom tests from barium enriched sea water, it is probable that the barium contents of these organisms are in the soft rather than hard parts. A more recent estimate of the particulate barium distribution is provided by Dehairs *et al.* (1979) and reproduced in Table 2 as percent of the total particulate barium load comprised of barite. It is seen that barite represents well over half of the total particulate barium, reaching a maximum in intermediate waters and decreasing with depth due to dissolution. It is estimated by Dehairs *et al.* (1979) that marine barite dissolution in the water column can contribute about 0.4μg Ba/cm^2-yr to the benthic flux of dissolved barium. By comparison, calcareous and siliceous biological debris contribute another 0.5μg Ba/cm^2-yr, yielding a total of 0.9μg Ba/cm^2-yr, of which about half comes from the fallout and *in situ* dissolution of marine barite. Some of the marine barite fallout must reach benthic sediments, and using an average value of about 1% by weight in carbonate-free sediments of the eastern equatorial Pacific depositing at about 1 mm/10^3 yrs, the flux of marine barite to deep-sea sediments is about 0.34μg Ba/cm^2-yr (Church, 1970). Increases in barite content with depth (1 to 3%) over an interval of about 0.2 m yield a sub-sediment barite growth rate of 0.04μg Ba/cm^2-yr.

Thus, it appears that of a total marine barite barium flux of about 0.8μg Ba/cm^2-yr to sediments, half dissolves back into deep water and the other half becomes inherited by barite-bearing deep-sea sediments.

Table 2. Contribution of barite and other barium phases to particulate ocean matter in low latitudes (after Dehairs, et al., 1979).

		Particulate Barium Contents (ng/kg)					
Waters	Total	SiO Tests	$CaCO_3$ Tests	Part. Org. Matter	Alumino-Silicates	Barite (difference)	% Barite
Surface	11	0.6	0.6	1.8	1.3	6.7	61
Intermediate	10	0.3	0.3	0.9	0.8	7.7	77
Bottom	14	0.3	0.3	0.9	2.7	9.8	70
Assumed Ba Concentration (ppm)		120	200	60	600		

The net result is that about one-third of marine barite appears pelagically generated from the water column, and about an additional two-thirds is diagenetically produced in the sediments. Both processes, however, appear to be basically the same; namely, the degradation or alteration of barium-enriched biological remains in restricted sulfate excess micro-environments.

Barite has been reported in deeply drilled cores of the central and western Pacific (Cook and Zemmels, 1972). Some of these were isolated, and from morphology and geochemistry appear indistinguishable from surface pelagic varieties (Church, unpublished).

Other Oceanic Barites

There appear to be isolated instances of barite found in the deep sea and coastal environments which appear to be non-marine in origin in that their barium supplies are not uniformly provided by open ocean water. These oceanic types plus those found on the continents can be readily distinguished from microcrystalline pelagic marine barites by their geochemistries (Goldberg *et al.*, 1969; Church and Bernat, 1972; Guichard *et al.*, 1979).

Church (1970) reports large barite panes associated with a friable manganese-palagonite concretion from the Indian Ocean which appears to have non-marine rare earth geochemistry (Guichard *et al.*, 1979). A rather dominant palygorskite deposit of the central Pacific also yielded spectacular barite panes (50 x 100μ), the palygorskite of which appears to have been formed out of equilibrium with sea water at elevated temperatures (Church and Velde, 1979). Bertine and Keene (1975) also report similar large barite crystals infilling opaline rocks with associated volcanic debris in the Lau basin. From isotope analyses, these barites formed at temperatures elevated well above the sea floor from the contact of hot barium and silica rich hydrothermal waters intruding cool sea water. The borderlands of southern California have yielded a number of dredge-hauls of rather pure barite slabs. From sulfur and oxygen isotope data (Church, 1970; Cortecci and Longinelli, 1972), these appear to be hydrothermal products of reoxidized barium-rich sulfide deposits. Other forms of barite largely in concretionary or rossette forms are found both in modern day estuarine varved clays of the southern Baltic

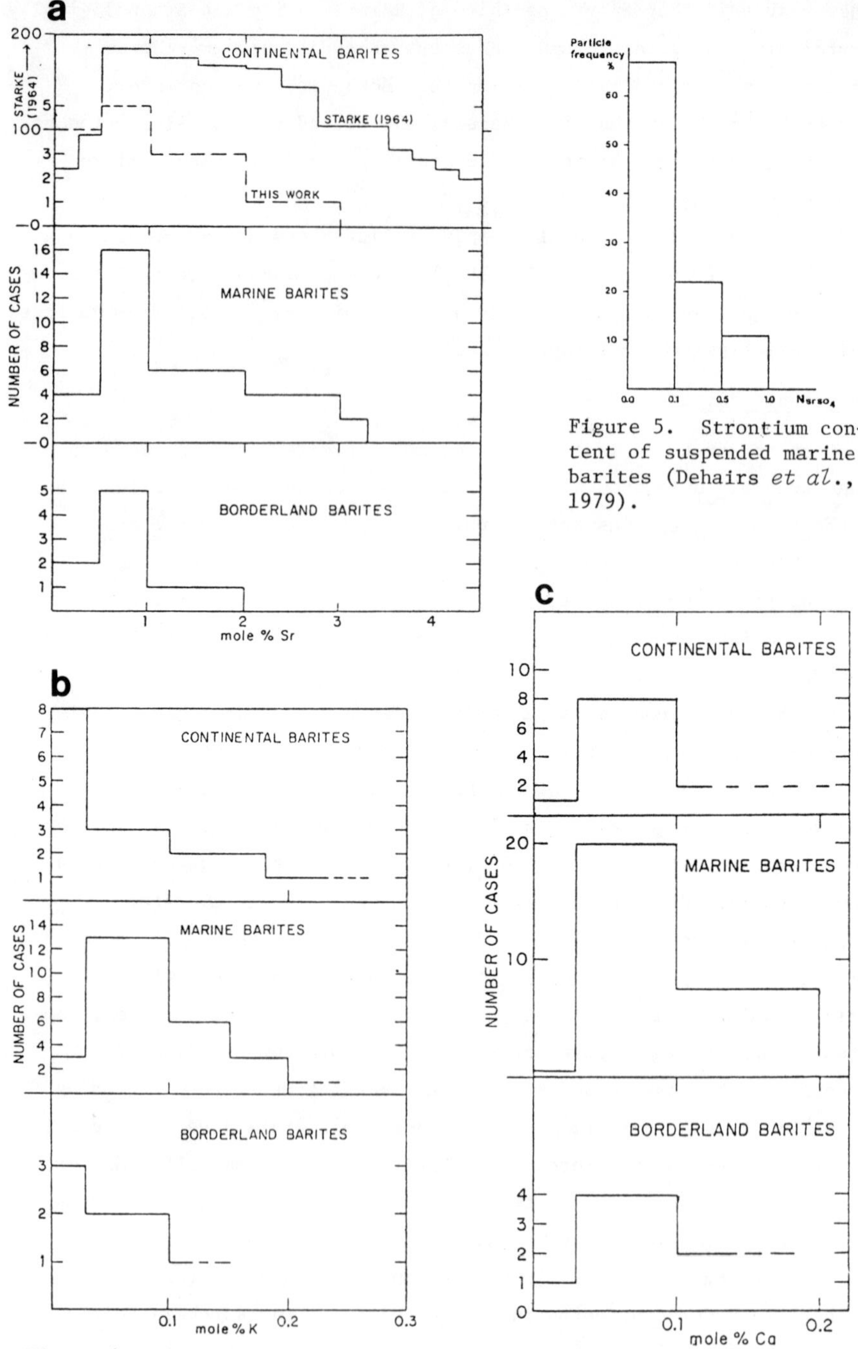

Figure 5. Strontium content of suspended marine barites (Dehairs *et al.*, 1979).

Figure 4. Histograms of (a) strontium, (b) potassium and (c) calcium concentrations in barite.

(Suess, 1979) and various continentally located ancient marine sediments (Guichard *et al.*, 1979). These appear to result from the post-depositional incursion of marine water (sulfate) into low-grade metamorphic or chalcopyrite-blende facies previously enriched in barium. Thus, most of the non-pelagic marine barites seem to be isolated as individual instances that do not largely participate in the barium biogeochemical cycles of the open sea.

GEOCHEMISTRY

The composition of marine barites is essential to understanding their locus and mode of precipitation, as well as distinguishing various oceanic and continental varieties. Marine barites generally are remarkably pure, containing mostly 1-3% strontium, lesser amounts (100-1000 ppm) of calcium and potassium and traces of several other elements including lanthanides and natural radionuclides. It is these trace elements, however, which are most useful to fingerprint origins and modes of formation of marine barite.

Major Elements

Church (1970) analyzed a variety of marine and continental barites for strontium, calcium, and potassium (Fig. 4). Strontium ranges from 0.2-3.4% with an everage of 1%. Barites formed in different environments seem indistinguishable on the basis of Sr content alone (Fig. 4a). However, most marine barites and borderland varieties are restricted to 1-3% Sr, while continental varieties generally have a greater range of Sr. Barites suspended in the water column have also mostly percent levels of Sr, but some can range the entire $BaSO_4$-$SrSO_4$ range (Dehairs *et al.*, 1979) as shown in Figure 5. For both water column and sedimentary marine barite, one might expect a rather constant Sr content if formed from a constant Sr/Ba (saturated) ion ratio in bulk sea water. The variability in Sr content suggests that, in fact, marine barites form in variable Sr micro-environments or at different rates rather than from open sea water.

Both Ca and K have concentrations between 0.01 and 0.1% with no apparent correlation to each other, the Sr content, or crystal size. Some portion of these low values might be due to slight contamination by trace feldspar inclusions. As shown in Fig. 4b, the distribution for Ca averages 0.01-0.1% and compares indistinguishably from continental varieties. As shown in Fig. 4c, the distribution for K shows a distinct clustering between 0.01-0.1%, while continental and borderland varieties tend to concentrate lower than the 0.01% level.

Minor Elements

Minor elements in three Pacific marine barites were analyzed semi-quantitatively by emission spectroscopy (Church, 1970). The results shown in Table 3 show most transition elements not detectable, with a

Table 3. Minor element survey of marine barites.
Estimated errors (in ppm) in square brackets.

Element	X-ray Fluorescence Detection Limit	ppm Dodo 8	Proa 101	Ris 60	Emission Spectrometry Proa 101 (E. Snooks)
Fe	100	330 [20]	180 [25]	275 [22]	100 [50]
Cr	100	0	0	0	10-100
Zn	150	250 [100]	200 [100]	0	-
Cu	400	0	0	0	1
Ni	200	0	0	-	-
Pb	100	0	0	-	10
Mn	200	0	0	-	10
Co	50	0	0	-	10
Sn	500	0	-	-	-
Zr	30	60 [15]	-	-	-
Na	-	-	-	-	100
Cl	100			200 [50]	-
Si					1000
P					10
Mg					< 10
Ag					1
Ti					10
Ga					10
Al					1000

few (Fe, Zn, Zr) less than several parts per million. No evidence was found for the high levels (100-1000 ppm) of Cr, Sn, Pb, and Zr reported by Arrhenius (1963). The amounts of Si and Al probably reflect mineral inclusions, while the Na and Cl would correspond to one percent of occluded sea water.

Rare Earths

There is only one study of rare earths in barites (Guichard *et al.*, 1979), and this lends information on the locus and mode of marine barite formation. As shown in Figure 6, the total lanthanide concentration of marine barites (like Th) are considerably more than terrestrial varieties. Also the chondrite-normalized patterns of marine barites show the generally concave upward "V" pattern of rare earths in sea water which seem indicative of a sea water lanthanide origin. One major

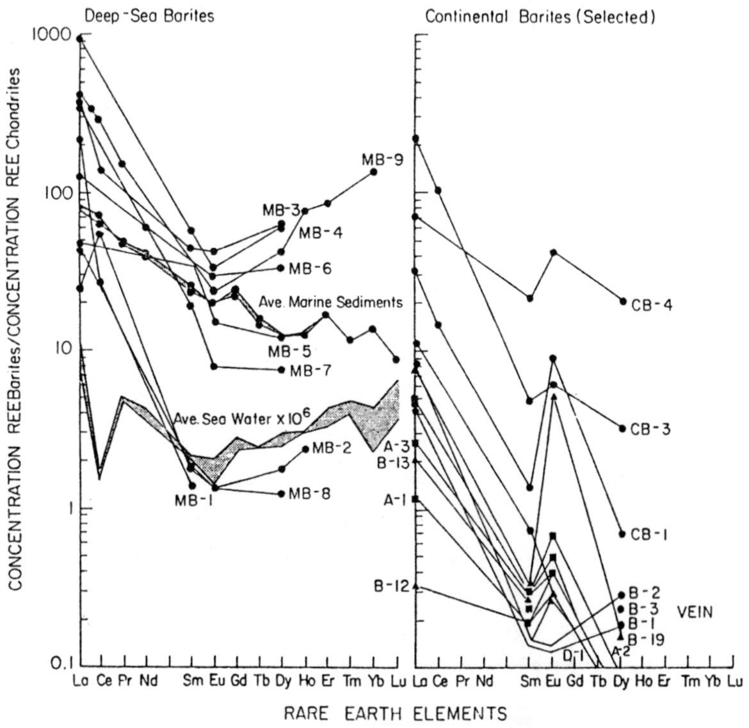

Figure 6. Distribution of rare-earths in deep-sea and continental barites (Guichard *et al.*, 1979).

exception is that marine barites lack the negative cerium anomaly of sea water. Unlike the other trivalent rare earths which should exist rather readily in sea water as poly-carbonate anion complexes, tetravalent Ce may exist as a more adsorbable neutral hydroxide species derived primarily from adjacent detrital supplies. However, for one marine barite isolated from a manganiferous concretion, the Ce anomaly was distinctly positive as is the case for most benthic manganese nodules. For another marine barite isolated from a strongly reducing environment, Eu displays a sharp positive anomaly. This would be expected for favorable co-crystallization of reduced Eu(II) cation, a ba(II) isomorph. The pattern of rare earths in barites can be interpreted on the basis of the combined effects of rare-earth substitution behavior and solution complexing. Marine barites under the normal Eh-pH conditions of sea water seem to reflect well and be good indicators of marine lanthanide sources. Only under marine or continental conditions of increased brine strengths and alkalinity do barites appear to reflect rare earths patterns fractionated by solution complexing.

Thorium, Uranium, and Radium

The first analyses for Th and U in marine barite come from the Atlantic (Somayajulu and Goldberg, 1966). These workers found lower $^{230}Th/^{232}Th$ activity ratios for barite than their bulk sediments. They attributed this difference to a more benthic formation of the barite than the sediments which derive their ^{230}Th isotope from surface productivity. Goldberg *et al.* (1969) showed marine barites contain significantly more Th (10-30 ppm) than continental varieties (<3 ppm).

Church and Bernat (1972) analyzed barite and bulk sediment in two Pacific cores for Th and U contents and isotopes (Fig. 7). They generally corroborated earlier results of higher Th contents and lower $^{230}Th/^{232}Th$ ratios relative to bulk sediments for equatorial Pacific pelagic marine barites. Also, they found that with depth the Th content of barites, while always greater than their bulk sediments, tended to decrease while the amounts of barite increased. They attributed this to marine barites growing diagenetically after burial with limited supplies of detritally provided ^{232}Th. Also, the identical $^{230}Th/^{232}Th$ chronologies for barites and their sediments suggest the bulk of marine barite formation is in

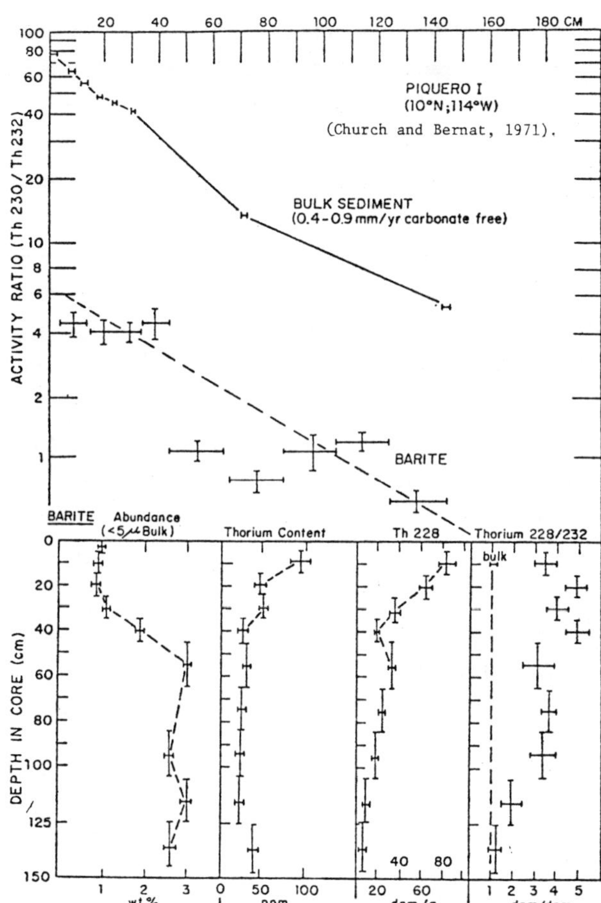

Figure 7. Thorium and its isotopes in a Pacific barite core - Piquero I.

surface sediments. The $^{230}Th/^{232}Th$ ratios of these marine barites are uniformly lower than bulk sediments, reflecting the differential surface and benthic supplies respectively for the two isotopes in the two phases as originally proposed by Goldberg *et al.* (1964). As noted in Fig. 6, the short-lived ^{228}Th ($t_{1/2}$ = 2 yrs), shows significant excesses in surface sediments. This suggests that the intermediate ^{228}Ra daughter ($t_{1/2}$ = 7 yrs) is probably efficiently scavenged and exchanged during diagenetic marine barite formation.

Radium in marine barite was later directly measured with depth in a central Pacific core analyzed by Borole and Somayajulu (1977) along with the daughter ^{210}Pb. Radium-226 was significantly enriched in marine barite at about a rather constant average $^{226}Ra/Ba$ ratio of 1.5×10^{14} atoms/mole. This ratio generally reflects the high $^{226}Ra/Ba$ pore water ratios of 8.5×10^{15} atom/mole reported by Somayajulu and Church (1973) for another Pacific core. This efficient and co-crystallized incorporation of ionically similar Ra for Ba also corresponds to the classical use of barite to scavenge radium from solution and behavior as a $BaSO_4$-$RaSO_4$ ideal solid solution. The constant $^{226}Ra/Ba$ ratio with depth (rather than a sharp decrease from radioactive decay of ^{226}Ra expected for a closed barite system) supports the rapid exchange of ^{226}Ra marine barite crystals and surrounding solutions on the order of its decay constant (5.78×10^{-3} yr^{-1}). Even the $^{228}Ra/^{226}Ra$ ratio of 0.02 ± 0.004 is about a factor of two below pore waters, suggests this *in situ* exchange of Ra must be quite rapid on the order of years. The isotope ^{210}Pb also analyzed by Borole and Somayajulu (1977) showed a range 20-66 dpm/g with an activity ratio to its ^{226}Ra parent only 3-12% of unity. Such a depletion is most likely due to the diffusive loss of the ^{222}Rn intermediate rare gas. Such a gaseous loss of radiogenic intermediates also limits the ^4He dating of marine barite. In one attempt by Church (unpublished, reported in Church and Velde, 1979) to ^4He date large marine barite panes with size correction for loss of alpha particles, ages of about 10^6 years were obtained which are at least consistent with their Tertiary sediments.

Stable Isotopes

The stable isotopes of sulfur have been analyzed on one borderland barite slab (Cortecci and Longinelli, 1972), which generally show the fractionation effects of sulfate reduction. Both sulfur and oxygen isotopes were run on a suite of marine, borderland and continental barites (Church, 1970). The results shown in Figure 8 show marine barites approximate marine sulfate more than continental varieties. Marine barites tend to fall into two classes largely on the basis of their oxygen isotopes. Large marine barites (>5μm) match or are slightly heavier in both isotopes than sea water sulfate, while smaller barites (<5μm) tend to be lighter in oxygen and perhaps also sulfur. These light values match in one case (PIQ I) a lighter value for coexisting pore water. Light sulfur isotopes in barite could arise from bacterial

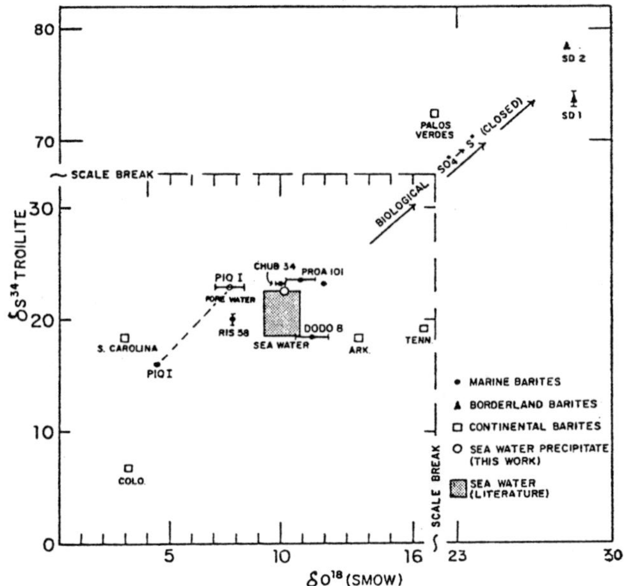

Figure 8. The fractionation factors of sulphur and oxygen isotopes in barites.

sulfate reduction, which removes the heavier sulfur isotope as pyrite with subsequent barite precipitation from the resulting net light sulfur reservoir. Barites thus probably grow close to the site of such reductions such as the interiors of tests or organic aggregates where the sulfate reservoir is closed and limited. Marine barites from more reducing environments, such as in the presence of pyrite, have lighter sulfur isotopes supporting this concept of closed system sulfate utilization and reoxidation within the restrictions of barite producing micro-environments. In sharp contrast, barites from borderland regions are very heavy in both sulfur and oxygen. This is best explained by an open system of continuously fractionated sulfate reduction (4 sulfur/1 oxygen). The light isotopes are removed by exchange or H_2S evolution followed by subsequent barite precipitation from a late stage residual sulfate reservoir. Such a fractionation sequence often occurs in oil brines or stagnant evaporite deposits. It is conceivable that borderland barites form under similar evaporite, brine-forming conditions, and not as a precipitate of normal sea water.

CRYSTALLOGRAPHIC AND THERMODYNAMIC PROPERTIES

Crystallographic Parameters

Barite can accommodate foreign ions by heterogeneous occlusion or homogeneous substitution (Gordon, 1954; Walton, 1967). A homogeneous solution forms when only a single substituted orthorhombic barite phase results. Barite is known to form continuous solid solutions with Sr, Pb, Ra cations or CrO_4 anions, while the substitution of Ca appears to form a discontinuous series (Deer *et al.*, 1965).

Marine barites contain between 0.5 and 3% strontium, and whether or not the Sr is in solid solutions with barite can be established if linear contractions in unit cell dimensions are observed with increasing Sr content (Vegard's Law). Church (1970) performed such an analysis on a suite of marine barites by measuring reflection spacings with a high precision Häag film camera (±0.003A). The cell dimensions and volume were calculated according to Boström (1968) using a computerized reiteration program. The data for both natural and artificial marine barites is summarized as the contraction of unit cell volume versus Sr content (Fig. 9). There is a definite contraction of all unit cell dimensions with increasing Sr content in marine barite, indicating that Sr is indeed in solid solution.

The rate of unit cell contraction with Sr content for marine barites is greater than that observed in high temperature barite precipitates (Boström, 1968), which obey Vegard's Law. This suggests an energy requirement greater than ideal to form the natural mixed solid at lower temperatures. Artificial precipitates undergo non-ideal contraction to a much greater extent, a portion of which is probably due to unusually enriched amounts of substituted calcium. The sea water equilibrated pure barite has similar spacings to pure barite which suggests strontium is taken up without permutation of the lattice.

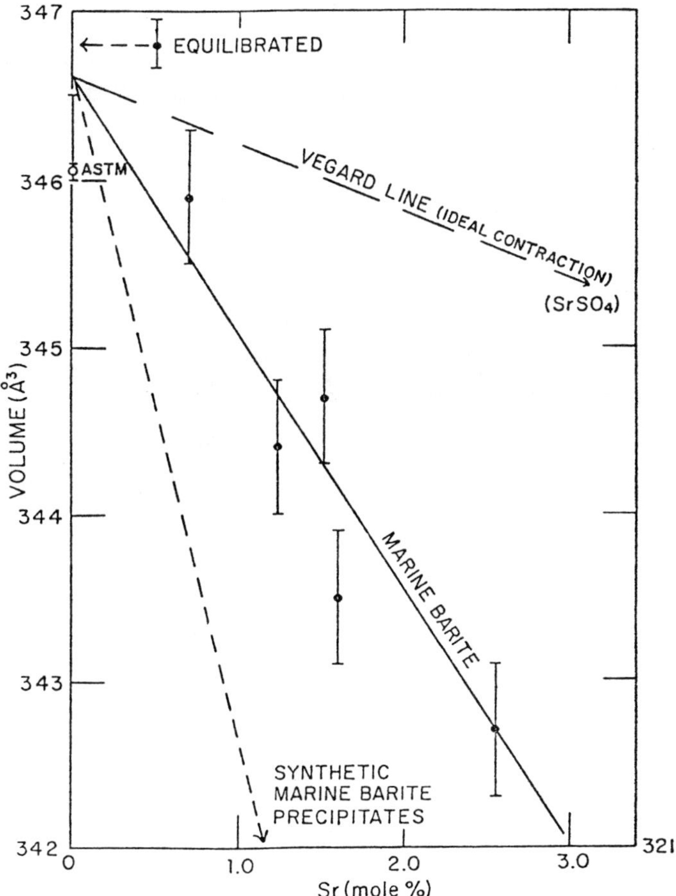

Figure 9. The contraction of unit cell volume with strontium content in marine barites.

Marine Barite Saturation in Sea Water

It is necessary to ask to what extent marine barite is formed in thermodynamic saturation with open and pore waters of the sea in order to evaluate their mode of formation and role in the marine geochemical cycle of barium. Chow and Goldberg (1960) first predicted barite should reach or approach saturated equilibrium in the sea. Church and Wolgemuth (1972) evaluated the saturation of marine barite as a function of benthic decreases in temperature and increases in pressure on the actual ion activity product of barite. In this analysis, the changes in ion activities of participating barium and sulfate were calculated from ion pairing models and compared to the ion activity solubility product of pure barite, also a function of temperature and pressure. The results of marine barite saturation are summarized in Figure 10 which plots the saturated molality of barite versus the actual barium molality with depth for the eastern Pacific. There is fair agreement with the barite saturation measurements of Burton *et al.* (1968) for surface English Channel water. The barium contents of barite-bearing sediment pore waters at depth is also compared to check the accuracy of the calculation. There is remarkable agreement between the calculated saturated barium molality and the pore waters of barite-bearing sediments (50μg/kg at the ocean floor). The overlying water column is undersaturated by a factor of

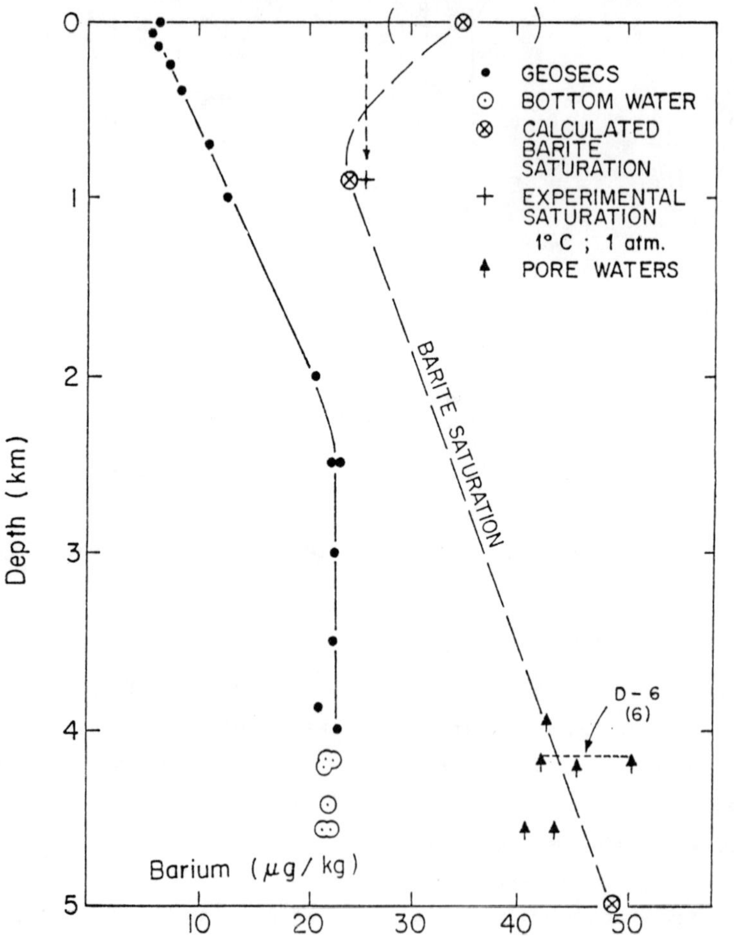

Figure 10. Concentration of barium in eastern Pacific waters versus calculated barite saturated barium concentrations (Church and Wolgemuth, 1972).

about two or more. Suspended marine barites must form in restricted microenvironments and being undersaturated should tend to dissolve with depth with disintegration of these biogenic environments. Such a dissolution of marine barite confirms the calculations of Dehairs *et al.* (1979) which shows significant dissolved barium fluxes to the deep sea from the fallout of pelagic barites. However, sedimentary marine barites in these Pacific sediments appear to be forming at equilibrium with their pore waters and thus stable after burial. In silicous sediments of the Antarctic, however, barium in pore waters appears supersaturated suggesting kinetic factors or organic coatings can support such thermodynamically unstable conditions (Li *et al.*, 1973). Desai (1961) also showed such supersaturation to be the case in organic rich culture solutions containing barite. The mixing regimes of river estuaries appears to create supersaturated conditions that could be forming marine barite (Hanor and Chan, 1978). This appears to result from the ion exchange of barium off riverine clays when they encounter sea water and sulfate. There needs to be a careful search for such a potential estuarine locus of marine barite formation and contribution to marine geochemical cycles.

Most saturation calculations are based on the assumption that the purity of marine barite renders them as ideal as pure barite in their saturation properties. Hanor (1969) demonstrated that if barites are assumed to form a non-ideal symmetric solid solution with Sr, assuming solid activity coefficients greater than unity, then lower saturated barium molalities could result in rendering surface sea water actually saturated with a $(Ba_{0.62}Sr_{0.38})SO_4$ marine barite. Church (unpublished) measured experimentally the barium contents of sea water after precipitating artificial marine barite and sea water equilibrated with pure barite. The pure barite equilibrations yielded saturated barium contents which matched well the saturation calculations for $1^{o}C$ (Fig. 11). However, artificial marine barites containing natural levels of Sr but elevated contents of Ca yield elevated, apparently supersaturated, barium in solution. Evidently the non-ideal substitution of cations as dissimilar as Ca to Ba significantly alters their solid activity coefficients. Unlike Hanor's (1969) prediction, these coefficients for Ca appear less than unity. A further enigma develops however if one calculates the theoretically saturated concentration of barium on the basis of solubility

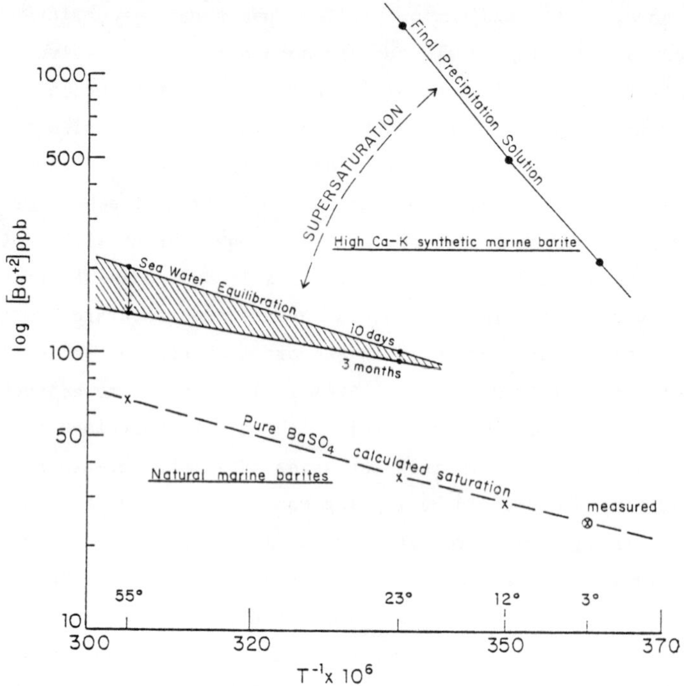

Figure 11. Gibbs-Helmholtz plate of barite saturated barium concentrations versus temperature. Displayed are precipitate and equilibrated pollutions for highly substituted synthetic marine barites; calculated and measured suration for pure barites in sea water.

---

constants estimated from mean alkaline earth sulfate activity coefficients other than $BaSO_4$ (unmeasured). The predicted saturation calculations are about half what one calculates for pure barites from ion activity product calculations or measured. This can only be explained if in fact the surfaces of marine barites equilibrated with sea water become more enriched in substituted alkaline earths than their bulk solid contents and thus yield apparent supersaturation. This same phenomena apparently alters the actual saturation properties of calcite and aragonite (Moorse, in press). Resolution of this enigma must await more careful equilibration experiments that monitor the surface chemistry of equilibrated marine barites using photo-electron spectroscopic techniques.

Substitution Properties of Marine Barites

The effects on solubility of substituting cations (e.g., Sr, Ca, K) into a marine barite can be evaluated by three approaches: (1) calculating the ideal composition of a marine barite; (2) the experimental precipitation of barite from sea water; and (3) the equilibration of pure barite in sea water. By comparing these compositions one can calculate thermodynamic values of enthalpy and entropy of substitution for a given cation and pass judgment on the accommodation of sea water ions by barite. These three approaches were taken by Church (1970).

The ideal composition of a marine barite can be computed by assuming that ions similar to barium form indistinguishable components of the barite structure, distributing themselves in chemical equilibrium between the solid and solution. This distribution is ideally proportional to their sulfate solubility products compared to $BaSO_4$. For dilute constituents the assumption is also made that the activity [ ] of the constituent in the solid is proportional to its mole fraction (X). The general distribution equation may be expressed for a substituent *M*

$$\left[\frac{X_S}{X_{Ba}}\right]_{Solid} = D_o \left[\frac{[M]}{[Ba]}\right]_{Solution} \qquad (1)$$

where $D_o = \dfrac{kd(BaSO_4)}{kd(MSO_4)}$ = ideal distribution coefficient.

The calculated composition of our ideal marine barite precipitating from surface sea water is listed in Table 4 using both the molal and activity ratios of component ions in sea water. The ideal Sr and Ca concentrations for a sea water precipitate of barium sulfate are one to two orders of magnitude higher than those of natural marine barites. This result is rather insensitive to the use of molal rather than activity ratios since most specific ion complexity effects tend to cancel. These effects are also assumed to largely cancel when the substituent calculation is repeated at lower temperatures and higher pressures of the deep sea. The results of an ideal marine barite composition at 0^oC; 500 atm are also shown in Table 4. The effect of lowering the temperature to 0^oC and raising the pressure to 500 atm decreases the substitution of cations in marine barite. The ideal concentration of K (assuming the bisulfate end member) agrees well with those found naturally in marine barite. The degree of non-ideal substitution into marine barite seems to be ion specific, and the order K, Sr, and Ca is the order of ion radius similarity to that of Ba.

Table 4. Ideal Sr, Ca, and K composition of marine barite precipitates as a function of temperature and pressure.

	Sea Water Molality	γ-free	f (frac. unassoc.)	Sea Water Activity	Ionic Radius (Pauling, 1953) r(Å)	Δr(Å), Ba-M
Sr	9.0×10^{-5}	0.24	0.95	2.1×10^{-5}	1.13	0.22
Ca	1.02×10^{-2}	0.24	0.91	2.9×10^{-6}	0.09	0.36
K	1.01×10^{-2}	0.62	0.99	6.2×10^{-3}	1.33	0.02
Ba	2.6×10^{-7}-1.8×10^{-7}(0°C)	0.24	0.93	5.7×10^{-8}	1.35	--

	25°; 1 atm.			0°C; 1 atm.		0°C; 500 atm.			Natural Marine Barite Compositions
	Kd(MSO_4)	Mole Percent (Activities)	Mole Percent (Molalities)	Kd(MSO_4)	Percent	$\Delta\bar{V}^*$ cc/mole	Kd(MSO_4)	Mole Percent	
Sr	2.6×10^{-7}	13%	12%	2.5×10^{-7}	9.5	-41.5	6.3×10^{-7}	5.0	0.3-3.0%
Ca	2.9×10^{-6}	13	12	6.0×10^{-5}	5.3	-41.3	1.5×10^{-4}	2.5	0.01-0.10
K	---	--	--	4.9×10^{-3}	0.062	-27.1	9.0×10^{-3}	0.043	0.01-0.10
Ba	1.0×10^{-11}	--	--	5.4×10^{-11}	---	-42.0	1.4×10^{-10}	---	---

Marine barite was synthesized under a variety of conditions which approximate those of the real sea water system in that dilute sulfate-free barium sea water solutions containing dilute barium were added slowly and homogeneously to large volumes of sea water. Such synthetic marine barites contain several precent Sr, one percent Ca, and tenths percent K, regardless of mode or order of reactant additions. Lowering the temperature to 0°C decreases the Sr and Ca levels, while raising the K concentration. A second set of syntheses using slower reaction rates and larger volumes produced even lower Sr, Ca, and K concentrations in the barite. More rapid addition apparently allows greater substitution, while longer digestion times allow ion exclusion and closer approach to an equilibrium distribution between marine barites and sea water. A comparison of ideal and synthetic composition for marine barites as a function of temperature are displayed in Figure 12.

The contents of Sr and Ca in marine barite show an increase with temperature, and reflect increases of barite solubility relative to the substituent and members. Celestite ($SrSO_4$) increases its solubility and anhydrite ($CaSO_4$) decreases its solubility with temperature relative to barite. Thus while the calculated and observed compositions for Sr and Ca display covariant temperature variations, artificial marine barites as a whole display substitution factors of five and ten times less than calculated ideally. Specifically, Sr concentration approximates that

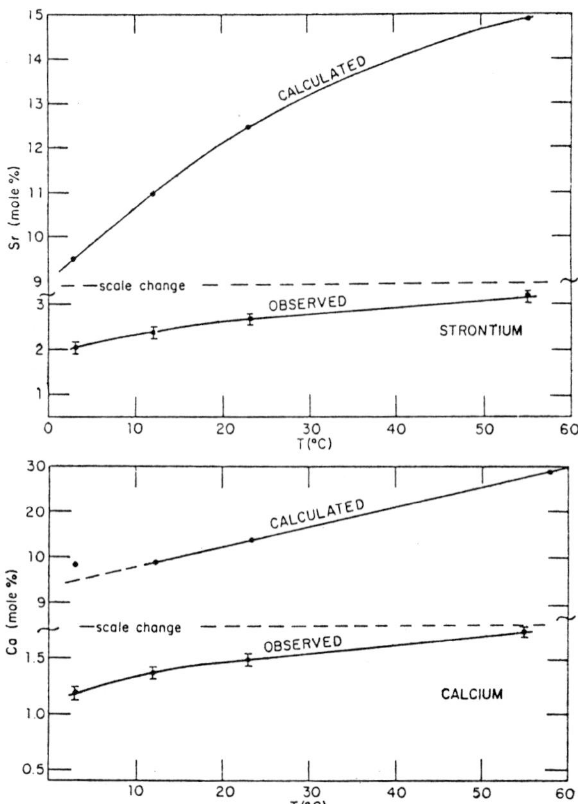

Figure 12. Composition of Sr and Ca calculated for ideal marine barites and observed in synthetic marine barites with temperature.

observed naturally (2.0 mole % at $3^{o}C$ vs 1.0 mole % naturally), while the Ca levels are still an order of magnitude greater than natural marine barites (1.2 mole % at $3^{o}C$ vs 0.01-0.1 mole % natural range). Potassium shows decreased substitution with temperature, and at $3^{o}C$ is several factors higher than observed in natural barites. Synthetic as well as natural marine barites seem to form with less Sr, Ca, and K than calculated for ideal solid solutions.

A third means to evaluate composition of marine barite from cation substitution is by digestion. Church (1970) equilibriated a pure barite and an artificial one enriched by Sr, Ca, and K in surface Pacific sea water at $3^{o}C$ for 17 months (Fig. 13). Strontium undergoes rejection from the enriched barite at about 40 ppm/day which yields a steady value of 1.7 ± 0.3%, agreeing reasonably well with the Sr levels found in natural marine barite. Calcium levels for the two barites converge

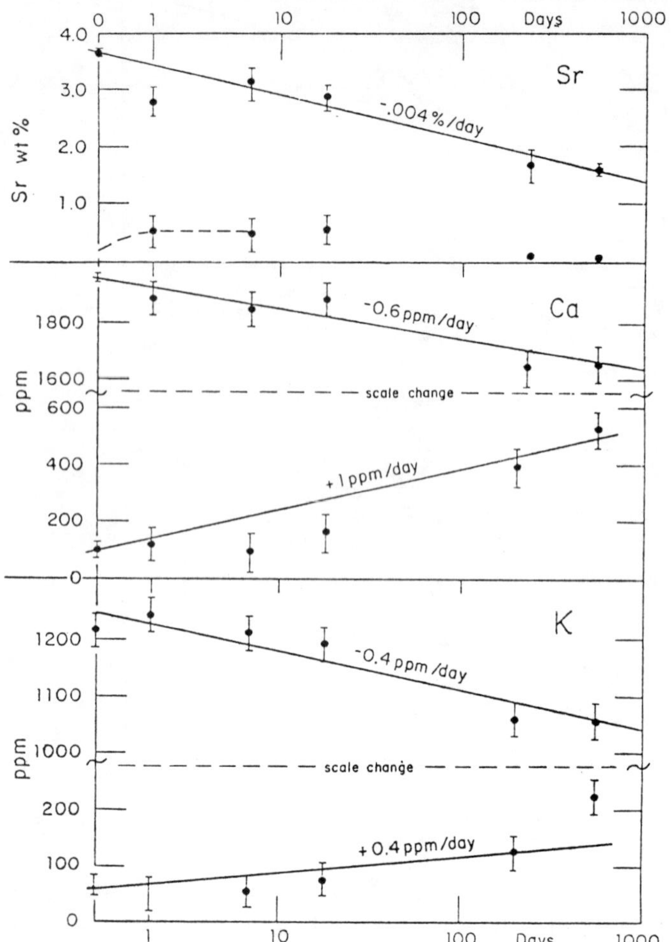

Figure 13. Digestion and equilibration of barite in sea water (3°C; 17 months).

linearly to 1060 ppm in about three years which agrees reasonably with the upper limits observed naturally (100-1000 ppm). Potassium behaves similarly, converging to 580 ppm in 55 months which agrees well with the K content of natural marine barites. Thus, barite is capable of undergoing exchange of homologous substituent cations with sea water over measurable and geologically short periods of time under natural deep-sea temperatures. Such equilibriated compositions resemble natural marine barites whose composition, too, may be surface controlled.

Thermodynamic Functions of Marine Barite Solid Solutions

The distribution function defined previously (equation 1) has a thermodynamic basis. Its ideal approximation fails to predict the empirical composition of marine barite because of the non-ideal nature of Sr, Ca, and K substitution in marine barite. In fact, the ideal and observed distribution coefficients can be related theoretically (McIntire, 1963) in a form that partitions ideal and non-ideal thermodynamic terms. Such an expression for the actual (observed) distribution coefficient D_{obs} for BC substituting in a crystal host AC is

$$D_{obs} = \frac{X_{BC}/X_{AC}}{M_{BC}/M_{AC}} = \underset{(a)}{\frac{Kd(AC)}{Kd(BC)}} \underset{(b)}{\left[\frac{\Gamma_{BC}}{\Gamma_{AC}}\right]} \underset{(c)}{\exp\,(-\Delta\mu/RT)} \qquad (2)$$

where

X and M are the concentrations with activity coefficients for λ and Γ in the solid and liquid respectively.

Three parts are apparent:

(a) The ratio of solubilities of the pure solid end members of the mixture; the ideal approximation to distribution (D_o).

(b) The ratio of the total interactions in solution for the component ions; the non-ideal term for components in the liquid phase.

(c) The free energy change associated with introducintg a foreign ion into a real lattice; the non-ideal term for the solute component in the solid solution phase.

The effects in term (b) largely cancel due to the chemical similarity of the substituent ions in solution. This leaves term (c) as the non-ideal expression for a solid solution. Thus, if the actual composition of the solid is known, the solute activity coefficient in the solid (λ_2) and the non-ideal free energy of solution ($\Delta\mu_2$) can be calculated as

$$\lambda_2 = \exp\,(\Delta\mu_2/RT) = \frac{D_o}{D_{obs}} \qquad (3)$$

and

$$\Delta\mu_2 = RT \ln \lambda_2. \qquad (4)$$

Thus values of λ_2 greater than unity, and positive values of $\Delta\mu$ are measures of the rejection of a substituent from a real solid solution relative to an ideal one.

Also since the distribution coefficients are expressed in thermodynamic forms which have as a basis temperature functions, they should obey log linearity with reciprocal absolute temperature consistent with the Gibbs-Helmholtz equation.

The molal free energy of non-ideal mixing, $\Delta\mu$, should also be a function of temperature described by the equation of definition

$$\Delta\mu = \Delta\overline{F}_M = \Delta\overline{H}_M - T\Delta\overline{S}_M \tag{5}$$

where $\Delta\overline{H}_M$ and $\Delta\overline{S}_M$ are the molal enthalpy and entropy of non-ideal mixing which can be calculated from this linear function.

Using the previous theory and data cited for substitution properties of marine barite, Church (1970) evaluated the thermodynamic behavior for the substituents Sr, Ca, and K in (1) artificial marine barites precipitated synthetically from sea water (0-50°C); (2) barites which have digested in sea water (3°C); and (3) a natural marine barite whose pore water was analyzed.

Distribution coefficients for these marine barite types are summarized and plotted in Figure 14 as pD versus $T(^\circ K)^{-1}$. Linearity suggests a real and thermodynamic basis for atomic substitution of Sr and Ca cations into marine barite. The ideal and observed separation of the two lines represent the non-ideal terms (λ_2 and $\Delta\mu_2$) for substitution. The larger and more positive terms for Ca relative to Sr represent greater exclusion of Ca. This would be predicted also on the basis of ion size dissimilarity.

The absolute values for the observed distribution coefficients in marine barite are anomalously low when compared to distribution coefficients found in the literature (Starke, 1964) extrapolated to the barite saturated, molal ratios of Sr and Ca to Ba in sea water at 23°C (Church, 1970). Thus marine barites seem to exclude Sr and Ca even compared to barites precipitated from simpler or less concentrated ionic solutions either in the laboratory or on land. A clue to this discrepancy may be in recognizing that marine barites are precipitating in large anion (sulfate) excess systems. Cambell and Nancollas (1969) have shown that the rate of barite precipitation is not only a function of supersaturation, but of the relative proportions of barium and sulfate ions as well. Excess sulfate systems retard the rate of barite precipitation, and follow second order kinetics which Nielson (1964) attributes to a

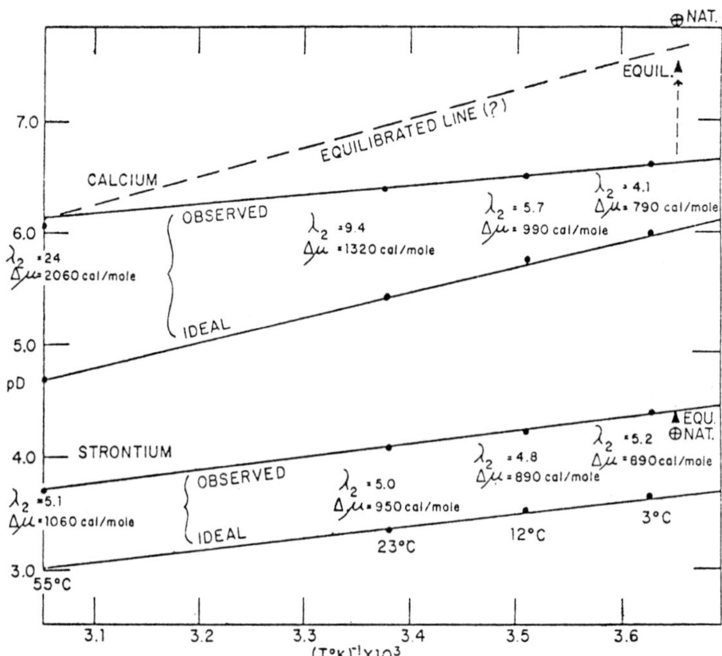

Figure 14. Tne negative logarithm of the distribution coefficient (pD) versus reciprocal temperature for Sr and Ca in ideal marine barites and observed in synthetic ones.

surface-growth rate limiting step. Thus ionic ratios of barium and sulfate and/or a surface mechanism of substitution might be important factors which explain the purity of marine barite.

When plots are made of $\Delta\mu$ versus temperature, a predicted linearity is observed (Fig. 15). The non-ideal enthalpy and entropy of mixing for Sr and Ca are also noted showing the greater energy involved for Ca substitution. When Sr substitutes into marine barite, the process is nearly isoenthalpic with a minimum of lattice distortion. Most of the free energy of substitution is involved in a small entropy decrease required for accommodaticn in the host structure. The requirement of a finite entropy of non-ideal substitution shows that marine barite does not in fact form a regular (symmetric) solid solution with strontium (Hanor, 1969). Calcium substitution appears to result in greater lattice distortions during substitution as was apparent in the large unit cell contractions of the high calcium marine barite precipitates in Figure 9.

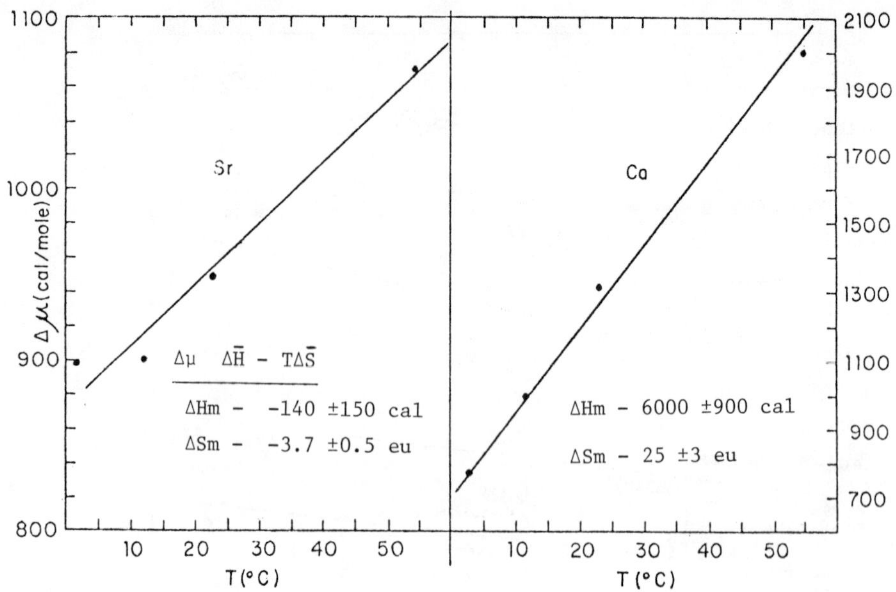

Figure 15. Free energy of non-ideal mixing and temperature for synthetic marine barites.

Thermodynamic Conclusions

Marine barite, as a thermodynamic precipitate of sea water, is a solid which not only discriminates against chemical contamination by solid solution with other species, but also whose degree of non-ideal substitution depends on the size of the contaminant ion. Potassium in marine barite precipitates or equilibriates with a composition nearly ideal. The equilibriated strontium composition in marine barites is less than predicted on an ideal basis. Calcium, however, equilibriates with difficulty to a composition which is an order of magnitude less than ideally predicted. This order of distinction (K, Sr, Ca) is the order of size similarity to the host, barium. Goldschmidt (1937) made the fundamental proposition that the incorporation of ionic substituents into minerals depends on the similarity of the charge and size of the substituent to the host ion. Marine barites as a sea water precipitate appear to adhere to this concept primarily on the basis of ion size.

SUMMARY

Marine barite is an authigenic precipitate of the sea and an important constituent of the marine geochemistry of barium useful in understanding marine diagenesis and precipitation.

Marine barite has often been recognized as a microcrystalline mineral of the ocean floor associated with biological debris. Distinct euhedral crystals normally 1-2μm can comprise up to two percent of sediments underlying productive areas of the equatorial oceans, particularly the eastern Pacific. Barite's occurrence in deposits of biogenic pelagic sediments is correlated with organic carbon and calcium carbonate, and the amount of barium conveyed by carbonate debris is sufficient for the amounts of barite observed. Occasionally large (25-100μm) barite panes are found in unusual deep-sea deposits characterized by altered volcanic debris (palygorskite) or manganiferrous phases.

Barite is found ubiquitously suspended in the ocean water column and appears to comprise the bulk of suspended barium phases. In both the water column and sediments, it is proposed that pelagic marine barite results from the concentration of barium in the marine biological cycle. Marine barite subsequently precipitates after degradation of biogenic organic phases and release of barium ion to sulfate rich micro-environments. In the water column, replacement of celestite tests of Acantharia (a planktonic order of foraminifera) may be an important precursor mechanism. It appears the flux of barite in the water column could contribute about half of the sedimentary barite. The remainder of marine barite appears to grow subsequently in surface sediments after burial as a consequence of further organic degradation in the initial 20-50 thousand years. In rarer deep sea or coastal environments, barite can form by other means such as barium release during volcanic alteration or ion exchange of barium during marine incursions of previously fresh aquatic environments organically enriched in barium.

Chemically, marine barites are quite pure with about a mole percent of strontium which is in solid solution from unit cell contraction X-ray data. The isotopes of Sr in barite are indistinguishable from those in sea water. Calcium and potassium range 100-1000 ppm, and from thermodynamic data may also be in solid solution. Rare earths and thorium are relatively concentrated in marine barites up to 100 ppm, considerably more for marine barite which distinguishes from barites found on land. Sulfur and oxygen isotopes in marine barites approximate those for sea water sulfate, or slightly enriched in the O^{16} under reducing conditions. Borderland marine barites, however, are distinct with heavy sulfate residuals after bacterial reduction. Natural thorium radionuclides show from ^{230}Th excesses that pelagic marine barites are forming at rates similar to their bulk sediments, and from ^{228}Th excesses that the radium daughter is being rapidly exchanged for barium on the order of years. The constant high $^{226}Ra/Ba$ ratios also suggest the rapid exchange of radium by barite over thousands of years. The ^{210}Pb daughter of the uranium series is about 90% deficient probably due to ^{222}Rn gas loss from the bulk 2-4μ microcrystals. A similar loss of ^{4}He restricts barites from being used for U-He chronology.

Saturation of sea water by marine barite is about 50μg/kg in the deep sea, and barite bearing sediments appear to be largely saturated at these levels. Barite in open sea water is at least a factor two under-saturated, suggesting suspended particulate barite must be forming in restricted bio-environments and not at equilibrium. In fact mass flux calculations show the increases of dissolved barium with depth in the ocean must be equally contributed by *in situ* dissolution of barite and release from remineralized biological debris. Estuarine regimes and organic rich siliceous pore waters often appear to be supersaturated. Thus, kinetic and organic surface effects can prevent formation of marine barites under conditions thermodynamically favorable for its formation.

Synthetic and calculated ideal marine barites should be more enriched in K, Sr, and Ca than observed. This is the order of decreasing size similarity to barium, increasing free energies of non-ideal substitution, and degree of increasing the solubility of a substituted marine barite. Thus, natural marin e barites are largely non-ideal in their purity and as a slowly forming precipitates of the deep-sea exhibit remarkable properties of crystallographic ion exclusion.

MARINE BARITE: REFERENCES

Arrhenius, G. (1959) Sedimentation of the ocean floor. In *Research in Geochemistry*, John Wiley & Sons, Inc., New York.

_______. (1963) Pelagic Sediments. In Hill (Ed.), *The Sea*. Wiley Interscience, New York, p. 655-727.

_______ and E. Bonatti (1965) Neptunism and vulcanism in the ocean," Progress Oceanogr., 3, 7-22.

Bertine, K.K. and J.B. Keene (1975) Submarine barite-opal rocks of hydrothermal origin. Science, 188, 150-152.

Borole, D.V. and B.L.K. Somayajulu (1977) Radium and Lead-210 in marine barite. Marine Chem., 5, 291-296.

Boström, K. (1968) Subsolidus phase relations and lattice constants in the system $BaSO_4$-$SrSO_4$-$PbSO_4$, Arkiv. Mineralogi Geologi, 4, 477.

Botazzi, E.M. and B. Schreiber (1971), Acantharia in the Atlantic Ocean, their abundance and preservation, Limnol. and Oceanog., 16, 677.

Burns, V.M. and Burns, R.G. (1978) Diagenetic features observed inside deep-sea manganese nodules from the north equatorial Pacific. Scanning Electron Microscopy, 78, 245-252.

Burton, J.D., N.J. Marshall, and A.J. Phillips (1968) Solubility of barium sulfate in sea water. Nature, 217, 834.

Cambell, J.R. and G.H. Nancollas (1969) Crystallization and dissolution of strontium sulphate in aqueous solution. J. Phys. Chem., 73, 1735-1740.

Chesselet, R., J. Jedwab, C. Darcourt, and F. Dehairs (1976) Barite as a discrete suspended particle in the Atlantic Ocean. EOS, Am. Geophys. Union, 57, 255 (abstr.).

Chow, T. and E. Goldberg (1960) On the marine geochemistry of barium, Geochim. Cosmichim. Acta, 20, 192-198.

Church, T.M. (1970) *Marine Barite*, Ph.D. Dissertation, Univ. California, San Diego, 100 p.

_______ and B. Velde (1979) Geochemistry and origin of a deep-sea Pacific polygorskite deposit. Chem. Geology, 25, 31-39.

_______ and K. Wolgemuth (1972) Marine barite saturation. Earth Planet. Sci. Lett., 15, 35-44.

Cook, H.E. and I. Zemmels (1971) X-ray mineralogy studies - Leg 8. In Tracy *et al.* (Eds.), *Initial Reports of the Deep-Sea Drilling Project*, Vol. VIII, pp. 901-955, U.S. Govt. Printing Office.

Cortecci, G. and A. Longinelli (1972) Oxygen-isotope variations in a barite slab from the sea bottom off California. Chem. Geology, 9, 113-117.

Cronnan, D. and J. Tooms (1969) The geochemistry of manganese nodules and associated pelagic deposits from the Pacific and Indian Oceans. Deep-Sea Research, 16, 335-359.

Deer, W., R. Howie, and J. Zussman (1965) *Rock Forming Minerals*, Vol. I, p. 4; Vol. V, p. 148. Longmans, London.

Dehairs, F. (1979) Discrete suspended particles of barite and the barium cycle in the open ocean. Doctor of Science Dissertation, Vrije Universiteit Brussel.

_______, R. Chesselet, and J. Jedwab (1979) Discrete suspended particles of barite and the barium cycle in the open ocean. Earth Planet. Sci. Lett., in press.

Desai, M.V.M., E. Koshy, and A.K. Ganguly (1969) Solubility of barium in sea water in presence of dissolved organic matter. Current Sci., 38, 107.

Goldberg, E.D., B.L.K. Somayajulu, J.N. Galloway, I.R. Kaplan, and G. Fauce (1969) Differences between barites of marine and continental origins. Geochim. Cosmochim. Acta, 33, 287.

_______, and G. Arrhenius (1958) Chemistry of Pacific pelagic sediments. Geochim. Acta, 13, 153-212.

_______ and J.J. Griffin (1964) Sedimentation rates and mineralogy in the south Atlantic. J. Geophys. Res., 69, 4293.

_______, M. Koide, J.J. Griffin, and M.N.A. Peterson (1964) A geochronological and sedimentary profile across the north Atlantic Ocean. In *Isotopic and Cosmic Chemistry*. North Holland, Amsterdam.

Goldschmidt, V.M. (1937) The principles of distribution of chemical elements in minerals and rocks. J. Chem. Soc., 1937, 655-672.

Gordon, L., C. Reimer and B. Burtt (1954) Coprecipitation from homogeneous solution. Anal. Chem., 26, 842.

Griffin, J.J. and E. Goldberg (1969) Clay mineral distribution in the Atlantic Ocean. J. Geophys. Res.

Guichard, F., T. Church, M. Treuil and H. Jaffrezic (1979) Rare earths in barites: distribution and effects on aqueous partitioning. Geochim. Cosmochim. Acta, 43, 983-997.

Hanor, J.S. (1966) *The origin of barite*. Ph.D. Thesis, Harvard Univ.

_______ (1969) Barite saturation in sea water. Geochim. Cosmochim. Acta, 33, 899.

_______ and G. Brass (1968) Stratigraphic variation in barium content in sediment cores from the East Pacific Rise. GSA Ann. Meeting.

_______ and L.-H. Chan (1977) Non-conservative behavior of barium during mixing of Mississippi River and Gulf of Mexico waters. Earth Planet. Sci. Lett., 37, 242-250.

Leoblich, A. (1966) *Treatise on Invertebrate Paleontology, Vol. C.* Geol. Soc. Amer., Univ. of Kansas Press, 789 p.

Li, Y.-H., T.-L. Ku, G.G. Mathieu, and K. Wolgemuth (1972) Barium in the Antarctic Ocean and implications regarding the marine geochemistry of Ba and Ra^{226}. Earth Planet. Sci. Lett., 19, 352-358.

Martin, J. and G. Knauer (1973) The chemical composition of plankton. Geochim. Cosmochim. Acta, 37, 1639.

McIntire, W.L. (1963) Trace element partition coefficients--a review of theory and applications to geology. Geochim. Cosmochim. Acta, 27, 1209-1264.

Murray, J. and A. Renard (1898) Deep Sea deposits, *Challenger Reports,* Longmans, London, 525 p.

Nielson, A.E. (1964) *Kinetics of Precipitation,* The MacMillan Company, New York, p. 80.

Pauling, L. (1960) *The Nature of the Chemical Bond,* 3rd ed. Cornell Univ. Press, New York, 644 p.

Puchelt, H. (1967) Zur Geochemie des Bariums in exogenen Zyklus. *Sitzungsberichte der Heidelberger Akademie des Wissenshalten,* Springer-Verland, Berlin, p. 175.

Somayajulu, B.L.K. and E.D. Goldberg (1966) Thorium and uranium isotopes in sea water and sediments. Earth Planet. Sci. Lett., 1, 102-106.

_______ and T.M. Church (1973) Ra, Th, and U isotopes in the interstitial water from the Pacific Ocean sediment. J. Geophys. Res., 78, 4529-4531.

Starke, R. (1964) Die Strontiumgehelte de Baryt. *Freiberger Forschungsh, C 150.*

Suess, E. (1979) Authigenic barite in varved clays; Result of marine transgression upon freshwater deposits and associated changes in interstitial water chemistry, In Fanning and Manheim (Eds.), *The Dynamic Environment of the Ocean Floor,* D.C. Heath & Co., Lexington, Mass.

Tendal, O.S. (1972) A monograph of the Xenophyophoria. Galathea Report, 12, 8.

Vinogradov, A.P. (1953) The elementary chemical composition of marine organisms. Sears Found. Marine Res., No. 2, p. 647.

_______ and V.V. Kovelisky (1962) Elemental composition of the Black Sea plankton. Dokl. Akad. Nauk. SSSR, 47, 217-219.

Walton, A.G. (1967) *The Formation and Properties of Precipitates,* Wiley Interscience, New York.

Chapter 8

MINERALOGY of EVAPORITES

William T. Holser

INTRODUCTION

The purpose of this chapter is to introduce briefly the chemistry and physics of seawater and its derivative brines, past and present, to summarize the mineralogical features of the evaporite facies, and to discuss the processes of crystallization that are important in generating them. In this limited space it is useful to emphasize the most common evaporite facies--calcium sulfate and halite--and consider in less detail the interesting but complex potash-magnesia facies, which is discussed in a variety of textbooks.

Additional chapters would be required to even outline the variety of geological settings in which evaporites are found today and have occurred in the past, and to review the controversies about their interpretations. However, it might be well at the outset to state, without documentation, a catholic but critical viewpoint as a context for the following discussions of mineralogy and geochemistry.

(1) Evaporites have been generated in a variety of physical situations: (a) coastal intertidal and supratidal zones, called sabkhas, (b) small lagoons on coasts and atolls, (c) large deep-water marine basins, (d) sub-sealevel basins with marine inflow, and (e) non-marine interior basins.

(2) The paleogeographic and tectonic settings for these physical situations are also of a wide spectrum: (a) continental margins and shelves, (b) interior cratonic basins, both shallow and deep, and (c) rifted continental margins.

(3) The generation of evaporites in these media has been pervasive in place and time: (a) nearly all sedimentary basins have some evaporites in the section, (b) nearly all periods of the past billion years have had some evaporites, (c) this distribution of evaporites spread wide in space and time has

been quantitatively very irregular--for example, huge amounts on several continents in the Devonian and Permian and insignificant and very localized deposition in the present time.

A word about the background of this work. The first version of Chapters 8 and 9 was essentially completed as a research report for Chevron Oil Field Research Company in 1970, and was revised in 1975 and again in 1979. I am grateful to the Standard Oil Company of California, and to Dr. William J. Plumley, for past encouragement and support in evaporite research, and for permission to publish these summaries. They draw heavily not only from my own work and that in the literature, but also from the extensive data and discussions provided by my colleagues in the corporation.

COMPOSITION OF SEAWATER

Today's Seawater

At the present time the main bodies of water in the oceans are very well mixed with respect to most components. The first list of ions in Table 1, which represents the major components, is constant in ratio (within analytical error) throughout the oceans of the world (Culkin, 1965). The actual total concentration of ions varies by 2 or $3^{o}/oo$ (parts per thousand or per mil) around the central average figure of $35.0^{o}/oo$ weight salinity, through slight concentration by evaporation of water or dilution by the inflow of water.

The principal driving force for mixing of the oceans at the present time is Antarctic polar water, slightly more dense by virtue of its low temperature, that drops down and flows northward, where it rises again to be heated. This process presently mixes the oceans on a time scale of less than a thousand years. This time is very short compared to the millions of years required to change most ions in seawater by input and output.

The trace constituents of modern seawater (Goldberg, 1965) are divided in Table 1 into those that are relatively constant throughout the sea, and those with evident variations. Only eight of these are present in concentrations greater than one part in 100 million. In a general way, most of the precipitation of minerals in evaporites is controlled by the main elements, whereas many of the minor elements are important in tracing geochemical processes. So while extremely small traces of practically

TABLE 1. COMPOSITION OF MODERN SEAWATER*

Principal Ions	*g/kg soln*	*g/kg H_2O*	*Residence Time, Yrs.*
Na^+	10.77	11.06	1.9×10^8
Mg^{2+}	1.30	1.34	1.4×10^7
Ca^{2+}	0.409	0.420	1.2×10^6
K^+	0.388	0.398	8.2×10^6
Cl^-	19.37	19.89	3.1×10^8
SO_4^{2-}	2.71	2.78	2.2×10^7

Constant Minor Constituents	*ppm, wt. (mg/kg soln)*	*Residence Time, Yrs.*
Br	65	Very long
Sr	8.0	5×10^6
B	4.5	1.6×10^7
F	1.3	-
Ar	0.6	-
Li	0.17	2.0×10^7
Rb	0.12	2.7×10^5
I	0.06	-
Al	0.01	100
Fe	0.01	140
Zn	0.01	2×10^3
Mo	0.01	1.4×10^5

Variable Minor Constituents	*Mean ppm*	*Range, ppm*	*Main Parameters*	*Residence Time, Yrs.*
C	26	10-50	Temp biol	8.8×10^4
N	13	-	Biol	1.2×10^4
O	8	0-10	Biol	-
Si	3.0	0-4.0	Biol	1.6×10^4
P	0.07	0.001-0.1	Biol	-
Ba	0.02	0.01-0.08	Pcptn ?	1.7×10^3

* Compiled from Wedepohl (1969-1978), Garrels, Mackenzie, and Hunt (1975), and Goldberg (1965).

any element may be found in seawater (including gold!), most elements beyond those in the table are very difficult to detect, much less to use in any way.

In terms of ordinary inorganic chemistry, seawater is not only relatively complex in its composition, as mentioned above, but it is relatively concentrated. This is important in the way that it affects the processes of interaction among the ions, including the process of precipitation. As brines become more concentrated during evaporation, deviations from ideal chemical behavior become greater. Only recently has it been fully appreciated that even in ordinary seawater, the constituents are not tied up in the simple manner given in the list of Table 1 and in most textbooks. This list of simple ions is a bit of fiction used by chemical analysts to express the amount of each *element* that has been determined in the solution. The real situation in ordinary seawater is shown in Table 2, after Garrels and Thompson (1962). Here we see that while Na^+ and K^+ represent accurately the ionic state of these elements in seawater, a significant fraction of Ca (8%) and an even larger fraction of Mg (10%) are present not as Ca^{2+} and Mg^{2+}, but as $(CaSO_4)^0$ and $(MgSO_4)^0$ un-ionized pairs. When we look at the "anions," such associations are even more important. While Cl^- characterizes practically all of the chlorine, half of the sulfate is divided between Mg- and Na-anion pairs. Thirty percent of the bicarbonate and 90 percent of the carbonate ions are also associated in this way.

Table 2. Distribution of Major Dissolved Species in Representative Seawater of Chlorinity 19°/oo, pH 8.1, at 25°C and One Atmosphere Total Pressure (Garrels and Thompson, 1962)

Ion	Molality (Total)	% Free Ion	% Me-SO_4 Pair	% Me-HCO_3 Pair	% Me-CO_3 Pair
Ca^{++}	0.0104	91	8	1	0.2
Mg^{++}	0.0540	87	11	1	0.3
Na	0.4752	99	1.2	0.01	-
K^+	0.0100	99	1	-	-

Ion	Molality (Total)	% Free Ion	% Ca-anion Pair	% Mg-anion Pair	% Na-anion Pair	% K-anion Pair
SO_4^{--}	0.0284	54	3	21.5	21	0.5
HCO_3^-	0.00238	69	4	19	8	-
CO_3	0.000269	9	7	67	17	-

In general as seawater concentrates during evaporation, the importance of these un-ionized pairs increases, and even larger complexes are also formed. A qualitative appreciation of these associations makes it easier to understand some of the wide variations in solubility that occur as evaporation proceeds. Even at ordinary concentrations, half of the sulfate in seawater is tied up with Mg and Na, and therefore is not so easily available to precipitate with Ca as gypsum or anhydrite. This effect competes with the rising concentrations to give the peculiar and important variation of solubility of the calcium sulfate minerals with seawater concentration, which will be discussed below.

Cycles of Elements in the Sea

While seawater is presently rather well mixed and of constant composition, in the long view this situation is really a dynamic balance. The balance of input against output of an element is important (1) to understand why seawater has the particular composition that it does have, and (2) to see on what time scale particular ions could possibly vary in concentration.

The dynamic situation concerning a particular ion is most succinctly expressed as its *residence time* in the sea (Goldberg, 1965). The residence time of an element is a measure of how long a given atom of that element might be expected to stay in the sea, between the time that it flowed in from a river and time that it was removed from the sea. Residence time is a vague idea, because it presumes an equilibrium (in an essentially dynamic situation) in which the inflow of an element over a geological interval is equaled by output from the sea. To the extent that dynamic equilibrium is real, the situation might be stated this way: Removal processes increase with concentration, and the concentration is built up by inflow to the point where the element is removed as fast as it is added. Things really may not be that constant, and in fact they probably are not, but residence time is still a convenient measure for comparing the stability of the concentrations of various ions in seawater. On the above assumptions, residence time has usually been calculated as:

$$\text{Residence time} = \frac{\text{(mass of the element in the sea)}}{\text{(net mass delivered to the sea by the rivers per year)}} \quad [1]$$

Table 3. Mean Composition World Rivers, ppm (Livingston, 1963)

Na^+	- 6.3		SO_4^{2-}	- 11
Mg^{2+}	- 4.1		HCO_3^-	- 58
Ca^{2+}	- 15		NO_3	- 1
K^+	- 2.3		SiO_2	- 13
Cl^-	- 7.8			

The "net" mass is the total river inflow corrected for the "cyclic" salts that the rivers derive from the atmospheric residues of sea spray. The residence times of various elements in today's seawater are indicated in the righthand column of Table 1. Table 3, taken from Livingston (1963) gives the mean composition of dissolved inflow from rivers in the various continents, and a weighted mean for the world. There are several important aspects of this table. Mean river water is dominantly a calcium bicarbonate water, although both magnesium and sulfate are very important. By contrast, seawater is a sodium chloride water with important amounts of magnesium and sulfate. Furthermore, the K/Na ratio in river water is several times higher than that in seawater. The second thing to realize about the table is that although it is stated as a mean, it is strongly dependent on certain very large rivers that may represent very special situations, and therefore inflow may vary *substantially* with both time and place.

The traditional geochemical viewpoint that emphasized inflow input to the ocean from dissolved components in river water has not been entirely successful in explaining the oceanic budgets of some elements. Recent work has focussed attention on interactions of ocean water with newly generated mid-ocean ridges, both through thermally induced systems of hydrothermal circulation, and through direct low-temperature weathering of the new basalt. Such interactions have been postulated to account for inputs of silica, calcium and metals, and output of magnesium, that are important relative to the river flux. Some estimates are given by Wolery and Sleep (1976); additional recent literature is too extensive to be listed here.

The processes by which elements are removed from the ocean are mainly: (1) precipitation, (2) evaporation, (3) participation in biological processes. An obvious example of the first process is the inorganic precipitation of calcium (and magnesium) carbonate, which is approximately saturated in present surface seawater. With extensive evaporation, calcium sulfate, sodium chloride and potassium minerals are precipitated. The importance of these processes in regulating the total amount of these elements in the ocean is not often appreciated. But other elements are also constantly being removed from seawater by precipitation. Manganese, zinc, cobalt and copper are all incorporated into the manganese nodules that cover large areas of the ocean floor (see Chapter 1, this volume). Other elements may engage in exchange equilibria with minerals on the sea floor: Potassium and boron may enter into clay minerals, magnesium into chlorite, and calcium into phillipsite. Some elements partition into solid solution, such as strontium in calcium carbonate and calcium sulfate minerals, bromide in halite, and rubidium in carnallite; these will be discussed in detail later.

Biological processes are extremely important in removing materials from the ocean, leading to wide variations in the amount of these constituents, depending upon the local biological situation. The elements shown in the third part of Table 2 are nutrients, particularly for the phytoplankton that make up most of the primary organic production in seawater. Typically, the nutrients are taken up by organic matter in the surface photic zone, and then sink into deep water. Most of such material is regenerated to the ocean by oxidation of the dead organic matter, which results in strong local variations of phosphate, nitrate, silica, etc. However, some of the organic matter is retained in deep-sea anoxic sediments, and the nutrients are thereby fixed in the sediments and removed from seawater. Other elements, such as barium and strontium, may be precipitated by certain organisms, and find their way to the sea bottom if they are protected by the organic matter. One of the main constituents of seawater, sulfate, also is profoundly affected by biological processes. It is removed into muds as sulfide by sulfur-reducing bacteria. As will be shown presently, this may have led to wide variations in the sulfate content of seawater during geological time.

All of these processes lead to a balance of some kind between input of elements to the sea and output from the sea, with a long time-scale for some elements and a very short time-scale for others. In general, we may expect the former to have relatively constant compositions with time, while the latter (such as organic nutrients) may vary widely in both time and place. In the following section we will examine the actual evidence as to whether the main and trace constituents that have long residence times *did* change very much during geological time.

Evidence for a Relatively Constant Composition of Ancient Seawater

A seawater constant in the composition of its main elements is not just a plausible assumption; we actually have quite a bit of evidence that *most* elements have remained relatively constant during at least the Phanerozoic. A constant composition for the starting material will be a big help in considering the chemistry and mineralogy of marine evaporites of all ages.

The best and most generally accepted model for the *origin* of the sea itself, while not direct evidence, at least provides a conceptual model to make such constancy plausible. In a presidential address to the Geological Society of America that has become one of the classics of modern geochemistry, W. W. Rubey (1951) restated a general geochemical balance to prove that weathering of primary igneous rocks could not account for certain elements in the combined product of sedimentary rocks, ocean, and atmosphere. These elements, which he called the "excess volatiles," are now found mostly in the ocean and atmosphere: H_2O, CO_2, S, N, Cl, Ar, F, B, Br. His conclusion was that these elements must have come, not from the primary igneous rocks, but as new or "juvenile" degassing of the earth's mantle via gases, hot springs, and other volcanic activity. This does not mean that all volcanic gases are juvenile; indeed, only a very small fraction need be (the rest being recycled through the burial of old waters and sediments), and we have not yet succeeded in definitely measuring this juvenile fraction. But the idea is a good one, and leads to two important conclusions: (1) We have no reason to expect changes with geological time of the *proportions* of material added from the mantle, so not only the relative concentrations of these elements in seawater, but also their actual concentrations relative to water (which is one of the things being added all the time), should remain constant. (2) The sea probably built up during 4.5 billion years of the Earth's history; consequently, the total volume of the sea probably increased less than 10 percent during the mere half billion years that we are talking about for known evaporites. If mantle degassing was *more* important earlier in the Earth's history, an even *smaller* fraction of the sea would have been added in the last half billion years.

A second good reason for thinking that the composition of seawater has not changed with geological time is the constancy of the mineralogy of evaporites, and the order in which these minerals form (Holland, 1972). Suppose that, at some time in the past, sulfate or carbonate had been more common than chloride in seawater, just as it is more common today in some playa lakes of interior basins. Then we might expect to find minerals such as mirabilite ($Na_2SO_4 \cdot 10H_2O$) or trona ($Na_3(CO_3)(HCO_3) \cdot 2H_2O$) as the common sodium-bearing mineral of the resulting evaporites, rather than halite. We do find sodium sulfates and carbonates in very large amounts in the Tertiary and Recent lake beds of the Great Basin and Central Asia. But we can look at cubic miles of evaporites of all ages and never find traces of those minerals. The few places (*very* few before the Eocene, surprisingly enough) where they are found, have other evidence labeling them as non-marine in origin (Holser *et al.*, 1979). The important minerals of marine evaporites are calcite and dolomite, gypsum and anhydrite, halite, and certain potash-magnesia minerals, and both the minerals on that list and their order of crystallization have remained constant since the earliest evaporites we can find, in the Proterozoic. The order of this crystallization puts even stricter limits on the possible past variations in seawater of elements such as Ca^{2+}. If Ca^{2+} and/or $(SO_4)^{2-}$ had ever been of much higher concentration, then gypsum would have been a ubiquitous companion of the carbonates at that time; if Ca had been of much smaller concentration, then halite would have preceded calcium sulfate in precipitation.

We have also uncovered a bit of rather direct evidence for the composition of ancient seas. Many halite crystals have microscopic cavities that trapped some of the brine from which they grew. In a few places such halite has been recrystallized and the cavities coalesced until they are large enough to be extracted and analyzed. We did this some years ago

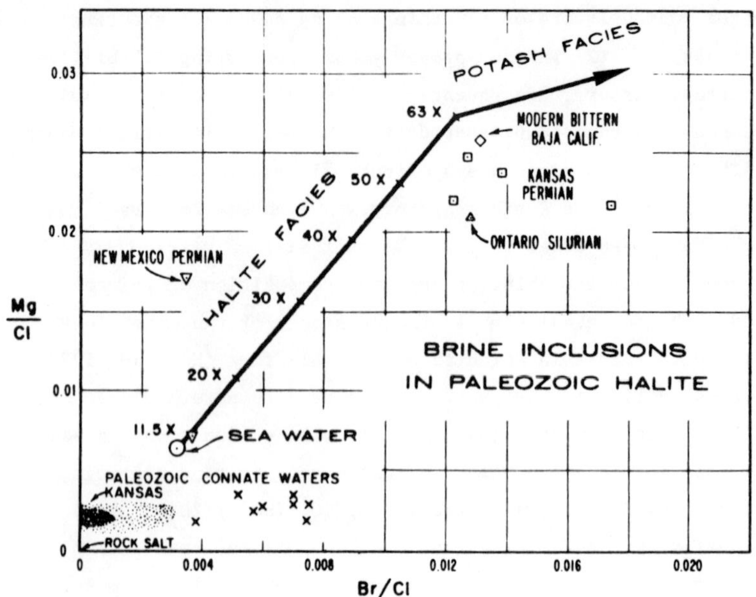

Figure 1. Analyses showing similarity of microscopic brine inclusions in Paleozoic halite to evaporation of modern seawater (Holser, 1963).

(Holser, 1963) and found that most such trapped brines had ionic ratios just like those we find when evaporating modern seawater (Fig. 1). When ancient seawater is not so carefully protected within a crystal but left to wander through the rocks, it undergoes a series of reactions that substantially change its composition (see Fig. 1) before we get around to analyzing it as a formation water. Nevertheless, it has proved possible to understand the general composition of formation waters as derived from ancient seawater, with no correlation between age and composition (Chave, 1960; White, 1965).

Even some trace elements may not have varied much through geological time. The way in which trace elements reflect brine composition will be discussed near the end of this review, but here it will be sufficient to say that bromide, which has a very long residence time in the sea (Table 1), shows no regular variation in basal halites of various geological ages (Fig. 2; a re-assessment by Herrmann *et al.* (1973) gives a similar result).

Isotopes act a lot like trace elements in their distribution between brine and crystals. Sulfur isotope measurements in sulfate minerals of marine evaporites of various ages do show a surprising variation (Holser and Kaplan, 1966) which will be discussed below. One reasonable interpretation of this variation is in terms of a redistribution, at various geological epochs, of the whole amount of sulfur between three important reservoirs: the sea, evaporites, and shales. On this basis I conclude that the sulfate content in the Permian sea may have been as little as 75 percent of its present level, and in the mid-Paleozoic as much as 50 percent higher than at present. As we will see, however, variations even as large as this do not change the evaporite minerals or the order of their facies.

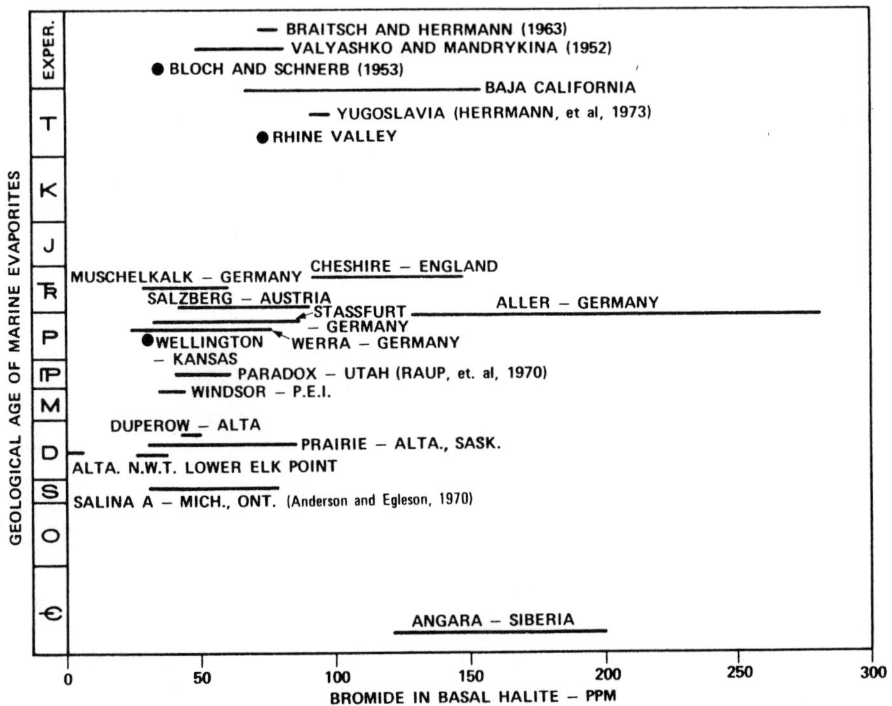

Figure 2. Bromide in the lowermost salt rock sections of various basins, as a measure of the Br/Cl ratio of ancient seawater. Data modified from tabulation of Kühn (1969). The upper section indicates the range of values expected from modern seawater, calculated from various experimental results.

Table 4. MEASURES OF CONCENTRATION

TOTAL SALTS

Salinity:	Mass of all salts, in grams, from 1 kg brine, dried to constant weight at 480 C. Standard seawater $\underline{s}_w$ = 35.00 o/oo ("per mil" = parts per thousand).
Volume Salinity:	Mass of salts per liter of brine. $\underline{s}_v = \rho\, \underline{s}_w$.
Chlorinity:	Mass of chlorine equivalent to all halides precipitated from 1 kg seawater. Standard seawater Cl = 19.37 o/oo.
Volume Ratio:	$\frac{\text{(Volume of original seawater)}}{\text{(Volume of brine)}}$
*Evaporation Ratio:	$\frac{\text{(Mass of water in original seawater)}}{\text{(Mass of water in brine)}}$

INDIVIDUAL COMPONENTS

Concentration (solution):	g component per liter brine.
*Concentration (wt solution):	g component per kg brine.
*Parts per million (ppm):	mg component per kg brine.
*Concentration (solvent):	g component per kg water.
*Molarity	g-moles component per liter brine.
Molality	g-moles component per kg water.
Normality:	g-equivalents component per liter brine (= molarity for univalent components).

PURE NaCl SOLUTION EQUIVALENT TO A BRINE

Depending on purpose may be equal to salinity, chlorinity, density, vapor pressure, etc.

*Preferred measures.

PROPERTIES OF SEAWATER AND ITS BRINES

Measures of Concentration

The "concentration" of seawater should be a pretty simple idea, but a dozen different measures have been used for it, often with confusion and sometimes in error. Some of this variety is displayed in Table 4.

Total salts are often stated in terms of *salinity*, following the lead of oceanographers. However, they did not have to deal with the precipitation of large amounts of salt from the brine; when that happens

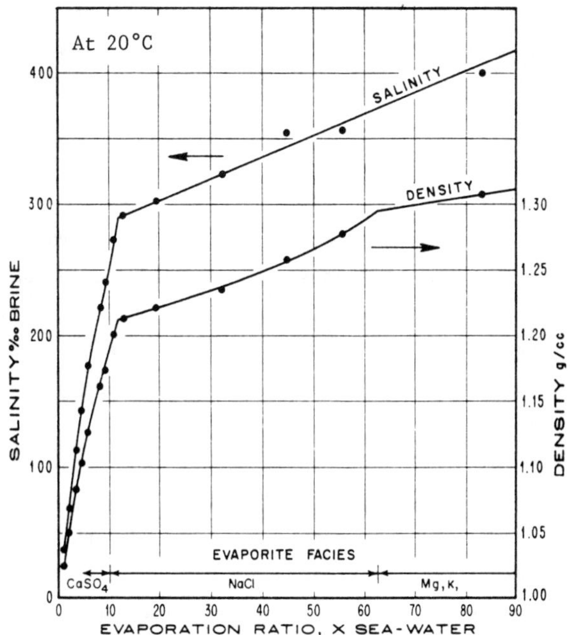

Figure 3. Density and salinity of concentrated seawater (Usiglio, in Clark, 1924; Rothbaum, 1958).

during evaporation, just about as much salt as water is removed. Now, if we would like to say how much the brine has undergone this process of concentration, rather than what concentration has resulted, we need a different measure. I like to use a concentration "seawater," that should properly be called the *evaporation ratio*. It is the weight of H_2O (*not* weight of brine) in the original seawater, divided by the weight of H_2O in the resulting evaporated brine. This simplifies calculations involving the essential process of removing water from the brine. The relation of salinity to evaporation ratio is shown on the left side of Figure 3, where it is evident that when evaporation goes on at a constant rate, salinity rises rapidly at first from its normal $35^{o}/oo$, but then only slowly beyond $300^{o}/oo$ as halite starts to drop out. Another measure is also often stated as "X seawater," but refers to the actual total *volume ratio* of seawater and brine, and it therefore differs from the evaporation ratio by also taking into account both the density of the brine and the different amount of salts remaining after evaporation.

An individual component (e.g., calcium) might best be stated in terms of weight concentration in the solution, as g/kg (kilograms of *solution*, not just kilograms of water) or as ppm. Many analysts state major concentrations in grams per liter. Minor concentrations are usually stated in "parts per million," which if unqualified should mean by weight. Watch out, though, as you will sometimes find that the analyst means milligrams per liter, and he may not have even measured the density, which is necessary to calculate weight concentrations. The difference can be as much as 30 percent in concentrated brines.

Normality, or equivalents, are useful for checking the ionic balance of a brine analysis; and molality, molarity or ionic strength are important for calculations in physical chemistry.

In geochemistry we often have to make use of data from synthetic systems rather than from natural materials. Seawater is a very complex system, not often measured by the chemists, but on the other hand simple sodium chloride is one of their favorites. But one has to be careful in applying data from sodium chloride solutions to seawater solutions: Any old measure won't do. For example, it is not correct to compare the transition temperature, gypsum → anhydrite, as measured in NaCl solutions and in seawater, at the same "chlorinity" as was done in one important published paper. The comparison must be made for two solutions having equal value of the particular property that controls the process of interest. In the case of gypsum → anhydrite, the controlling property is not the amount of chloride, but the water vapor pressure. Consequently, this transformation is expected at the same temperature in solutions of the same water vapor pressure, not of the same chlorinity.

Density

The density of seawater brines rises sharply from the 1.03 $g \cdot cm^{-3}$ of normal seawater to 1.29 at the beginning of halite deposition, then more slowly to a maximum of 1.31 in the potash-magnesia facies, as shown on the right side of Figure 3 (Usiglio, in Clark, 1924).

The density of brines is most important for its control over the circulation pattern in evaporite basins. When evaporation has raised the density of a surface brine sufficiently to make it mechanically unstable, the heavy concentrated brine will overturn to a lower level (probably the bottom), displacing the lighter brine back towards the surface. The threshold of mechanical instability that must be reached for this convection event is a subtle function of the viscosity and the thickness of the brine. There is also a question as to whether the convection set up by evaporation from the surface will stabilize as a set of compatible cells that are nearly equidimensional, as shown in Figure 4a, or as one large cell shown in Figure 4b. The former would be expected in a simple symmetrical system, but either the asymmetry of continual inflow of new seawater from one side, or constant wind-driven currents in either direction, or higher evaporation in shallows, would probably modify the pattern towards the latter form (Briggs and Pollack, 1967). Another unanswered question is whether convection would be expected to be periodic, perhaps after a certain amount of evaporation, or whether it would be continuous, everything else being constant. Can layers of distinct brine concentration result from certain convection situations, or will a convection always reach the bottom to make a uniformly dense column? And what about the velocity of the overturn: Can it generate density currents

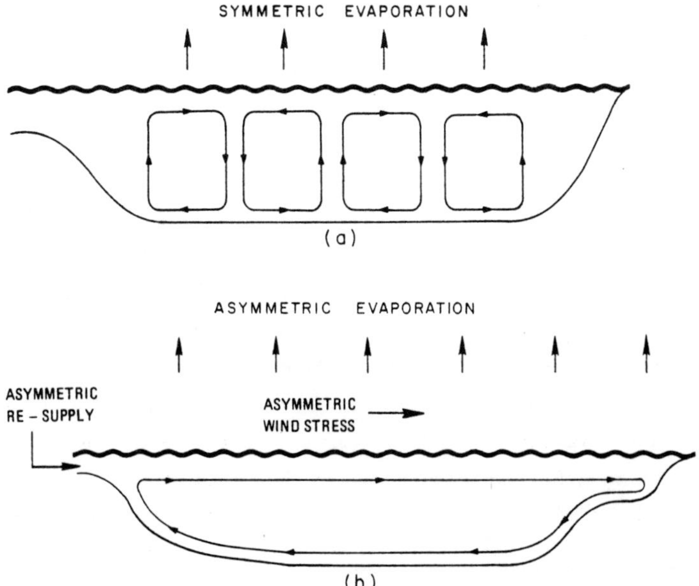

Figure 4. Types of evaporation-density convection in an evaporite basin.

of sufficient strength to move precipitated sediments across the bottom of deep basins? Actually, we know less about circulation patterns than about anything else in evaporite basins. Some aspects of circulation have been considered by Brongersma-Sanders and Groen (1970).

There are other aspects of circulation in which the density of brines can be overwhelmingly important. If the formations beneath the evaporite brine are sufficiently permeable, or if (as in a sabkha environment) brine depth is insignificant, then overturn and convective displacement may extend into these sediments as well, although with a much greater friction involved. A similar circulation of dense brines into sedimentary formations has been postulated by Bloch and Picard (1970) to be active in the erosion of salt domes. Whether such downward migration of brines will take place depends largely on the characteristics of the sediment. Practical experience with artificial brine ponds has demonstrated that once a few inches of *halite* has grown uniformly over the smooth bottom surface of one of these ponds, brine losses are negligible. On the other hand, subsurface brine storage has been calculated to be an important factor in the salt budget of some natural salt lakes where it has been studied (Langbein, 1961; Eardley, 1970).

Dense brines will also flow outward through permeable sides of the basin; this is probably an important method of reflux that also accounts for dolomitization of the barrier rock, as postulated by Adams and Rhoades (1960) and later calculated for a practical case by Deffeyes *et al.* (1965). In this connection the head driving such reflux flow depends not only on the densities of the columns in the basin and in the neighboring sea, but also on the relative level of their surfaces: If the basin is at a sufficiently lower level the flow may stop or even reverse and flow inward through the barrier. For example, even a brine precipitating halite with a density of 1.3 can maintain a reflux flow through the barrier only if its surface is not more than 30 percent (of its depth) below the sea surface. Whether such flow is *significant* either into or out of an evaporite basin is an interesting question. K. S. Hsü (personal communication, 1972) has calculated that if the Mediterranean Sea were cut off from the Atlantic Ocean at the Straits of Gibralter, as he asserts happened in Miocene times, then flow through the surrounding rocks even with a permeability as high as a Darcy, could not feed in Atlantic Ocean water fast enough to keep up with the evaporation in the Mediterranean basin. Maiklen (1971) has postulated that evaporation was sufficient on the Middle Devonian Elk Point Basin to draw the surface of the basin down far below sealevel; he did not consider the contribution of flow in either direction through the barrier reef.

Density layering of brines has overwhelming consequences for the organic cycle of the basin. With even moderate concentrations of brine that prevent convection for an appreciable time, the bottom brine will be highly deficient in oxygen, which will tend to both accumulate and preserve organic remains and sulfides. And as the density of the deep brines rises above 1.29, the organic material drifting down from organic

Table 5. Density Changes in Brines at 20°C

	Change in density with:		
	Concentration	*Temperature*	*Depth (Pressure)*
Concentration, C (times seawater)	$\left[\frac{\partial\rho}{\partial C}\right]_{t,d}$ *(dimensionless)*	$\left[\frac{\partial\rho}{\partial t}\right]_{c,d}$, °C^{-1} (a)	$\left[\frac{\partial\rho}{\partial d}\right]_{c,t}$, M^{-1}
[Pure water]		-22×10^{-5} (b)	49×10^{-7} (b)
1.0	26×10^{-3} (b)	-27 (b)	46 (b)
3.5	25 (c)	-40 (c)	43 (d)
11	1.1 (e)	no data	30 (d)
63	0.4 (e)	no data	no data

(a) With decreasing temperature, and with no precipitation during temperature change.

(b) International Critical Tables 3, 100 (1928).

(c) Hara, Ryosaburo, Nakamura, and Higashi (1931); Fabuss, Korosi and Huq (1966).

(d) Holser (unpublished).

(e) Usiglio, in Clarke (1924, p. 220). See Fig. 7-3 of this report.

production near the surface will not even sink to the bottom, but will accumulate as an interbrine layer.

In the ocean and in most lakes, the main driving force for convection is the change of density due to temperature changes, while concentration by evaporation is only a local modifying factor. By contrast, if a brine is concentrated sufficiently to deposit even gypsum, its increase in density is very much greater than would ever happen with even the greatest range of temperature. The two effects, as contrasted in Table 5, usually oppose one another because the brine will ordinarily be heating up at the same time it is evaporating. A change of temperature as great as 5°C would change the density only 10 percent as much as a change of concentration from say 3X to 4X seawater.

The effect of pressure on density will be less important for even the deepest evaporite basins with which we have to deal. For example, each 100 m of depth increases the density of X1 seawater as much as lowering the temperature by less than 2°C, and the effect of pressure is even less at high salinities, as shown in the last column of Table 5.

Vapor Pressure

The pressure of water vapor over solutions decreases with increasing concentration, approximately in proportion to the increase in density of the solution. Figure 5 shows the measured decrease at various temperatures; because of this relation, vapor pressure will mirror density in Figure 3. The vapor pressure rises sharply with temperature, as does also the rate of decrease ($\partial \rho / \partial C$), with increasing concentration.

The primary importance of vapor pressure is its control of the rate of evaporation (Myers and Bonython, 1958). Aerodynamically, evaporation rate is a product of wind velocity and the difference of water vapor pressure between that in the main mass of air and that near the surface of the brine. In looking at Figure 5, it is therefore evident that not only will evaporation decrease as the brine concentration increases, but at high concentrations it could easily reach the point where it equals the water content of the air itself, so that evaporation stops altogether. For example, Figure 5 indicates that if the water vapor pressure of the air was 13 mm Hg, evaporation of brine at 20°C would not continue beyond the point where halite began to precipitate. Compared to the water vapor pressure

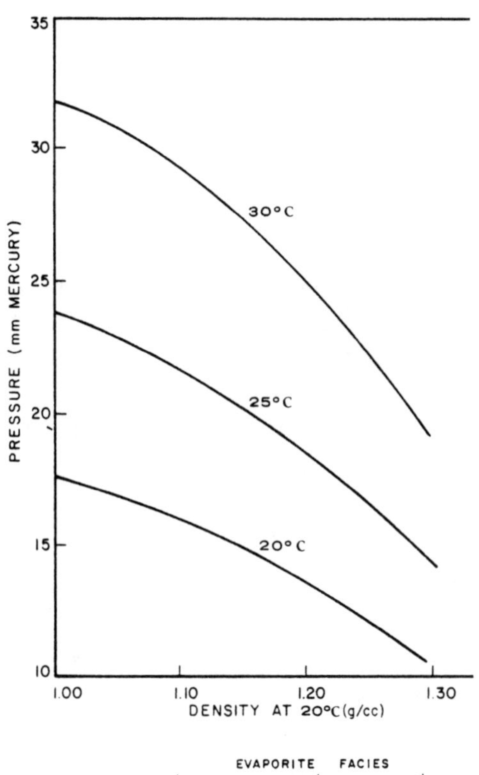

Figure 5. Vapor pressure of seawater concentrates and sodium chloride solutions. (Rothbaum, 1958).

over pure water at the same temperature, this level is usually expressed as a *relative humidity* of 75 percent. The above is only an immediate view of the mechanism of evaporation. Actually for the large bodies of brine with which we are mostly concerned, the air humidity is *determined* to a great extent by the evaporation rate in the surrounding region. The overall process depends on the energy balance between heat input from the sun, the heat used in evaporation itself, and the re-radiation (itself a strong function of water temperature) (Kohler and Parmele, 1967). Consequently, we might expect that in an evaporite basin whose climate was controlled by marine air would have less tendency toward higher evaporite facies, perhaps even being restricted to the calcium sulfate facies (Kinsman, 1976). Probably for this reason an evaporite formed in the middle of a continent along a rift zone is more likely to evaporate seawater to the highest level of the potash-magnesia facies, as is evident in the occurrence of tachyhydrite ($CaMg_2Cl_5 \cdot 12H_2O$) in the Cretaceous rift evaporites of Brazil/Congo (Wardlaw, 1972).

Other Properties of Brines

Concentrated brines have a substantially lower heat capacity and slightly lower thermal conductivity than seawater (Kellogg Co., 1965). They are also less transparent to light so that 10 cm of brine absorbs most of the solar energy (Garrett, 1966). For both of these reasons, incoming solar radiation raises the temperature of brines much faster than it does seawater, particularly if the brine is overlain by a layer of fresh water that is more transparent in the ultraviolet but still absorptive in the far infrared ("greenhouse effect") (Hude and Sonnenfeld, 1974). Temperature cycles of up to 10°C between day and night become possible in brines only a few centimeters deep, with consequent variations of vapor pressure (Fig. 5) and evaporation.

The viscosity of brines increases rather strongly with an increase in concentration (Meleshko, 1959). By the time a brine has been concentrated sufficiently to precipitate minerals in the potash-magnesia facies, it is not only dense but syrupy in consistency.

IMPORTANT EVAPORITE MINERALS

Evaporites have a distinct and characteristic mineralogy, and this mineralogy is probably the most important characteristic in the study of their geochemistry, origin, and geological setting. It is usually more important to know that a rock contains sylvite than merely that it contains traces of potassium, or to know whether the chlorine obtained by chemical analysis is present as original halite or as brine in microscopic fluid inclusions. This section lists most of the minerals that will be encountered in evaporite rocks. The characteristics of these minerals that are most pertinent to the geochemical discussion, such as chemical composition and density, are given. This is a minimum basis for the reviews of deposition, and trace element and isotope chemistry of these minerals. Here, an attempt is made to describe the minerals in a practical way that combines downhole logs and visual field observation with whatever laboratory methods may be necessary to make a final determination. The methods used for identification are given in a rough order of complexity and accessibility. In this respect, while x-ray diffraction is nearly always the really definitive characteristic, in most cases an x-ray laboratory analysis will only be necessary to confirm a mineral determination now and then during work on a given evaporite sequence.

More complete tables listing all the physical and chemical characteristics of evaporite minerals are available elsewhere, such as in Braitsch (1971, p. 8-15) or Borchert and Muir (1964, p. 4-7). Logging is described in detail by Nurmi (1978).

No attempt is made in this review to describe all the textural features of evaporite minerals. In anticipation of a fundamental fact to be discussed further on, the minerals are listed here in the order of marine evaporite chemical facies. As emphasized below, minerals of one facies may also be found in facies of succeedingly higher evaporation stage. In line with this viewpoint, the list begins with a group of minerals that are for the most part introduced from outside the evaporite basin as detrital constituents. They might therefore be considered accessory minerals in evaporites, but in practice will be found in all proportions with pure evaporite minerals.

The following abbreviations and conventions are adopted for this listing: ρ is density in $g \cdot cm^{-3}$, V is acoustic velocity in ft/sec, H is Mohs' hardness, n, ω, ε, α, β, γ are indices of refraction, and main x-ray diffraction lines are denoted by d-spacing in Angstroms followed by intensity (on a scale of 100) and Miller indices.

Silicates

Clay-size minerals, mainly silicates, are often an important component of evaporite sections, where they may occur as thick beds of "salt-clay" or shale, as thin layers separating laminations of evaporite minerals, or dispersed between the crystals in evaporite beds. Kühn (1968, p. 481-487) has summarized the moderate amount of literature on clay minerals in the Zechstein deposits; the only other extensive study is that of Jones (1965) on the Wellington Formation of Kansas.

Clay-size minerals in evaporites may be considered under three headings: (1) authigenic minerals crystallized from solution in the evaporite basin; (2) "semisalinar" minerals formed by reaction between incoming detrital minerals and the concentrated brines of the evaporite basin; and (3) purely detrital components. Considering the last one first, detrital contributions to an evaporite sequence are somewhat limited by two factors. Firstly, many evaporites, particularly salts, accumulate at such a high rate that normal inflows of clastic minerals are highly diluted by the salts themselves. Secondly, the very conditions of evaporite geography severely limit river inflow, which otherwise would dilute the basin below the concentrations necessary for evaporite precipitation. In some evaporites, there is evidence suggesting that practically the only clastic material is the wind-blown dust that is also characteristic of many deep-sea deposits. Within these limitations, practically any clastic minerals may be found in small amounts.

The usual small amount of clay minerals in evaporites requires their concentration for study. Depending upon the method of obtaining insoluble residues, fine-grained anhydrite and carbonates may constitute a large fraction of the residue. Some detailed studies in the Zechstein and a preliminary survey in the Prairie Evaporite of Alberta indicate that, at least in some cases, insoluble clay residues have significant variations through a salt section and are potentially useful for correlation. In addition, components of such residues that are uniquely or dominantly a product of the evaporite environment may be used to infer the presence of an evaporite in a sedimentary section where it has since been removed by solution. Such interpretations have not yet been practiced, but in view of the important structural consequences of evaporite solution in a sedimentary section, they might well be pursued with profit in the future.

Some "clay" beds associated with evaporites (German *Salzton*) actually include large proportions of fine-grained but soluble evaporite minerals such as halite.

Quartz. SiO_2. $\rho = 2.65$; V = 19,400. Hardness of seven distinguishes it from nearly all other elements encountered in evaporites. Insoluble. In thin section low birefringence (0.009) and low refractive index (ω = 1.544) are characteristic.

In evaporites rounded, wind-transported grains are common; sandy beds may be very widespread over an evaporite basin.

Authigenic quartz is often found (see also Chapter 3, this volume), and evaporite conditions are apparently very favorable for its formation.

The authigenic quartz is typically idiomorphic, in short equant crystals dominated by pyramids rather than prisms (Grimm, 1962); different form ratios may have stratigraphical significance (Schettler, 1972). The crystals can sometimes be found with inclusions of evaporite minerals such as anhydrite, which not only demonstrate the authigenic origin of the quartz, but prove the early crystallization of anhydrite. The finely fibrous variety of quartz, chalcedony, is said to be always length-slow in evaporites (elongated along *c*) (Pittman and Folk, 1970).

Feldspars. These, particularly albite, also appear as authigenic constituents of evaporites. Sedimentary authigenic feldspar is always very pure (>99.9%!) albite or microcline, and shows cathodoluminescence, both characteristics distinguishing it sharply from detrital feldspar of igneous or metamorphic origin (Kastner, 1971).

Clay Minerals. Generally ρ = 2.2 to 2.6, V = 12,000. Clay-like insoluable grains and softness (H = 1.5) are characteristic. High gamma activity on a log is characteristic of illite, but may be confused with low concentrations of the even more gamma-active potash minerals. Confirmation of, and distinguishing among, the clay minerals is difficult in thin section. X-ray diffraction is recommended, including a parallel determination in which the clay layers are expanded by an organic compound such as glycol. Characteristic basal spacings are given in Table 6. This procedure may also pick up some species that are rarer, but are probably authigenic and indicative of evaporite conditions (even in the

TABLE 6. CHARACTERISTICS OF CLAY MINERALS

Mineral	Typical Composition	Basal Spacing, Å	
		Non-expanded	Glycol
Kaolinite group	$Al_4Si_4O_{10}(OH)_8$	7.2	7.2
Talc	$Mg_3Si_4O_{10}(OH)_2$	9.3	9.3
Montmorillonite group	$(Al,Mg)_2Si_4O_{10}(OH)_2 \cdot (H_2O)_4$	9.6	17.0
Muscovite 1M(d) (illite)	$KAl_2(AlSi_3)O_{10}(OH)_2$	10.0	10.0
Chlorite group	$Mg_3Si_4O_{10}(OH)_2 \cdot Mg_3(OH)_6$	14.2	13.2
Corrensite	(1:1 chlorite-vermiculite)	29.	32.
Koenenite	$Mg_7Al_4(OH)_{22} \cdot Na_4Mg_2Cl_{12}$	11.?	11.?

absence of evaporite minerals that may have been dissolved); talc, and corrensite.

The most common clay minerals in evaporites are muscovite $1M_d$ (illite), mixed-layer clays (including the one-to-one regular alternation of chlorite and vermiculite layers known as corrensite), chlorite group minerals, talc, and the clay-like magnesium-aluminum hydroxychloride koenenite. Montmorillonite is less common, and where it does occur, volcanic input should be suspected (see also Chapter 5, this volume).

Origins of clay minerals range from purely clastic to purely authigenic but an appreciable part probably results from reactions between an input of clay material and the brine constituents, particularly magnesium. This implies that talc may be an indicator of the potash-magnesia facies (Kolosov *et al.*, 1969). A general review is given by Braitsch (1971, p. 230).

Carbonates

Both detrital and locally precipitated or replacement carbonates are common in evaporite regimes.

Calcite. $CaCO_3$. $\rho = 2.71$, $V = 21{,}000$, $H = 3$. Where visible in either hand specimen or thin section, rhombohedral cleavage distinguishes the calcite group from other minerals. Easy solubility in 1*N* HCl with effervescence, is characteristic, but depends strongly on the temperature, grain size and porosity of the sample, and effervescence of a small amount of calcite may obscure identification of major amounts of other minerals with which it is commonly associated, such as dolomite, clays, or anhydrite. The Mg/Ca ratio of acid-soluble carbonate can be determined by atomic absorpiton analysis or a small titration kit. Alternatively, carbonates can be identified as a group, in thin section, by their cleavage, refractive index (calcite $\omega = 1.658$, $\varepsilon' = 1.566$) and extreme birefringence, and calcite distinguished from the others by routine alizarin stain of a half of the thin section (Friedman, 1971).

Tertiary to Recent carbonates may hold an appreciable amount of Mg in non-equilibrium solid solution. Such magnesian calcite will have the optics and staining properties of calcite, but show an appropriate amount of Mg on chemical analysis; they may be distinguished from calcite by small increase of unit-cell dimensions, as determined by x-ray diffraction.

Calcite and dolomite of evaporite origin are characteristically colored gray with bitumen (Shearman, 1974).

Aragonite. $CaCO_3$. $\rho = 2.94$, $H = 4$. Much aragonite is formed in shell material, but in evaporite terranes pisolitic caliche is an important site of aragonite growth (Scholle and Kinsman, 1974). Index of refraction, birefringence, and chemistry (but no Mg in solid solution) are similar to calcite, but cleavage is rectangular instead of rhombohedral.

Although aragonite is unstable and either dissolved or transformed to calcite in nearly all rocks older than the Tertiary period, relics of pisolitic texture may be indicative of the sabkha type of environment where much aragonite is deposited.

Dolomite. $(CaMg)(CO_3)_2$. $\rho = 2.87$, $V = 23{,}000$, $H = 4$. White, gray from organic matter, often brown from surface oxidation of small amounts of Fe in solid solution. Euhedral rhombs seen easily in thin section and often by hand lens are characteristic of dolomite, although by no means all dolomite is of this form. Does not effervesce easily in HCl. In thin section optics similar to calcite, but not stained by alizarin. The carbonate associated with evaporites is usually (but not always) dolomite.

Magnesite. $MgCO_3$. $\rho = 2.96$, $H = 4$. Fine grained, dead white. Rare, but probably more common in evaporites, especially of the potash facies, than it is usually given credit for. No effervescence with HCl, or stain with alizarin. Optics similar to dolomite, but refringence ($\omega = 1.700$, $\varepsilon = 1.509$) may be distinguishably higher. Determination should be checked by x-ray diffraction [*d*: 2.14(100)(104), 2.102(43)(113), 1.700(34)(116)].

Sulfates

Anhydrite. $CaSO_4$. $\rho = 2.96$, $V = 20{,}000$, $H = 3$. White, gray, light brown. Alteration of the surface to a dead-white, fine-grained adherent coating of gypsum, by exposure of a core or drilling chips to water for several hours, is a clear determinant for anhydrite. This characteristic "stain" can be generated more quickly by immersion for 15 or 20 minutes in a 4M solution of K_2SO_4.

In thin section moderately high refractive indices ($\alpha = 1.570$; $\gamma = 1.614$), high birefringence, and parallel extinction, are characteristic. Brick-shaped sharp euhedra are common, sometimes as isolated crystals several millimeters long replacing a fine-grained matrix, more often as a silt-size aggregate seen under the microscope. Acicular crystals also common, and are ascribed to the replacement of original gypsum (Ogniben, 1957).

On a larger scale, a variety of nodular, flaser ("chicken-wire") and enterolitic structures in anhydrites are ascribed to a supratidal regime of crystallization, either as primary precipitation of anhydrite (Kinsman, 1974a) or as an early diagenetic replacement of primary gypsum (Butler, 1970); but other, subaqueous, regimes of formation are not excluded (Richter-Bernberg, 1955). Many other structures, lamellar and coarsely bladed, are certainly derived from original gypsum (see detailed discussion in a later section).

Gypsum. $CaSO_4 \cdot 2H_2O$. $\rho = 2.32$, $V = 19,000$. Softness ($H = 2$) is characteristic. In hand samples and under the microscope the cleavages that break into plates, and break further with other cleavages at an angle of 67°, are a unique feature of gypsum. Refractive index ($\omega = 1.521$, $\varepsilon = 1.530$) and birefringence (0.009) are both low.

Gypsum from modern sabkhas takes a variety of forms: (1) a disorganized mush of small discoidal crystals flattened perpendicular to the *c* axis; (2) very large "selenite" crystals, also flattened perpendicular to *c*; (3) both small and large crystals flattened in the *a*-*c* plane (Butler, 1970; Schreiber and Kinsman, 1975; Shearman, 1978). The morphology perpendicular to *c* may be caused by high salinity (Cody, 1976), or by the adsorption of organic molecules to the {111} form (Barcelona and Atwood, 1978). Replacements of the large gypsum crystals by anhydrite or by halite can be recognized in older rocks.

In rocks older than Tertiary nearly all gypsum is an alteration of anhydrite by near-surface ground water. While occurrences below a couple of thousand feet are not unknown, such determinations should be recorded as questionable unless confirmed by optics or x-ray. Bassanite, ($CaSO_4 \cdot \frac{1}{2}H_2O$), is very rare, always an alteration product and must be confirmed by detailed x-ray examination.

Celestite. $SrSO_4$. $\rho = 3.00$, $H = 3.5$. Ubiquitous but rarely in large amounts, its presence may be suggested by Sr analyses. Then it may be distinguished from Sr in solid solution in sulfates or carbonates by searching thin sections or insoluble residues for crystals of high refractiv index (ω = 1.622, ε = 1.631), low birefringence (0.009), and prismatic cleavage; or by optical or x-ray examination of a gravity separate. That extremely small inclusions of celestite account for most of the Sr detected in many gypsums is confirmed by electron microprobe studies (Link and Ottemann, 1968).

Chlorides

Halite. NaCl. ρ = 2.17, V = 15,000, H = 2.5. Usually colorless to gray, from reducing conditions in evaporite brines (Kinsman *et al.*, 1974), but rarely (Canning Basin, West Australia) oxidized and stained dark enough brown by finely disseminated goethite so that even in thin section it resembles sphalerite. Blue from color centers (reviewed in Kirchheimer, 1976). Soluble in water, so that in common drilling practice it may be dissolved in cuttings and from the surface of cores, which then look highly porous. Deliquesces at 75 percent humidity, so samples must be protected from humid climates. Thin sections must be prepared in kerosene to preserve halite or potash minerals. In thin section the zero birefringence and low refractive index (n = 1.544) allows halite to be sometimes overlooked or regarded as open porosity: Look closely for rectangular (100) cleavage and growth zones and for material slightly less refractive than the mounting medium. Commonly in gravel-size anhedral grains, whose boundaries under the microscope may only be evident by irregular surfaces outlined by microscopic fluid inclusions. "Chevron" type upward-pointing (100) growth zones have important genetic significance (see a following section). Under high deformation, as in salt domes, crystals may exhibit preferred shape orientation (and even structural orientation, which is difficult to detect), as well as parting (that resembles cleavage) along (110) glide planes at an angle of 45° to the real (100) cleavage. *Hydrohalite* ($NaCl \cdot 2H_2O$) is a possible precursor of halite in the crystallization of some winter layers (Fiveg, 1948), but this has not been confirmed.

Polyhalite. $K_2MgCa_2(SO_4)_4 \cdot 2H_2O$. ρ = 2.78, V = 17,400. In some potash rocks polyhalite is the counterpart, and often a replacement of, the pure calcium sulfate, anhydrite or gypsum. Gamma ray activity moderately high. Hardness (H = 3) and slight solubility are more like anhydrite than like other potash minerals. Color opaque white, orange, or pink. Fine-grained in acicular or anhedral crystals. In thin section the low refractive index (ω = 1.548, ε = 1.567), moderate birefringence (0.019), and polysynthetic twinning superficially resemble plagioclase, but the extinction is asymmetrical to the twin composition plane. Identification may be confirmed by x-ray diffraction, in which the lines at 6.00(8)($1\bar{1}1$) and 5.82(7)(200) stand out from nearly all other minerals. The occurrence of polyhalite may extend farther and be less subject to subsurface solution than other minerals of the potash-magnesia facies, and for that reason it is important to recognize.

Sylvite. KCl. $\rho = 1.98$, $V = 13{,}500$. Gamma ray activity very high. Slightly more soluble, with a bitter taste compared to the strictly salty taste of halite. Deliquesces at 80 percent humidity. Softer ($H = 2$) and sectile when cut with a knife, compared with the inevitable cleavage fracture of halite. White, transparent to translucent, often red with microscopic or nearly amorphous disseminations of hematite. Optics similar to halite.

Carnallite. $KMgCl_3 \cdot 6H_2O$. $\rho = 1.67$, $V = 12{,}800$, $H = 2.5$. Gamma ray activity high. Very soluble with a bitter taste. Waters other than residual brines will decompose it incongruently to sylvite; consequently, some cuttings and cores will show as evidence for carnallite only surface porosity lined with shiny pink plates of sylvite, resembling mica. Deliquesces at 53 percent humidity. Often idiomorphic in barrel-shaped orthorhombic (pseudo-hexagonal) crystals. Dull, with pink to red colors common as a consequence of precipitation of hematite from an original solid solution between Mg- and Fe-carnallites. Optics: $\alpha = 1.466$, $\gamma = 1.494$, no cleavage. Plastic deformation gives polysynthetic twinning on (110). Carnallite is soluble in alcohol with precipation of sylvite, whereas halite, sylvite and most other evaporite minerals are only slightly soluble if at all.

Kainite. $4[K_4Mg_4Cl_4(SO_4)_4] \cdot 11H_2O$. $\rho = 2.13$, $H = 3$. Vitreous, easily soluble, colorless to gray. Confirm by x-ray diffraction: 3.08(100)(402, $\bar{5}31$); 7.78(80)($\bar{1}11$); 7.39(40)(111).

Kieserite. $MgSO_4 \cdot H_2O$. $\rho = 2.57$, $H = 3.5$. Massive with $\alpha = 1.520$, $\gamma = 1.582$. This is the most common magnesium sulfate in evaporites, an important constituent of the rock type *Hartsalz*. Often a decomposition product of kainite. The dead white efflorescence to epsomite and other higher hydrates of $MgSO_4$ is indicative, although this also occurs on other magnesium minerals.

Other Minerals

Other minerals sometimes encountered in the potash-magnesia facies are langbeinite, $K_2Mg_2(SO_4)_2$; schoenite, $K_2Mg(SO_4)_2$; bloedite (astrakhanite), $Na_2Mg(SO_4)_2 \cdot 4H_2O$; bischoffite, $MgCl_2 \cdot 6H_2O$, and tachyhydrite, $CaMg_2Cl_6 \cdot 12H_2O$. A number of borates and phosphates are found as accessories. In non-marine evaporites the range of mineralogy is much greater, but trona, $Na_3CO_3(HCO_3) \cdot 2H_2O$; glauberite, $Na_2Ca(SO_4)_2$; and mirabilite, $Na_2SO_4 \cdot 10H_2O$ (or thenardite, Na_2SO_4) are often found, with silicates such as the zeolite group as important accessories.

EQUILIBRIA OF SOLUBLE MINERALS

Evaporite minerals are products of chemical crystallization; consequently, relations at equilibrium play an important part in their generation. Although the seawater system is a complex one, with at least six controlling components, the important principles of equilibrium crystallization can be understood by starting with some of the simpler systems of which it is composed.

Simple Binary System--$NaCl-H_2O$

The solubility of sodium chloride in water at 20°C is 36 g NaCl/100 g H_2O. To understand the process of evaporation and crystallization, suppose that we start with a solution that is something like seawater, with only 3 g NaCl/100 g H_2O. If we remove by evaporation 91 grams of water, the resulting brine is 100/(100-91) = 11X concentrated compared to the original solution. The solution has [3/(100-91)] x 100 = 33 g NaCl/100 g water, just about at saturation with respect to sodium chloride. Up to this point, the concentration has risen in direct relation to the amount of water remaining. Once saturation is reached, however, for each 100 grams of water that is removed, 36 grams of salt must precipitate so that the remaining brine will remain at saturation.

It is very important to clearly understand that the amount of salt deposited from the solution is 36 grams for *every* 100 grams of water removed, *regardless of the size of the body of brine*. That is, once saturation has been reached, the rate of precipitation in grams per square centimeter of the basin is simply related to the rate of evaporation of water in grams per square centimeter, regardless of the depth of the brine. This is not to overlook the fact that, at a given rate of evaporation, it would have taken proportionately longer to *reach* saturation with a deep body of brine than it would with a shallow body of brine.

Temperature and Solubility

In binary systems, solubility varies with temperature, but the amount of this variation is very different in different systems. On the one hand, in the NaCl-H_2O system just considered, the temperature coefficient of solubility is relatively small within the range of temperatures usually found in brine. On the other hand, sylvite, KCl, has a much greater change in solubility with temperature. This contrast is displayed for the simple two component systems in Figure 6 (this is only an approximation to the corresponding variations of halite and sylvite solubility in seawater brines).

The variation of solubility of salts with temperature is sometimes called upon to explain crystallization of salts at the bottom of a

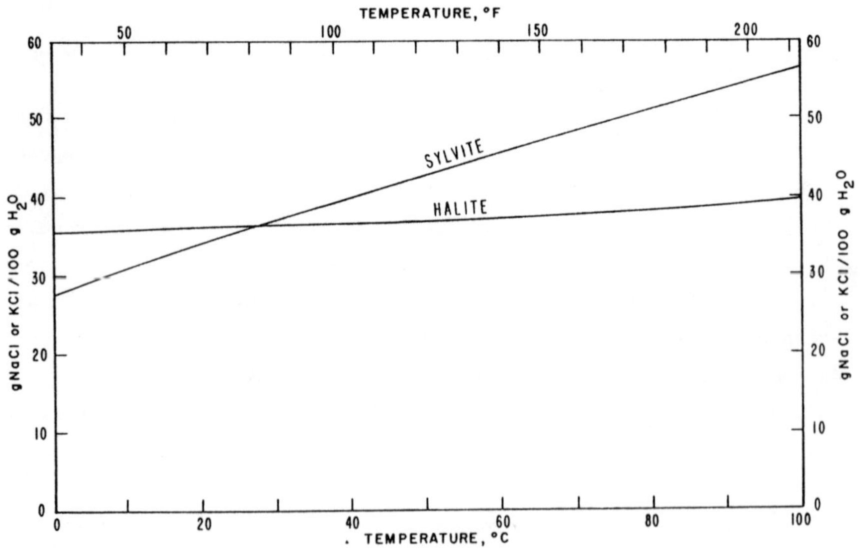

Figure 6. Solubilities of halite and sylvite in pure water as a function of temperature.

brine basin. Brine is heated while being evaporated by the sun on the surface; then the dense brine may circulate to the bottom of the basin and precipitate salt if it loses its heat there by conduction or radiation. Also, brines saturated in the summer may precipitate in the winter, as can be observed today in temperate interior basins where the range of summer-winter temperatures is large.

Two things are important in evaluating the effectiveness of this system of crystallization: (1) crystallization ultimately depends upon evaporation, and the change of temperature may only shift the time and place of crystallization; (2) the amount of salt precipitated depends on the *difference* of solubility at the two temperatures, not on the solubility itself. Thus in the simple system that was considered above, whereas evaporation of 100 grams of water necessarily led to the crystallization of 36 grams of salt at 20°C, a lowering of the temperature by as much as 10°C only precipitates 0.4 grams of halite. Of course, this could only happen if the brine had already been brought to saturation by extensive evaporation. Sylvite is more effectively precipitated by this mechanism, precipitating 6.7 grams during the same temperature drop. In the more complex system of seawater brines a similar variation of solubility may explain some halite-sylvite layering (Fig. 7) as an annual summer-winter cycle of temperature change in continually saturated brines.

That the change of solubility with temperature is a function of the mineral itself as well as of the solution is dramatically illustrated in Figure 8. In the system $CaSO_4$-H_2O, gypsum, the phase stable at lower temperatures (see below) has only a very little rise in solubility with increasing temperature. But anhydrite and the metastable bassanite decrease sharply in solubility as the temperature is increased.

Braitsch (1963) has published a very readable survey of the various ways that the temperature of evaporite deposition may be determined.

Pressure and Solubility

The effect of pressure on solubility varies widely depending on the system being considered. In many systems of concentrated brines, the effect of pressure on solubility is unimportant. In the system NaCl-H_2O considered above, an increase of pressure to 700 bars, corresponding to hydrostatic pressure at a depth of 5000 meters, the solubility of sodium

Figure 7. Seasonal layering of halite (white) and sylvite (gray: red in original); Unteren Kalilager, Oligocene of the Rhine Valley, Büggingen, Germany (BRD). Photograph by H. Roth.

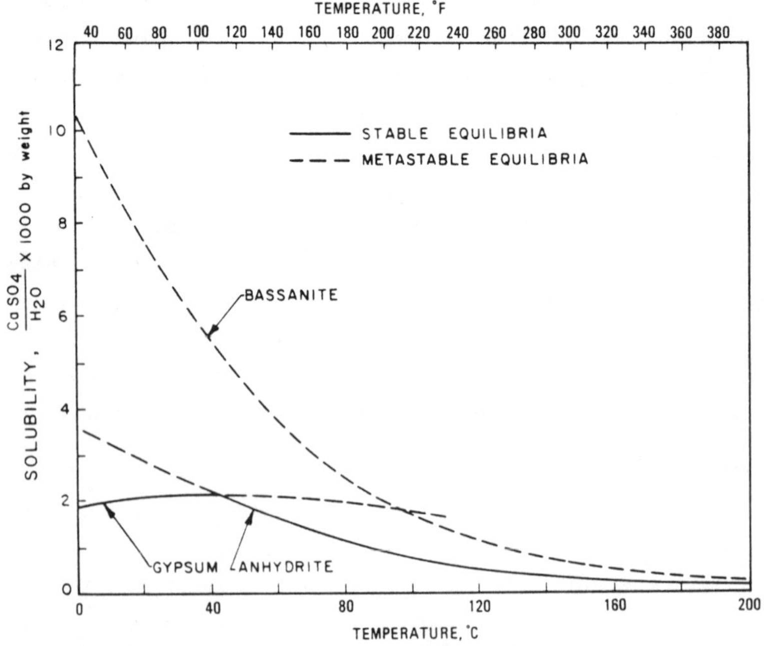

Figure 8. Temperature variation of solubility of calcium sulfate minerals in pure water (after D'Ans *et al.*, 1955).

chloride increases by only 10 percent. This factor is therefore of little consequence in such systems.

In contrast, as shown in Figure 9, the solubility of anhydrite at the boiling point of water more than doubles under this pressure increase. A diagram such as Figure 9 brings out a number of interesting and practical relations. For example, the decrease in solubility of anhydrite with temperature is only partially compensated by the usual increase in pressure down a well. Consequently, water on the surface that has already been saturated by contact with gypsum at about two parts per thousand (Fig. 8) might rise in temperature to perhaps 180°C in being pumped down a well to a depth of 3,000 meters where the hydrostatic pressure would be 300 bars.

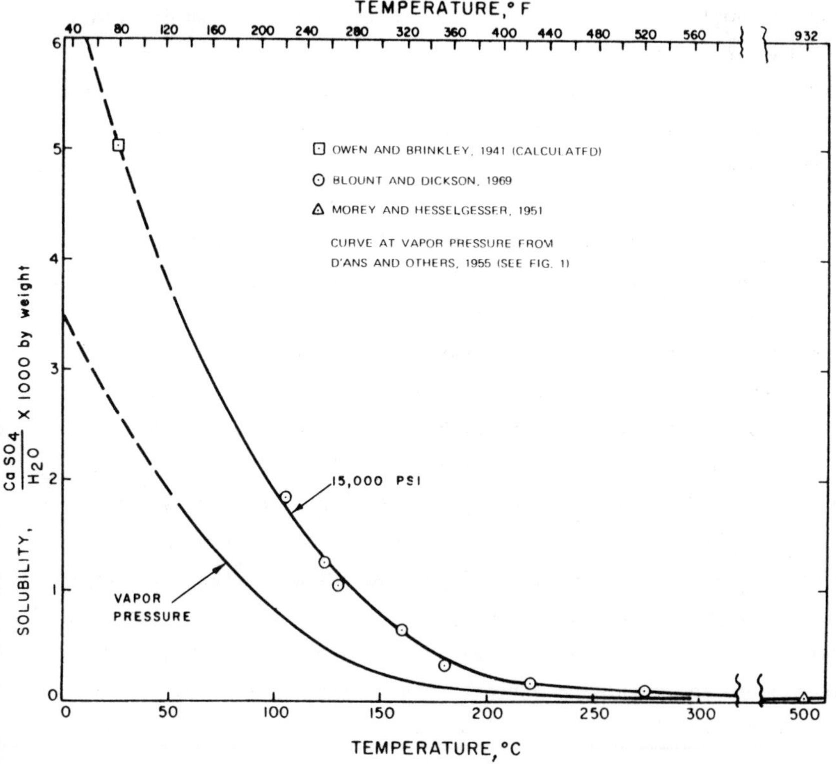

Figure 9. Effect of pressure on the solubility of anhydrite in pure water.

By interpolation between the two upper curves in Figure 9, its solubility there would be only 0.2°/oo so that the difference of 1.8 parts $CaSO_4$ per thousand parts of water would scale out on the way down. In another case, formation water at high pressure saturated with calcium sulfate might precipitate as much as half of its load of calcium sulfate, at a constant temperature, if the pressure were decreased radically on entering a well bore because of a shift from lithostatic to hydrostatic head, or simply because of a reduction of pressure during pumping.

Multicomponent Equilibria

Of the components of seawater listed in Table 1, the ones that have a major effect on the crystallization of salt minerals make up a six-component system K-Mg-Na-Cl-SO_4-H_2O. The details of such a system are necessarily complex and particularly difficult to display in diagrams, although the principal features of the system were worked out experimentally 70-80 years ago by Van't Hoff and his colleagues. We will consider here a few of the important results. For further details see Braitsch (1971) and Rowe *et al.* (1972).

As new components are added to a system, the important *common-ion effect* may operate. This is because the solubility, to the extent that it is a constant, is actually a solubility product; that is, equilibrium is established for a certain value of $[Na^+][Cl^-]$ where the values in brackets are the activities (to a first approximation the concentrations) of the designated ions. Consequently, all of the Cl in a solution, regardless of its origin and whether it may be formally counted as part of another component such as $MgCl_2$, is effective in precipitating halite. In the ternary system NaCl-$MgCl_2$-H_2O the solubility of halite decreases from 36 g/100 g H_2O in pure water to about 26 g/100 g H_2O in a brine with 1.3 g $MgCl_2$/100 g H_2O--roughly the amount in a seawater brine that is just beginning to crystallize. In the complete and more complex brine of real seawater, as shown in Figure 10, the solubility of halite shows a decrease of this kind as the magnesium and potassium chlorides accumulate during evaporation through the halite facies. Contact between brines of varying degrees of concentration could lead to halite precipitation, as discussed in a following section.

As emphasized in Table 2, ions are tied up in various complexes in seawater. This complexing may increase tremendously in the brine facies. Referring to the bottom curve on Figure 10 (see also Kinsman, 1974) the very sharp rise in solubility of calcium sulfate with seawater concentration increasing up to 2X, is probably the result of complexing of sulfate in $(MgSO_4)^o$ and $(Na_2SO_4)^o$ neutral pairs in the brine. Solubility continues to increase until the common-ion effect takes over and solubility drops to a very low but constant level at high brine concentrations. Because of this very low solubility in concentrated brines, nearly all of the calcium has been precipitated as gypsum before the halite facies, and any new seawater mixed with such brine will precipitate its own gypsum.

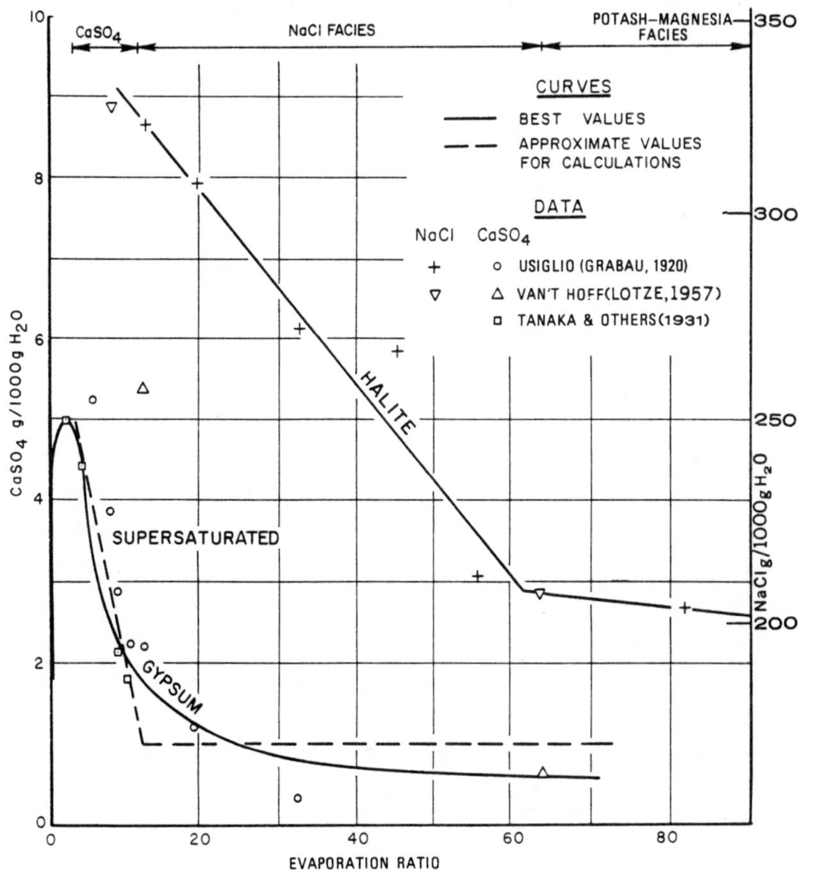

Figure 10. Solubilities of gypsum and halite in concentrated seawater.

For the same reason, seawater concentrated to gypsum saturation will precipitate *more* gypsum if it also dissolves halite.*

As more components are added to form this complete seawater system, it becomes more difficult to indicate the relations on simple diagrams. It becomes necessary to extract into such diagrams only the essential features that are new as higher concentrations are attained. We have just seen that nearly all of the calcium will be precipitated as sulfate at an early stage of the evaporation, just as carbonate was precipitated at an even earlier stage. Furthermore, halite precipitates through a long period before the complex potash-magnesia minerals begin to appear, while a little halite *continues* to crystallize throughout all of the variations of that complex system. These constant elements allow us to simplify display of the equilibrium relations in the seawater system, as shown in Figure 11, by neglecting calcium completely and assuming that halite is always present. In this triangular diagram, necessarily drawn for only a single temperature, the proportions of the remaining three components (in moles) are shown: Mg, K_2, and SO_4. This drawing shows the fields in which particular minerals of the potash-magnesia facies are in equilibrium with a brine of the composition indicated by position on the diagram. For example, seawater, which has $Mg:K_2:SO_4$ in the mole proportion 70:7:23, is barely inside the field of equilibrium with the mineral bloedite, $Na_2Mg(SO_4)_2 \cdot 4H_2O$. This means that after gypsum and then halite, the next mineral to precipitate at equilibrium "should be" bloedite. As bloedite and then the magnesium sulfate epsomite are precipitated, the composition of the remaining brine moves toward the upper right, away from the lower-left sulfate corner of the diagram. As the brine is further concentrated kainite is added to the minerals in equilibrium, and through successive reactions, as shown by the arrows on the diagram, kieserite and carnallite become equilibrium phases. Finally, the last amount of water is removed (at the end of the last arrow) from a mixture of kieserite, carnallite and bischoffite. We are not concerned here with the details of this sequence, which also varies widely with temperature, although such diagrams play a very important role in the geochemistry of crystallization of the potash minerals, in their subsequent alteration, and in the chemical processing of brines and minerals.

----

*Contrary to a statement used in a model for halite solution and gypsum growth by Schreiber and Schreiber (1977).

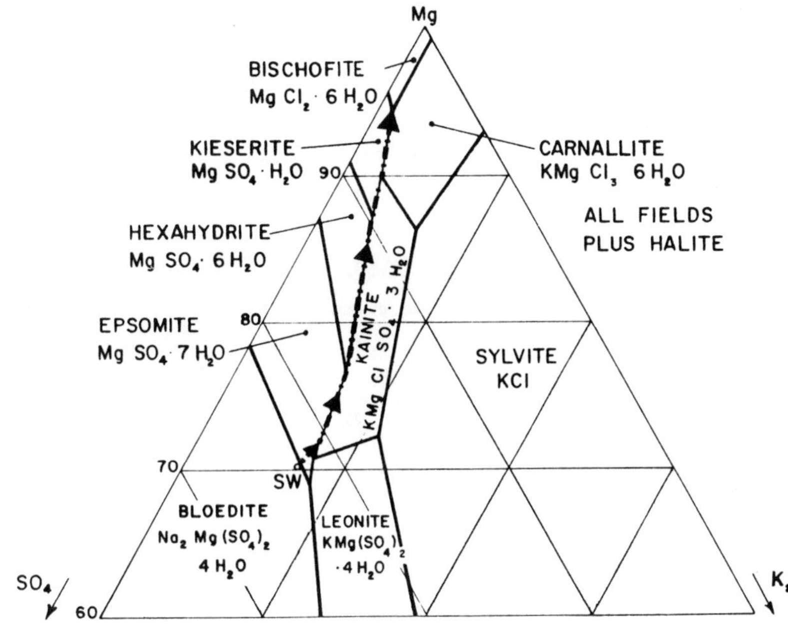

Figure 11. Stable equilibria in the system Na-K-Mg-SO_4-Cl-H_2O at 25°C. Scale in moles. (After Borchert and Muir, 1964, p. 99; for corrections see Braitsch, 1971, p. 67).

Here it is sufficient to make a couple of points: (1) potash-magnesia mineralogy is complicated and includes many minerals; and (2) the common mineral sylvite is not reached during this sequence of equilibrium crystallization as studied in the laboratory, unless sulfate is strongly depleted. Another possible reason for the latter discrepancy is considered in the following section.

METASTABLE SYSTEMS

Principles

A stable chemical system under particular conditions of composition, temperature, and pressure can best be understood as the *most* stable set of mineral (and brine) phases that can exist. The problem is that other minerals do in fact exist under identical conditions, and this very existence implies a degree of stability which leads to the label *metastable minerals*. Metastable minerals exist in spite of the fact that another mineral may be of lower energy and, hence, intrinsically more stable, because of a certain energy hump that must be got over in order to reach that most stable situation. The energy barrier can be overcome in some cases by a rise in temperature, or by the internal energy of plastic deformation. Time certainly works in the direction of obtaining the most stable mineral. Two things should be clearly understood, however: (1) transformations can go from a metastable phase to a stable phase, or even from a metastable phase to another metastable phase, but never from a stable phase to a metastable phase; (2) in some cases, particularly at low temperatures, metastable phases have persisted for geological time without transforming to the stable state.

Consequently, the stable equilibria that we have discussed above are only a starting point for the understanding of the realities of geochemical mineralogy. The mineral relations discussed there are the end products that should finally result if all the reactions have run that are possible. In this section we discuss some important examples of minerals which, in fact, do exist metastably in significant amounts.

Calcium Sulfate Mineral Relations

The calcium sulfate minerals are an important and often puzzling example of metastable relations. The following summary can be supplemented by details in Kinsman (1974). In Figure 8 it was already evident that at low temperatures gypsum really had the edge over anhydrite in having a low solubility, while at higher temperatures anhydrite was the stable phase. Bassanite, or "plaster-of-Paris," $CaSO_4 \cdot \frac{1}{2}H_2O$, is not stable at any temperature. However, if gypsum is heated in the laboratory in the presence of water, the gypsum will continue to exist metastably up to temperatures of nearly 93°C, and then transform to bassanite. Although continued dehydration at high temperatures can yield anydrite in the laboratory, it is nearly impossible to grow from solution. However, some modern sediments and older rocks composed of anhydrite have textural evidence indicating that they grew as anhydrite directly from the brine (Kinsman, 1974). In this respect the situation is similar to that of primary dolomite, which is easier to find than to make. Figure 12 suggests the stability relations among the calcium sulfate minerals, and diagrams in a qualitative way some of these stable and metastable relations. The thicker arrows represent easier, quicker transformations. For experimental variation, see Cruft and Chao (1970).

The equilibrium relations for calcium sulfate in seawater brine are shown in Figure 13 as a function of temperature and concentration. Increasing sea salt concentration decreases the temperature at which gypsum can transform to anydrite, until at the beginning of halite precipitation it is at normal surface temperature. In the high salinites and high temperatures

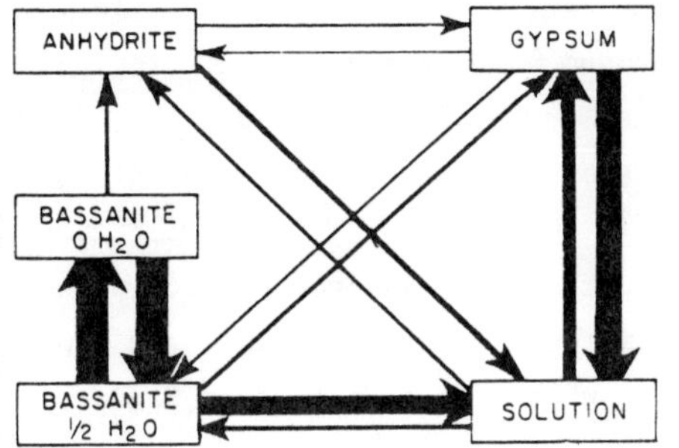

Figure 12. Relative ease of changes between various forms of calcium sulfate. Wider lines are qualitatively easier transformations.

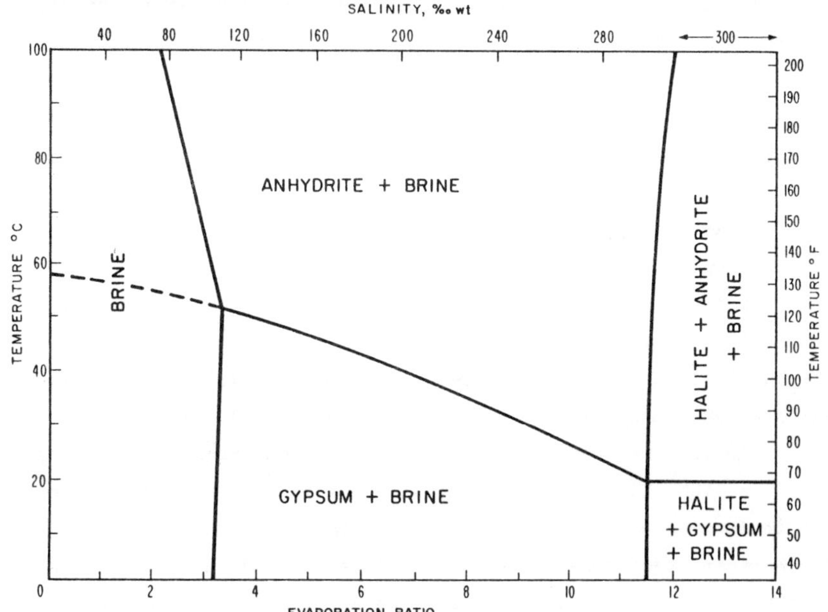

Figure 13. Equilibrium relations for gypsum and anhydrite with varying temperature and brine concentration (calculated from data of Hardie (1967) and many other sources).

of some sabkha muds, anhydrite does, in fact, grow from brines by capillary evaporation. But in many situations, gypsum not only grows, but persists, at temperatures and brine concentrations where anhydrite is the stable phase.

As sediments of this sort are buried and the temperature rises above about 100°C by virtue of the geothermal gradient, the transformation inevitably goes to anhydrite. These relations are indicated on Figure 14, in which the effect of pressure with depth is also taken into consideration. What this diagram says is that even if gypsum was deposited in its region of stability in relatively fresh waters, as in the lower left part of Figure 13, burial will soon bring it along one of the geothermal gradients into the transformation region, as shown in the upper right part of Figure 14. As a buried anhydrite rock, formed in this way, is brought back within some tens of meters of the surface by uplift and erosion, it enters a zone of lower temperature and fresher ground in which, as indicated in Figure 14, it tends to transform back to gypsum.

Several important consequences follow upon these gypsum → anhydrite → gypsum transformations. If the water released by gypsum as it dehydrates to anhydrite after burial is retained in the rock, it adds (to the porosity) at least 40 percent of the volume of the original gypsum, with a distribution likely to greatly decrease the rock's strength. Rocks of this sort, made into quicksand by the internal release of this water, may be responsible for (1) lubrication of faults, and (2) extremely deformed textures and structures of some anhydrites (Heard and Rubey, 1966). If, on the other hand, such dehydration water is expelled from the rock, it will become an important internal source of water. Such water, being undersaturated with respect to other minerals of the evaporite sequence, including carbonates as well as halite and potash minerals, may lead to important solution and/or diagenetic changes in those rocks (Borchert and Muir, 1964, p. 108, 158). Such waters should also be considered as possible media for the migration and transformation of hydrocarbons.

Another important consequence of the loss of such water is the decrease in volume of the sedimentary section after burial. Succeeding layers of sediments will show anomalies similar to those caused by compaction, but reference to Figure 14 suggests that the changes in sediment

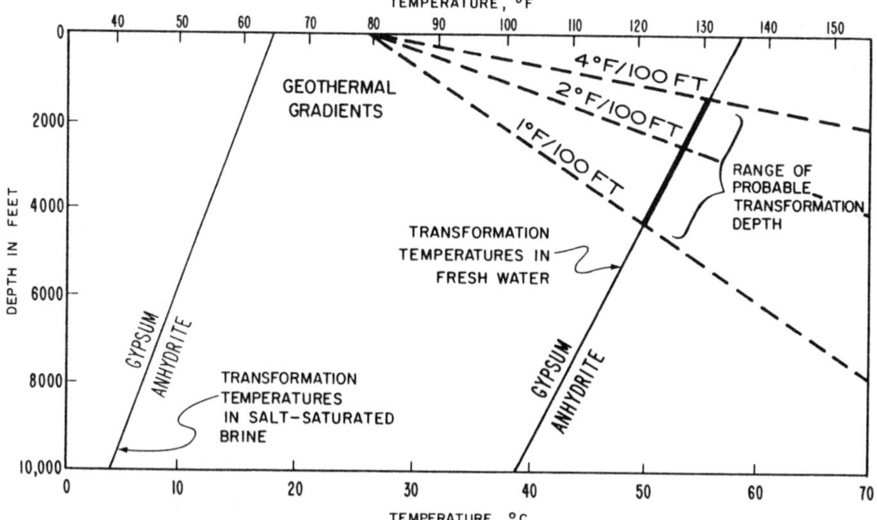

Figure 14. Gypsum-anhydrite transformation under hydrostatic pressure, showing normal range of geothermal graidents (calculated after McDonald, 1953).

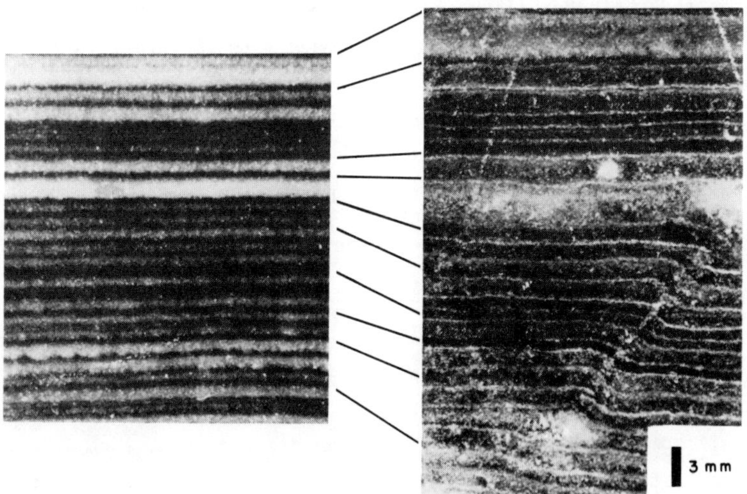

Figure 15. Expansion of buried anhydrite (A) to gypsum on the outcrop 5 km away (B), as shown by correlation of varves of the Permian Castile Formation, Texas. Reproduced by permission of R. Y. Anderson and D. W. Kirkland and the *Bulletin of the Geological Soceity of America* (1966).

thickness will be generated at some shallow (rather critical) depth, rather than gradually over any wide range of depth as in normal compaction of shales. For example, dehydration of gypsum may generate a characteristic drape of succeeding sediments in an evaporite basin surrounding a more solid reef.

The rehydration of anhydrite to gypsum, as uplift and erosion bring the rock back into the zone of cooler fresh ground waters, is accompanied by a corresponding increase of 64 percent in volume, assuming no initial porosity in the anhydrite rock. A volume increase has been definitely proved for the Castile Formation by the varve correlations of Anderson and Kirkland (1966), as shown in Figure 15. It is important to note that, at least in this particular case, the expansion during gypsification has been accommodated by a thickening of the rock without folding or distortion of the sedimentary layers. Inasmuch as gypsification ordinarily takes place under a load of only a few hundred feet of overlaying sediment, this vertical style of expansion may be expected. Although folding is often attributed to rock pressures generated during gypsification, it would be well to reserve the possibility that such rocks were deformed before gypsification. Such deformation might be much more important in interpretation of the geological history than is the relatively local process of gypsification itself. Close observation has established that generally, although not inevitably, gypsification preserves the general features of textures and structures of the previously buried anhydrite rock.

Potash-Magnesia Minerals

When the stable potash-magnesia minerals were discussed in the previous section, it was pointed out that sylvite, KCl, cannot be formed during the simple evaporation of ordinary seawater, although sylvite is actually the most common potash mineral in evaporite rocks. This discrepancy could be explained if some of the complex potash-magnesia minerals, such as kainite, did not precipitate very easily (reviewed in Valiashko, 1972). That would allow the compositions of the residual brine to lose further magnesium sulfate, and move across the triangle to the edge of the field where sylvite is stable. Then as shown in Figure 16, first sylvite and then carnallite could be precipitated as primary minerals. Other explanations for primary sylvite involve removal of magnesium or sulfate from the brine,

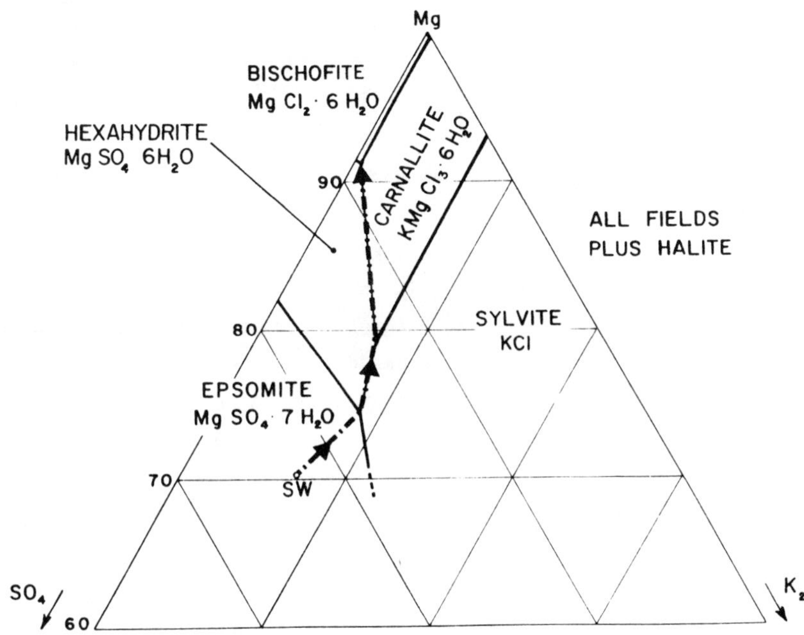

Figure 16. Metastable equilibria in the system Na-K-Mg-SO_4-Cl-H_2O at 25°C (after Valiashko, 1972).

such as the biological reduction of sulfate (Braitsch, 1971, p. 246-250). Further evidence for sulfate deficient brine is the occurrence of thick beds of tachyhydrite in the Cretaceous of Brazil (Wardlaw, 1972) and Thailand (Hite and Japakasetr, 1979).

The distinction between primarily-crystallized and secondarily-formed potash minerals is important to the geological interpretation of the evaporite sequence, with consequences that are crucial to the orderly development of a potash mine and which may even affect petroleum exploration. The most important alteration path leading to secondary formation of sylvite can be most easily understood by recalling (Fig. 11 or 16) that carnallite is originally precipitated in equilibrium with a brine that is extremely rich in $MgCl_2$. Now if the composition of the fluid in contact with carnallite is changed to one with less or even no $MgCl_2$ (such as the water derived from the dehydration of nearby rocks in the calcium sulfate facies), some of the carnallite will transform to sylvite according to the reaction:

$$\underset{\text{carnallite}}{KMgCl_3 \cdot 6H_2O} \rightarrow \underset{\text{sylvite}}{KCl} + \underset{\text{brine}}{MgCl_2 + 6H_2O} \qquad [2]$$

The reaction of carnallite to sylvite proceeds with a change in volume of +13 percent if the resulting new magnesium chloride brine is retained in the porosity, or −78 percent if it is expelled. Particularly in the latter case, the decrease in volume of the rock may lead to drape in the overlying formation superficially similar to that discussed above for the gypsum → anhydrite transformation. However, the carnallite → sylvite transformation depends not on a certain temperature of burial, but on a new supply of fresh water. Consequently, thickening in the overlying sedimentary section could result at any level as a consequence of contraction of the potash bed at a later time period. Wardlaw (1968) has demonstrated that drape due to carnallite → sylvite transformation can be recognized in seismic sections and isopach maps of the overlying formations. Thickenings may be important not only in the interpretation of the potash deposit itself, but should be recognized in the interpretation of geophysical exploration.

In addition to recognizing this transformation by direct or indirect observation of thickness changes, the distinction between primary and secondary carnallite or sylvite can also be based on petrography or on the distribution of trace elements such as bromide and rubidium (see below).

FACIES OF MARINE EVAPORITES

Principles

The word *facies* is used to describe critical variations in the mineralogy of evaporite rocks, and in this respect, its use is somewhat different from other uses of the word, such as the temperature-pressure facies of metamorphic rocks, or the textural-paleontological facies of carbonate rocks. In evaporite rocks facies characterize in a general way the stage of evaporation of the brine from which the minerals were deposited. They represent practical subdivisions of the evaporite process recognizable through the appearance of new minerals.

General Facies Sequence

The general facies sequence is outlined in Table 7, along with some of the important characteristics of the rocks in each facies. Each new facies is characterized by the appearance of a new mineral: a calcium carbonate (aragonite, calcite, or dolomite), a calcium sulfate (gypsum or

TABLE 7. SUMMARY OF EVAPORITE FACIES

Facies[a]		Mineralogy	Measures of Evaporation Stage				
			Salinity Wt %	Density g/cm^3	Fraction Evaporated Wt % H_2O	Concentration X Seawater (Wt H_2O)	Concentration X Seawater (Volume of Brine)
	Bittern subfacies	Bischoffite, tachyhydrite					
					99.2	120	
Potash-magnesia facies (supersaline)	Potash subfacies	Carnallite, sylvite, kainite, kieserite, halite; anhydrite or polyhalite					
			380	1.31	98.7	75	78.0
	$MgSO_4$[b] subfacies	Epsomite, bloedite, halite; polyhalite or anhydrite					
			375	1.29	98.4	65	68.0
Halite facies (saline)		Halite; anhydrite or gypsum					
			300	1.20	91.0	11.5	12.2
$CaSO_4$ facies (penesaline)		Gypsum or anhydrite; dolomite					
			150	1.10	72.0	3.5	3.6
Dolomite facies		Dolomite, calcite					
			35	1.02	0	1.0	1.0
$CaCO_3$ facies (normal marine)		Calcite, aragonite	10[c]	1.01	-		
(brackish)			1[c]	1.00	-		
(terrestrial)			0	1.00	-		

[a] Designations for environments according to Sloss (1953) are shown in parentheses.

[b] Usually altered to potash subfacies.

[c] Subdivisions of salinity according to Gorrell (1958).

anhydrite), halite, or a potash-magnesia mineral (usually sylvite or carnallite). These appearances are the natural consequence of the process of crystallization (either equilibrium or metastable) as water is removed by evaporation, and depend on a balance between the proportions of various components in the original seawater and the solubilities of the corresponding minerals in brine.

It seems necessary to emphasize that a facies in evaporites is distinguished not by the proportion of the minerals, but by the appearance, in *any* proportion, of a critical new mineral. To be in the halite facies, a rock is required to contain not 50 percent, but any amount of halite. The very fact that halite could precipitate at all is the critical factor in the appearance of the new facies.

For some purposes evaporite geologists have found it useful to subdivide the potash-magnesia facies into subfacies characterized by various minerals of the evaporation or even of the alteration sequence; these are reviewed by Braitsch (1971, p. 84-97).

Facies with Seawater Input

The evaporite facies are most directly characterized by the sequence of minerals crystallized during simple evaporation of a given volume of seawater. However, virtually the same sequence of minerals is found if seawater is added to the brine during crystallization. Sometimes the amount of water added is less than that lost by evaporation and outflow, so that the brine and its precipitated minerals will move forward through the normal sequence of facies. But in some situations the amount of seawater added could more than compensate for that lost by evaporation, in which case the brine and its precipitating minerals will regress towards a lower facies. In the important and practical case where the total volume of brine in the basin remains virtually constant, such a regressive phase would *only* be possible with excess reflux back to the sea, to provide room for new diluting seawater.

Inasmuch as evaporites would not ordinarily reach any appreciable thickness without the addition of new seawater, it is useful to consider in some detail the chemical relations that apply to evaporite facies when new seawater is added. The process is most easily appreciated by considering the effect of adding the new seawater in a batch that represents

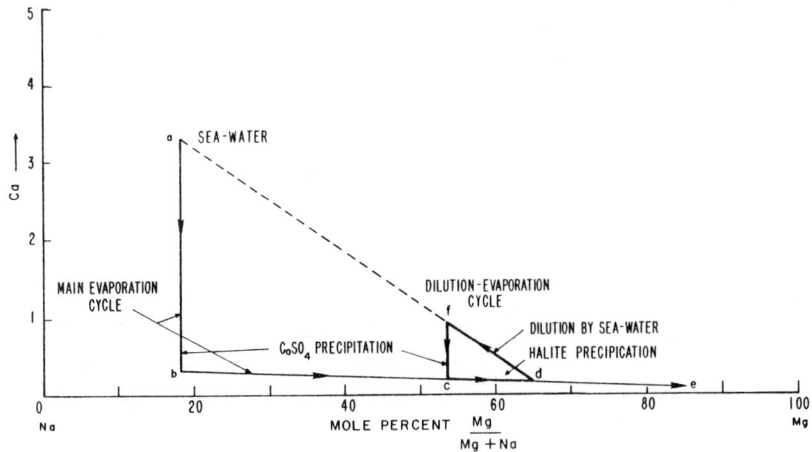

Figure 17. Dilution-evaporation cycle in seawater.

a small fraction of the total volume of brine. This batch might by a yearly volume of inflow, or inflow following isolation for a longer period of time. Continual inflow during evaporation is equivalent to such a model, if the added increments are made smaller and smaller at shorter and shorter intervals. The important thing is not the manner of seawater addition, but the overall, long time balance of inflow and reflux.

Figure 17 is a diagram designed to show what will happen when seawater is added to a brine that has already reached a high degree of concentration. The diagram is drawn in relation to three important components in the composition of the brine: The proportions of the main cations vary across the diagram from 100 percent Mg on the right to 100 percent Na on the left, while the small amount of Ca in the brine increases upward. This diagram considers only the cations in solution; anions follow along from the composition of the original seawater. It is often useful to think of this diagram with an added third dimension in which the proportion of water in the brine is plotted upward above the level of the page.

The composition of normal seawater with a mole ratio of about 4:1::Mg:Na and about two percent Ca lies at point *a*. Precipitation of a small amount of $CaCO_3$ is neglected. Consider first of all how the brine changes composition as seawater undergoes simple evaporation *without additions*. Until calcium sulfate begins to precipitate, the proportions Ca:Na:Mg remain those of *seawater* at point *a*. The composition starts far above the page along the imaginary H_2O axis and moves vertically towards *a* until gypsum is saturated and begins to precipitate. Then as further evaporation takes place, the removal of Ca as gypsum makes the brine composition move southward on the page, away from the northern pure Ca edge of the diagram. In terms of water content, the composition can be considered as sliding down along the saturation surface for gypsum. When point *b* is reached, halite is saturated (at about 11X evaporation ratio). Now most of the material precipitating will be halite, because as shown in Figure 10 the solubility of gypsum (also anhydrite) is very low and unchanging, so only a little gypsum drops out as water is removed. Consequently, the composition of the brine moves mostly to the right away from the pure Na corner, and only slightly downward away from the top Ca edge. If such evaporation continued indefinitely, point *e* would be reached at which the Mg concentration would be high enough to precipitate the first minerals of the potash-magnesia facies.

Along this line *b-e* in the halite facies it is important to see clearly the effect of adding new seawater that has not yet been evaporated through a previous calcium sulfate facies. Consider, for example, a brine that has crystallized a considerable volume of halite rock to reach point *d*. If now a small amount of new seawater is to be added to make up for evaporation, there are two ways that this might happen: (1) The new seawater could mix with the old brine to form a new brine of slightly shifted composition; or (2) the new seawater could float as a separate body of density 1.02 on top of the brine that is in the halite facies with a density perhaps 1.25. Figure 17 goes on to consider in detail what would happen in the first case. The diluted brine would have Ca:Na:Mg proportions along a line between *d* and *a*, say at *f*. In its water content it would also be diluted, perhaps lying a little above the saturation surface for calcium sulfate (but note discussion in following section). As new evaporation proceeds, the first thing that happens is removal of sufficient water to saturate the brine with calcium sulfate, this action being projected on the diagram at point *f*. Next an amount of calcium sulfate will precipitate along the line *f-c* of the diagram to bring the entire brine to saturation again with halite. Further evaporation will move along the line to the right of *c*, precipitating halite and a very small amount of calcium sulfate.

In many cases evaporation will go a little farther to the right of *d* to bring the final concentration for this dilution-evaporation cycle a little higher in the halite facies than when the cycle began. The net result of the dilution-evaporation cycle (which might, for example, represent a year's precipitation) is *mainly the fallout of calcium sulfate and halite in the amount they were brought in by the added seawater*. To this may be added a small increment of halite which serves to advance the general stage of brine concentration as stated above; or a small amount may be subtracted if in fact the general trend of brine concentration at this stage is regressive (the level *d* for the previous year is not reached).

Figure 18a indicates in a similar manner the relations for the second case listed above, where the inflowing seawater does not mix with the older concentrated brine. Now the upper thin layer of new seawater will of course go through the sequence of concentration: precipitation of calcium sulfate and precipitation of halite, that the original large mass of seawater went through. Eventually it may be expected to reach

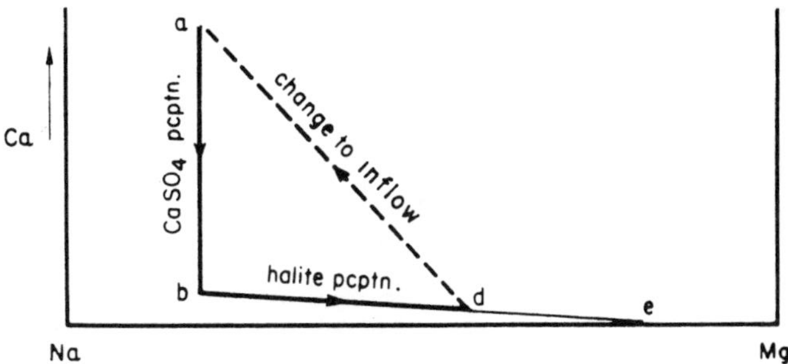

(a) Inflow - evaporation cycle

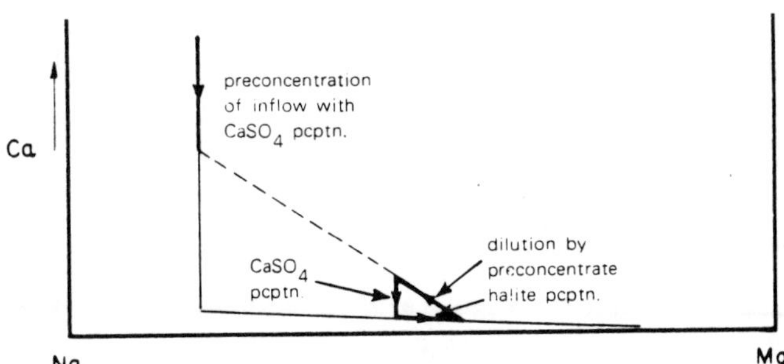

(b) Dilution - evaporation cycle — Inflow preconcentrated

Figure 18. Other evaporation cycles.

point *d*, at which time its density will be the same as the main mass of brine, and mixing will be inevitable. In this case the main deposition is again a layer of calcium sulfate and halite, nearly in the proportions of the inflowing seawater. It should be noted that although *f-c* is a shorter line on the diagram than *a-b* (and *c-d* is shorter than *b-d*), the amount of minerals is the same in the two cases, because the volumes of brine being evaporated are correspondingly different.

It is important to realize in either situation, mixing or top-flow, that a lot of calcium sulfate is precipitated, although the average composition of the brine in the entire basin has remained essentially in the middle of the halite facies. In case (1) the brine suffered only a slight change in composition by mixing with the inflowing seawater; in case (2) the surface water was actually of greatly different composition, but still the average for the entire basin was highly concentrated. Thus minerals of a lower evaporation facies will usually be produced in bulk in a basin filled with brine of a higher evaporation facies, as an addition to the main minerals of that latter facies.

In some cases the mineralogy may be slightly altered in this process. For example, although it is inevitable that calcium sulfate will precipitate in some form or other from fresh seawater flowing into a basin that is already filled with concentrated brine, it will tend to precipitate as anhydrite or even polyhalite, $K_2MgCa_2(SO_4)_4 \cdot 2H_2O$, in brines of the potash-magnesia facies. These are the minerals stable in that facies, so that even if gypsum formed at the locale of crystallization, it would likely transform either while falling through the concentrated brine, or soon after settling on the bottom. Similarly, dolomite or even magnesite (Kinsman, 1966) may take the place of calcite and aragonite in brines of higher magnesium concentration. For example, polyhalite and magnesite are the common sulfate and carbonate minerals in the Salado Formation of the Delaware Basin, New Mexico and Texas, not only in the rocks of the actual potash-magnesia facies in the Carlsbad district, but through much of the formations surrounding the district.

The geographic situation of evaporite formation may include a certain amount of isolation between a succession of evaporite basins, between the various parts of an evaporite basin, or even between the various parts of a relatively shallow shelf area of evaporation. In such circumstances, it may be useful to consider the main volume of more concentrated brine as a system in itself, whose inflow is not likely pure seawater, but is perhaps seawater that has been preconcentrated by a certain amount of evaporation and precipitation of minerals of lower evaporite facies. The degrees of isolation involved in real circumstances are, of course, highly variable. At present, it is sufficient to indicate that *any* model involving the inflow of preconcentrated brine into a basin can be evaluated by a slight variation of the same principles as described above. In Figure 18b the inflowing brine has been preconcentrated until it is halfway through the calcium sulfate facies, perhaps 7X seawater. Figure 18b indicates a dilution-evaporation cycle for such a model, involving for each additional increment of inflowing brine the precipitation of its *remaining* calcium sulfate, along with an appropriate fraction of its halite load. An analogous diagram could be drawn, similar to Figure 18a, for a model in which the preconcentrated inflow did not mix with the previous brine. In either case the principles are the same as those governing the inflow of ordinary seawater, the only difference being that part of the calcium sulfate was deposited elsewhere, reducing the amount of calcium sulfate deposited in the basin being considered.

Precipitation Due to Mixing

Raup (1970) has shown that some mixtures of brines can be, without further evaporation, supersaturated with respect to halite or sylvite. This is easy to show for a simple ternary system, such as $NaCl$-$MgCl_2$-H_2O, where the solubility of sodium chloride is markedly reduced by the common-ion effect when substantial amounts of $MgCl_2$ are added with another brine. Raup has made further experiments, the results of which are indicated in Figure 19. Mixing two brines, one highly concentrated in $MgCl_2$, results in immediate precipitation of halite. Raup has shown by model experiments that the halite crystals may grow not only during thorough mixing of two solutions, but also along a diffusion layer between a light overlying dilute brine and a heavy underlying concentrated brine analogous to the "case (2)" (Fig. 18a) of the previous section. We are thus provided with the mechanism for a third locale of mineral crystallization, at the boundary of two brines, in addition to the previously recognized sites of crystallization: (1) the surface of the brine where the evaporation is actually taking place; and (2) on the floor of the basin, where crystals may grow,

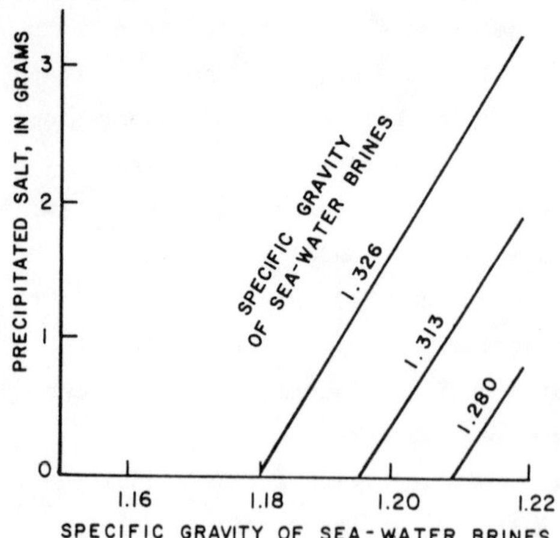

Figure 19. Precipitation of halite (in grams) from mixing 50 ml each of two seawater brines in various concentrations at 25°C. Reproduced with permission of O. M. Raup and the *American Association of Petroleum Geologists Bulletin* (1970).

perhaps on seeds already fallen there, either by a lowering of temperature or possibly by the overturn of previously supersaturated brines from the surface.

The chemistry of precipitation during mixing may be appreciated by referring back to Figure 10. Brines that lie along any straight solubility line on this diagram will still only be saturated when they are mixed. However, where the solubility line is concave upward, the mixing of two brines leads to supersaturation. Although the data available do not yet allow the exact configuration of the solubility limits in highly concentrated seawater, Figure 10 indicates that, in general, mixing-precipitation will take place only when brines of one facies are mixed with those of a previous facies. Thus, halite precipitates when brines of the potash-magnesia facies (that have already precipitated some potash minerals) are mixed with brines of the halite facies; analogously, mixing of brines far along in the halite facies with those of the calcium sulfate facies is likely to precipitate gypsum or anhydrite. The data in Figure 10, in fact, suggest that a small amount of halite might be precipitated by mixing brines of the potash-magnesia facies with brines as low as 6X in the calcium sulfate facies. More dilute brines probably could not reach supersaturation by mixing.

The relative importance of these three sites of evaporite mineral crystallization have yet to be evaluated, on the basis of petrographic and geochemical evidence, for actual evaporite sequences. Such investigations will undoubtedly make a great contribution to our understanding of the dynamic model of circulation and crystallization in real evaporite basins.

Mass and Volume Relations

The large volume of seawater required to make an evaporite deposit is dramatically illustrated on the left side of Figure 20. The first two columns dramatize the large volume of original seawater required to deposit a few feet of salts, even if *all* the salts in the seawater were extracted by evaporation. The third profile divides the constituents of seawater into components roughly related to the successive mineral facies although, as discussed in a previous section, there is not a strict correspondence, e.g., some of the NaCl precipitates as halite with the various minerals of the potash-magnesia facies, and also the components of the latter facies precipitate in various combinations as particular minerals of that facies. Anyway, it is evident that the complete theoretical evaporation of seawater would give a rock that is dominantly halite.

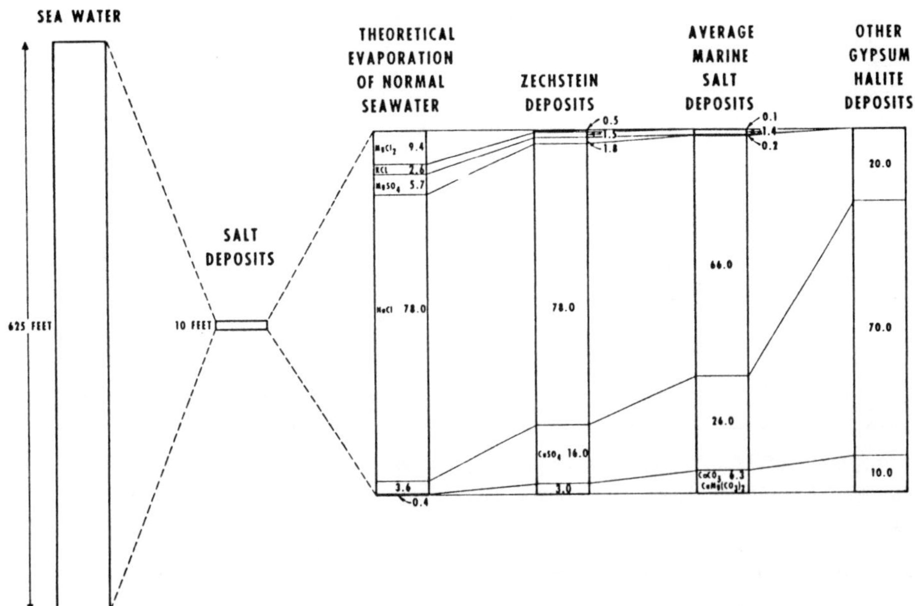

Figure 20. Comparative precipitation profiles of theoretical evaporation of seawater and precipitation profiles of evaporite rocks. These profiles sum components in the order of their first appearance, but are not actual compositional profiles. Data from Borchert and Muir (1964, p. 74), and other sources.

In real evaporite rocks, the relative volumes of the various minerals are distorted, largely as a natural result of the fact that evaporation only proceeds so far before it is limited by either atmospheric humidity (Kinsman, 1976), the extent of communication of the evaporite basin with the sea (Dean, 1979a), or a change of evaporite regime. Consequently, in average marine salt deposits, the ratios of calcium carbonate to calcium sulfate, of calcium sulfate to halite, and of halite to potash-magnesia minerals, are all higher than in normal seawater, as shown on the right side of Figure 20 (see also Richter-Bernburg, 1955). Another way of looking at this is that inevitably some basins at some times reach only a lower facies, or part way through one of the higher facies, so that much of the more concentrated brines escape back to the sea. The final product of evaporation of seawater, bischofite, $MgCl_2 \cdot 6H_2O$, is in fact deposited only very rarely (Hite and Jakapasetr, 1979).

In another way these volume relations severely limit some aspects of real evaporite precipitation. Suppose that an evaporite basin were filled with as much as 500 m of brine (which is not out of the question), and the whole column of brine was just beginning to precipitate calcium sulfate. As the brine continues to evaporate and be refilled from the sea, it will very soon reach the concentration of 11 times that allows the precipitation of halite--a new facies--*unless there is not only inflow but also outflow of residual brine.* In the example, at the beginning of halite precipitation the brine will still be nearly 500 m deep. At a seawater concentration factor of 11X that means that a total of about 5500 m of seawater has evaporated. But a back-of-the-envelope calculation from the numbers on Figure 20 will show that this amount of seawater can crystallize only 3.2 m of calcium sulfate! Similarly, by a slightly more complex calculation (Holser, unpublished), concentration of the same 500 m of brine, to the beginning of the potash-magnesia facies (Table 7) can only half fill the basin with halite. The important consequence is that evaporites of a single continuous facies thick compared to the depth of the brine from which they were deposited, require a large fraction of reflux. Otherwise they will quickly proceed to the next facies. Thus many--perhaps even most--of the evaporite series that are thick enough in a given facies to attract the attention of geologists, require some system to get rid of the residual brines. Of course a shallow basin (or a "zero-depth" sabkha) compounds this problem. These residual brines must either be returned to the sea (which is nearly impossible if the basin is sub-sea level) or they must be stored as connate water in the underlying sediments.

A corollary to this thesis is that the important thick sections of uni-facies evaporites that we witness in the sedimentary column must have generated tremendous volumes of brine. This brine was so dense (Fig. 3, Table 7) that it cannot have been mixed by a temperature overturn (Table 5) like that which presently accounts for vertical mixing of the oceans. Consequently, these large volumes of brines are likely to have accumulated in the deeps of oceans or mediterraneans, where they may have been responsible for the deposition of black shales containing extraordinary concentrations of sulfide and organic carbon (Holser, 1977).

The time rate at which a volume of evaporites accumulates also has important geological consequences. A nominal rate of evaporation of seawater in the present subtropics is about two m/yr, although, as discussed in a previous section, this may be substantially reduced as the vapor pressure of a concentrated seawater brine (Fig. 5) decreases toward the ambient humidity. This amount of evaporation sets a maximum on the fallout of new crystals each year, giving a consequent thickness for a yearly deposit. Table 8 shows a calculated maximum of the rate of evaporite deposition in each facies, for an evaporation rate of two m/yr, and compares these with rates of subsidence in sedimentary basins. This comparison has two important consequences: (1) To anticipate the discussion of a following section, the calculated thicknesses are compatible with those of the laminations in certain evaporite sediments of comparable facies (Table 8), adding to other evidence that the laminations are yearly varves; and (2) a previously "starved" sedimentary basin--even a fairly deep one--can be filled with such evaporites (especially salt) at a rate that is geologically instantaneous.

Another volume relation is important. Valiashko (1951a, 1972) illustrated the relative volumes of salts and residual brines, as shown in Figure 21. This diagram indicates the volume of brine remaining (assuming no reflux) and the volume of salt deposited (assuming no porosity) from 100 g of seawater. He emphasized that, by the time the potash-magnesia facies is reached at 60 or 70X seawater, the volume of brine has been so

Table 8. Evaporite Deposition Rates Compared to Subsidence mm/yr

A. EVAPORITE DEPOSITION RATES

Basin	*Carbonate*	*Anhydrite*	*Salt*	*Reference*
Calculated maximum for evaporation of 2 m/yr	0.13	1.2	25	Holser
Jurassic Todilto, NM	0.13	7		Anderson and Kirkland, 1966
Permian Zechstein 2, Germany	0.04-0.08	0.2-1.2	50-100	Richter-Bernburg, 1955
Permian Upper Kama, USSR		1-2	30-100	Fiveg, 1948
Permian Castile, TX	0.4	1.8	28	Anderson *et al.*, 1972
Pennsylvania Paradox, UT			25-150	Raup, 1966
Devonian Elk Point, Canada		1-10	10-250 20-100	Fuller and Porter, 1969a Wardlaw and Schwerdtner, 1966
Silurian Salina, MI			30-75	Dellwig, 1955

B. SUBSIDENCE OF SEDIMENTARY BASINS

Method	*Rate*	*Reference*
Geodetic measurements	0.1-1.5	International Union Geodesy and Geophysics, 1966
Recent geological features	0.2-1.5	Coleman, 1966
Thickness of ancient sediments	0.03- 0.6	Kay, 1955; Hudson, 1964

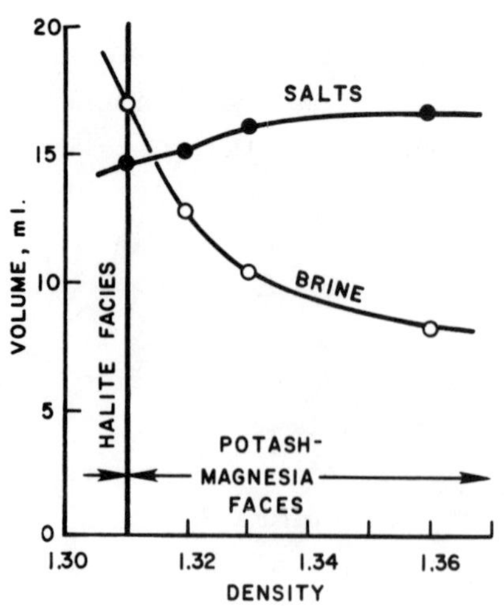

Figure 21. Volume relations in brines concentrated from 1000 g of seawater. After Valiashko (1951).

much reduced, and the volume of evaporites (mainly halite) has increased so greatly, that the two volumes are nearly equal.*

Valiashko goes on (1956) to point out that if the salts have crystallized with a porosity of 40 to 50 percent, most of the remaining brine of the potash-magnesia facies would then be found filling such porosity. This is quite a different physical model than we have usually been used to thinking about in evaporite deposition and has a number of important consequences. In the first place, as soon as the brine level goes below the surface of the salt bed, evaporation is hindered and considerably slowed, and capillary action is dominant. This may be another reason, in addition to the requirements of humidity discussed above, that accounts for the relative scarcity of deposits high in the potash-magnesia facies. In modern sabkhas and in many modern playa lakes, most of the concentrated brine surfaces are interstitial, and capillary evaporation is the rule. Another consequence is that subsequent crystallization may take place by alteration and replacement of earlier-deposited evaporite rocks and by porosity filling, rather than by a simple sequence of sediments deposited on the bottom of a brine-filled basin. Furthermore, such brines might migrate through the salt porosity for considerable distances under changing hydrodynamic and tectonic conditions.

Irregular deposits of potash rocks, with a dominant factor of replacement, are common in some districts such as the Permian of Carlsbad, New Mexico; but in other basins such as the Permain Zechstein of Germany, the Permian Kungur of the pre-Urals, and the Oligocene of the Rhine Valley (Fig. 7), beautifully banded and possibly varved potash rocks must have been formed by sedimentation. Perhaps in these areas the underlying salt was originally deposited with rather low porosity. Garrett (1970) states on the basis of his experience with modern artificial salt ponds that we may expect an initial porosity in halite of 40 percent, decreasing to 15 to 20 percent at a depth of 60 cm of salt, and 5 to 10 percent under 6 to 12 m of salt. The relative importance of Valiashko's model of interstitial brines must be evaluated for each geological case.

----

*In view of our discussion above, which indicates that the required inflow of seawater would add lower facies salts to the precipitate, the contention of Kopnin (1964), that only late-stage precipitates should be counted, cannot be accepted.

Even in the calcium sulfate facies, a sabkha environment exhibits the pervasive influence of diagenetic reactions taking place in the porosity of the primary sediment (Kinsman, 1974).

Relation to Nonmarine Evaporite Facies

River waters and their related ground waters can also be a source of evaporite deposition, as a look at the present Great Basin of the western United States vividly demonstrates. Table 3 lists the mean composition of world rivers, with the ions in the order of their importance in seawater, for easy comparison with Table 1. It is immediately evident that while rivers are probably the long-term source of the elements in the sea, the vastly differing residence times (Table 1) have radically distorted the composition of the sea relative to this supply composition. In particular, river waters are generally rich in calcium bicarbonate and calcium sulfate, instead of sodium chloride. Furthermore, we see in Table 3 only the mean composition, depending on the source of elements in solution. The sources of dissolved ions in such waters include: atmospheric sea salt and other constituents of rain and snowfall, weathering and dissolution of earlier rocks (including evaporites), and seepage of ground waters (including connate waters). It is therefore not surprising that a very wide range of compositions is found.

As discussed in a previous section, a common result of the differing composition of river waters and seawater is that the resulting nonmarine evaporite facies contain considerable calcium carbonate, sodium carbonates (such as trona, $Na_3CO_3(HCO_3)\cdot 2H_2O$), or sodium sulfates (such as mirabilite, $Na_2SO_4\cdot 10H_2O$). The occurrence of these minerals, which are outside any variations of seawater crystallization, is certain evidence that nonmarine waters were important in an evaporite basin.

Nonmarine waters can be precipitated in nearly as wide a range of situations as can marine evaporites. River waters may feed lakes of interior basins, some of which, as in the Green River of the Eocene in Wyoming and the Owens River system of the Pleistocene-Recent of eastern California, may form extensive systems with thick deposits of lake-sedimented evaporites. In other situations, evaporation of ground water may deposit large amounts of material, ranging from simple caliche cement to rather extensive evaporite accumulations. Patterson and Kinsman (1975;

see also Kinsman, 1969a) have demonstrated the importance of continental sources for the salts along the continental edge of the Persian Gulf sabkhas. There calcium sulfate and carbonate minerals are equally important in both the marine and nonmarine sections of the sabkha, although the residual brines have differing compositions in the two parts of the area.

Evaporites with mineralogy of the nonmarine facies sequence are surprisingly rare in rocks of pre-Tertiary age, although other evidence can, in some cases, be used to attribute earlier evaporites to a nonmarine origin. The distinction may be of considerable importance in developing the geological history of an area, and of course in guiding the search for reefs and other exploration objectives.

STRUCTURES AND TEXTURES

Calcium Sulfate Facies

Two descriptive varieties of structure are quite distinctive in the calcium sulfate rocks: lamellar and nodular. These have had a central importance in interpretations of the history of these rocks, and many aspects of these interpretations continue to be controversial. Table 9 takes these two descriptive varieties, subdivides them into some genetic subtypes, and adds some other and less controversial structures and textures to round out the picture. The discussion will by only cursory--more details and illustrations should be studied for example in Jung (1958), Murray (1964), Maiklem *et al.* (1969), Shearman and Fuller (1969), Hardie and

TABLE 9

SOME IMPORTANT STRUCTURES IN ROCKS OF THE CALCIUM SULFATE FACIES

Descriptive Types	*Genetic Subtypes*
Lamellar	Varve-lamellar
	Clastic-lamellar
	Algal-lamellar
Nodular	
Pegmatitic	
Massive	

Eugster (1971), Anderson *et al.* (1972), Kinsman (1974a), and especially Schreiber (1978).

A previous section has emphasized the convertibility of gypsum to anhydrite, and vice-versa. Apparently such conversions take place with little change of structure or even of finer details of texture, as discussed in a previous section. Hence in the following discussion one can generally also understand gypsum when anhydrite is cited.

A typical lamellar anhydrite rock (Syn.: laminated, rhythmic, Liniena., Gebanderta., balatino) consists of alternate lamellae of white,

Figure 22. Varve lamellar gypsum rock, Jurassic Todilto Formation, New Mexico. See Anderson and Kirkland (1960).

nearly pure anhydrite, and gray to black lamellae rich in dolomite and organic carbon. The lamellar pair is commonly 0.2 to 2 mm thick, but may be as much as 10 mm, and the dark portion may be a fifth to a tenth of the couplet thickness. The thickness of lamellae is usually within a range feasible for precipitation in one year (Table 8). Thinner sub-laminations also occur. Some individual lamallae show variations in structure: wavy, enterolithic, or micro-nodular. Figures 15, 22, and 23 illustrate some varieties of lamellar gypsum/anhydrite. Detailed petrography is illustrated by Anderson *et al.* (1972).

Many lamellar anhydrites are remarkably uniform. Individual lamellae have sharp contacts, are perfectly flat, and are continuous across a core, a mine face, or an outcrop. The vertical sequence of thicknesses seems to be cyclic, and detailed time-series analysis of long sequences in the Todilto Formation of New Mexico and the Zechstein Series of Germany have indicated the dominance of 6, 11, 35, 60, 170, and 400 unit cycles (Anderson and Kirkland, 1960; Richter-Bernburg, 1964). Most remarkable of all, sequences can be accurately correlated for tens and even hundreds of kilometers across these sedimentary basins. In these and other basins such lamellar gypsum sequences are underlain by more finely lamellar clastic and carbonate rocks, and overlain by nodular anhydrite and banded (coarsely lamellar) halite rock (Anderson and Kirkland, 1966; Anderson *et al.*, 1972).

These characteristics have led to two conclusions: (1) that the lamellar sulfate-carbonate was sedimented in an extremely protected "zero-energy" environment usually characterized as a "deep-water" basin; and (2) that the lamellae are yearly varves. The remarkable uniformity of the lamellae certainly attests to deposition below wave base, perhaps some hundreds of meters, but it should be mentioned that sharp brine stratification in an evaporite basin might damp wave motion at depths less than would be normal for a more uniform water column. Paleotopography of bounding older reefs and other structures corroborates depths of over 500 m for several basins that develop these varved sequences.

The lamellae of some anhydrite rocks may be much more irregular. Their wrinkled form (Fig. 24) is indicative of an origin in algal stromatolites of the intertidal sabkha zone (Nurmi and Friedman, 1974; Gill, 1977), a situation in which they are also found in modern sediments.

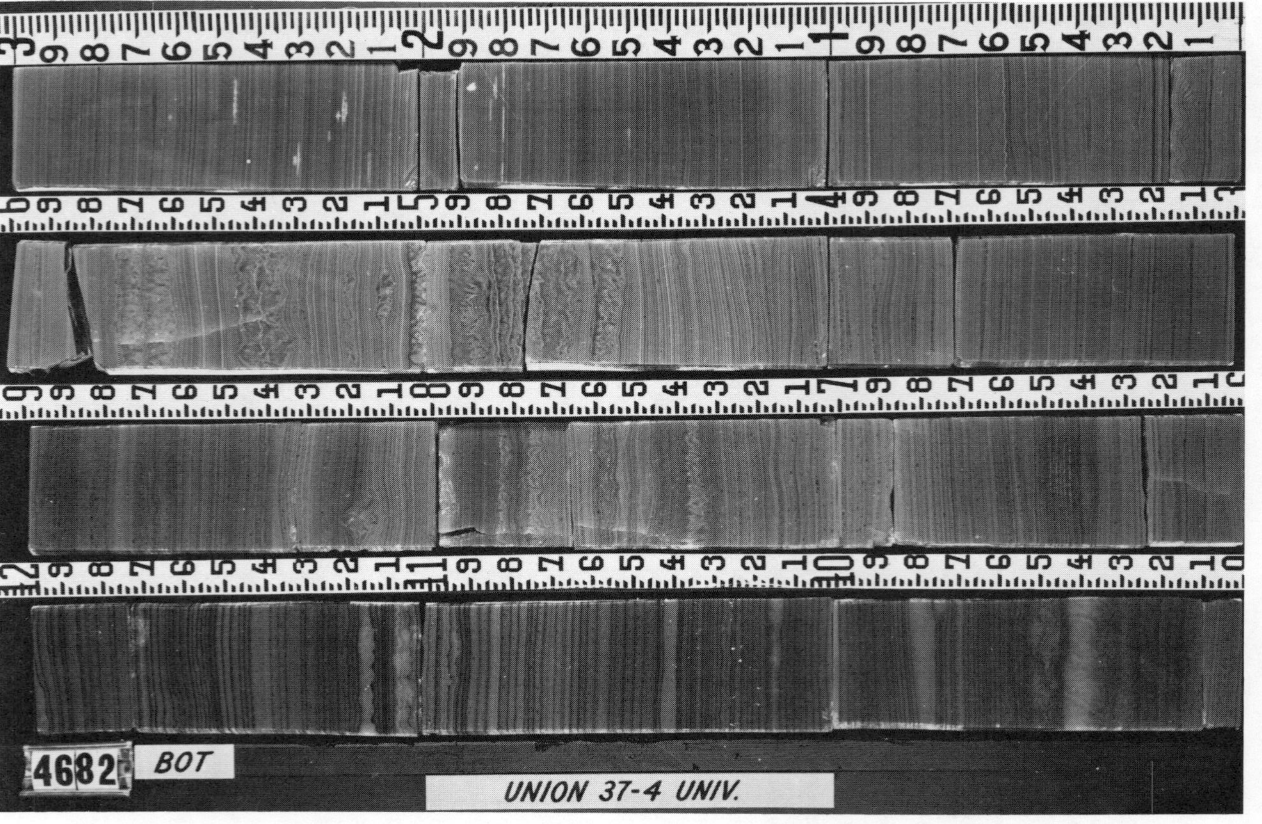

Figure 23. Varve lamellar anhydrite, rock, and some variations. Anhydrite I Unit of Permian Castile Formation, 4670+ feet (top to 4682- feet (bottom) in Union 37-4 University, Winkler County, Texas. Scale is in feet, tenths, and fiftieths. See Anderson *et al.* (1972).

Figure 24. Algal lamellar gypsum rock, unnamed Pleistocene formation, Puertocitos, B.C., Mexico.

Corroborative evidence for such an origin of ancient anhydrite rock may be provided by discontinuous lamellae, mound structure, polygonal dessication cracks, ripple marks and cross bedding; the lamellar thicknesses will not be correlatable over any appreciable distance.

An intermediate style of lamellar anhydrite displays the irregularities characteristic of clastic sedimentation, such as ripples, cross bedding, and graded bedding, but with little or no evidence of algal structures. Clastic sedimentation has apparently been a dominant feature in the deposition of such materials, in the shallower, higher energy environment of intertidal or subtidal-lagoonal regimes, with the layers due to tide or wind influx of seawater (Schreiber and Kinsman, 1975; Von der Haar and Gorsline, 1975). Figures 25 and 26 illustrate a modern example from Laguna Ojo de Liebre, B.C., Mexico, and a wide variety of such structures are illustrated by Hardie and Eugster (1971) from the Miocene Solfifera Series of Sicily. There is also evidence for turbidite deposition of thick banks of anhydrite (Schlager and Bolz, 1977).

Apparently lamellar anhydrites may be formed in the same wide variety of environments as carbonate rocks. In many cases the structural evidence is ambiguous or conflicting, in which case the options of varve, clastic, algal, or other origins should be kept open.

Judging from modern examples, the carbonate portion of the algal lamellae and some inorganic lamellae, was probably initially aragonite. Petrography of ancient varved lamellae suggest that their carbonate was either original or early diagenetic dolomite. But typically both kinds of lamellae in ancient anhydrite rocks are presently dolomite, in 0.02 to 1-mm rhombs. In some cases (Permian Castile Formation of the Delaware Basin of Texas; Devonian Upper Elk Point Formation of western Canada) the dark lamellae of the varve couplets are (unaccountably) calcite, while in varve anhydrites closely associated with rocks of the potash-magnesia facies (potash zone of the Permian Salado Formation of the Delaware Basin,

Figure 25. Clastic lamellar gypsum sediment, Laguna Ojo de Liebre, B.C., Mexico. In the section exposed below the present evaporite flat, by this hole, the white layers are gypsum and the black layers are accumulations of organic material, including both salt grass and algal debris (but without algal constructs). Scale is 15 cm long. See Holser (1966a).

Figure 26. Banded salt and clastic lamellar gypsum sediment, Laguno Ojo de Liebre, B.C., Mexico. Scale gives depth of core in inches below present evaporite flat. Mixed salt and gypsum at top, with decreasing salt and increasing organic matter downward.

Figure 27. Isolated nodules of gypsum in red-bed siltstone, Sewemup Member of the Triassic Moenkopi Formation, Pariott Mesa, Grand Co., Utah. See Shoemaker and Newman (1959).

New Mexico; Permian Wellington Formation of Kansas: Jones, 1965) the carbonate may be altered to magnesite, in that case often dead white.

The other dominant and distinctive structure is nodular anhydrite (Syn.: pearly, flaser, chicken-wire, mosaic, mottled, authiclastic). Typically white nodular aggregates of finely crystalline (0.02 mm) anhydrite may range from 20 to 50 mm in size, set off by dark streaks of dolomitic material. The shape of the anhydrite nodules is typically lens-like (flaser), probably derived by diagenetic compression from originally more spherical aggregates. The dark outlining material often is anastamosing, leading to the name "chicken-wire." Such textures are commonly observed as displacive growths in the carbonate muds of modern supratidal sabkhas (review in Shearman, 1979). For these reasons such structures are widely acclaimed as evidence for sabkha deposition. Figure 27 shows isolated gypsum nodules in a red-bed sabkha sequence. However, they have also been found in deep-water evaporites (Permian Zechstein Series of Germany: Richter-Bernberg, 1959; Permian Castile Formation of Texas: Anderson *et al.*, 1972) overlain by varve-banded halite, as illustrated in Figure 28. As Dean (Dean *et al.*, 1975; Dean and Anderson, 1978) has pointed out, the main requirement for the formation of gypsum or anhydrite nodules is growth under a mud in contact with high-salinity brines. Consequently, nodular structure must be used with extreme caution in drawing environmental conclusions.

The mineral of the modern nodules may be either original anhydrite (Shearman, 1966), original gypsum (West *et al.*, 1979) or possibly anhydrite after gypsum (Butler, 1969). Of course, subsequent transformations are usual.

Some thickly lamellar anhydrite may grade into a flaser structure, giving the impression that the latter has formed by deformation. In other cases, the anhydrite aggregates are much less sharp and regular than usual (reticulated, Wolkena., Knollena.), grading into structures that are more obviously breccias, like those described in a subsequent section.

An extraordinary structure of the calcium sulfate facies is what might be termed *pegmatitic* (Syn.: macrocrystalline, selenitic): giant crystals of gypsum (or pseudomorphs after gypsum) centimeters to meters in length, standing vertically or in near-vertical sheaves. Figure 29 illustrates a bed of such gypsum from the Pleistocene of Baja California; similar

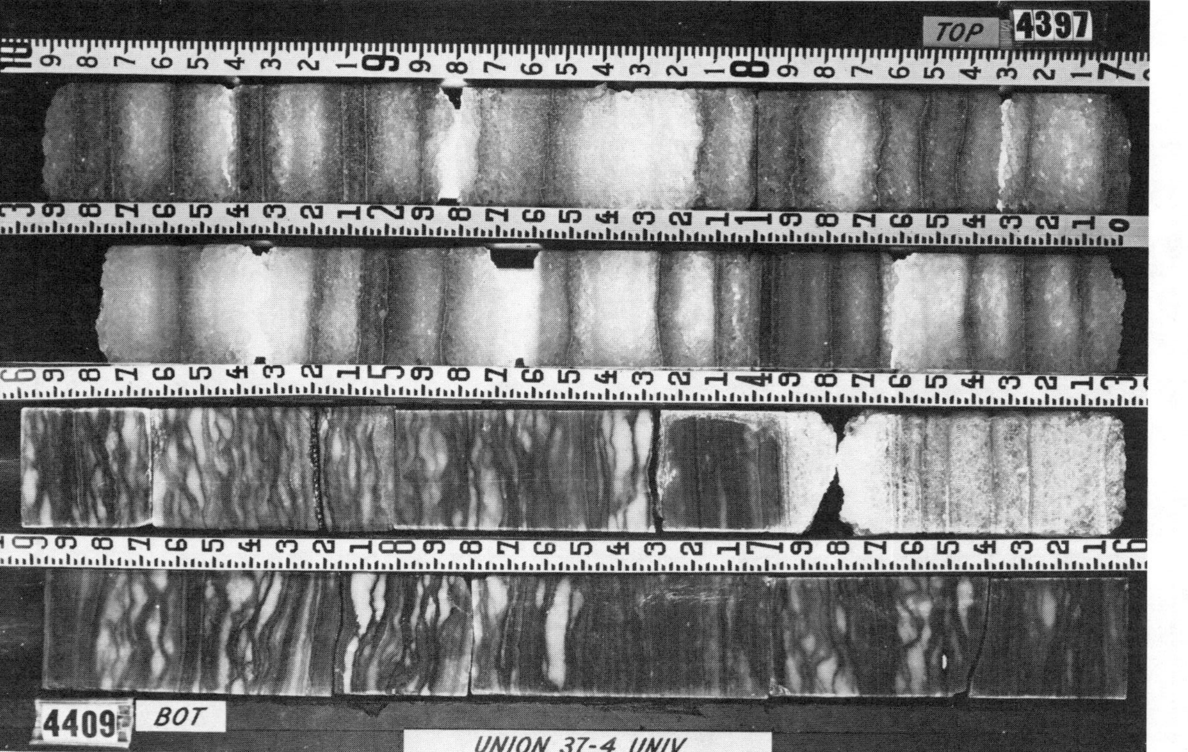

Figure 28. Nodular (flaser) anhydrite rock underlying banded halite rock of Halite II Unit of Permian Castile Formation, at 4397+ feet in Union 37-4 University, Winkler Co., Texas. The top half of the core--all of which is halite rock--is illuminated by both reflected and transmitted light; the lower half of the core--flaser anhydrite below 4404 feet--is illuminated only by reflected light. In the halite section note the typical triplet composed (upward) of a thin anhydrite lamella, cloudy halite, and clear halite. In the anhydrite section note that the nodular anhydrite now and then alternates with sequences of 10 or 20 flat or slightly wavy varve lamellae. The bottom of this piece of core, at 4408.8 feet, is the top of at least 50 feet of varve lamellar anhydrite of Anhydrite II Unit. See Anderson *et al.* (1972).

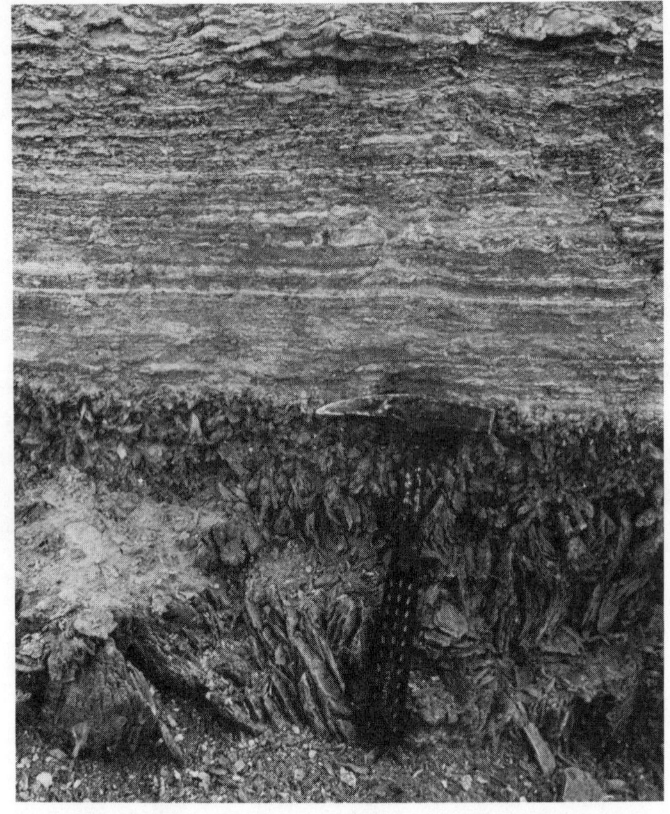

Figure 29. Pegmatitic (selenitic) gypsum rock overlain by clastic lamellar gypsum, unnamed Pleistocene formation, Puertocitos, B.C., Mexico.

gypsum is common in the Eocene of the Paris Basin and the Miocene of Sicily, and pseudomorphs by anhydrite, halite, or polyhalite are found in many older evaporites. Very detailed studies have been done on the Sicilian material, as reviewed by Schreiber (1978; see also Hardie and Eugster, 1971) Typical material shows (100) fishtail twins, fins up, crossed by (120) growth zones marked by clay or other inclusions. The gypsum crystals evidently grew free under brine of a depth between tens of centimeters (Von der Haar, 1975) and tens of meters (Beets and Roep, 1978), but the difference of environment of this texture compared with others has not been determined.

Another common form of anhydrite rock is dense and massive, structures being either subtle or totally absent. The analog in modern sabkha sediments may be the commonly observed "mush" composed of small discoidal gypsum crystals.

In addition to pervasive replacement of gypsum by anhydrite and vice-versa, replacement of carbonates by sulfates and of sulfates by carbonates are found in many evaporite rocks. Small-scale, mostly microscopic evidence of the replacement of carbonates by anhydrite has been described from many areas. Large-scale replacement has been claimed by several geologists, but on much poorer evidence.

The most convincing evidence of at least local replacement of carbonates by anhydrite is the presence of fossils or oölites that are now partly or entirely anhydrite (e.g., Adams, 1932; Dunham, 1948). It should be noted that some such fossils, in their normal mode of preservation, are casts that may be either empty or partly filled with dolomite crystals. Patch and vein anhydrite in bituminous dolomite from the Permian of England is well illustrated by Stewart (1951), who has also documented replacement of carbonate by massive anhydrite on a scale of hundreds of meters in the same formation. Many other descriptions of large-scale replacement are extrapolations of microscopic evidence, without affirmation from either relict structures or a believable source of sulfate.

A particular transformation that is widespread in the surface alteration of anhydrite rock is dedolomitization by reaction between the $CaSO_4$-rich waters generated by solution of the anhydrite, and the dolomite of the anhydrite rock, resulting in a coarse calcitic limestone with or without residual gypsum (Braddock and Bowles, 1963; Shearman and Fuller, 1969a; Smith, 1972; Vaughan *et al.*, 1977; West, 1964).

Halite and Potash-magnesia Facies

Both textures and structures are important in the interpretation of the history of these rocks. Compared with rocks of the calcium sulfate facies, preservation of these features is enhanced by their larger scale and (in the case of halite) by their lack of polymorphic transformations, and debased by the easy recrystallization of the salt minerals. Consequently, most features are best observed at low magnification in core slabs, except for the most opaque cloudy halite for which thin sections may be advantageous. A. E. Kliske (Chevron Standard, Calgary, pers. comm., 1967) recommends the following procedure. Any original core or rough-sawed surface suffices for starting material. Smooth a flat on the surface with 12- to 16-inch flat machinist's file , graded from coarse rasp cut to fine cut. Cover the surface with a 2 x 2-inch slide mount or other thin glass sheet, affixed with glycerine (n = 1.57). Observe with a stereo microscope illuminated at grazing incidence from a point-source light. Zonal concentrations of inclusions become more clear as the specimen is rotated so that they are parallel to the microscope axis.

Some of the most important textures and structures of halite rocks are listed in Table 10. Cloudy halite is translucent white from myriad microscopic cubic fluid inclusions. As illustrated in Figure 30A, these inclusions are typically concentrated along (100) cubic zones in an alternation of inclusion-rich and inclusion-poor zones. The zonal laminae are 0.1 to 0.3 mm thick, about what would be expected from a daily alternation of fast crystallization during daytime evaporation with incorporation of a cloud of fluid inclusions, and slow nighttime evaporation giving clear salt (A. E. Kliske, pers. comm., 1967). That the laminae incorporate defects other than fluid inclusions is demonstrated by the natural or artificial development of blue salt or other color centers in similar laminae (P. J. Schlichta, pers. comm., 1965).

One important manifestation of cloudy halite is in the hopper-shaped crystals that can be observed floating on the evaporating surface of modern salt pans (*novosadka*: Valiashko, 1951). The typical uni-directional stepped surface of these crystals, due to preferential growth along (110) and (111) compared to (100), is outlined by cloudy laminae. The hopper crystals usually aggregate into "rafts" of crystals, and within a few days fall to the bottom of the brine, but their form and surface origin can often still be recognized in the salt rock deposit that may eventually result (Dellwig, 1955).

A second style of cloudy halite is formed on the bottom of the brine, as the fallen hopper crystals or broken fragments of them continue to grow. Growth continues to prefer (111) or (110), but now is mainly upward from the bottom of the brine. This leads to preferred growth of those crystals oriented with (111) up, in which the cloudy laminae form a sequence of upward-pointing chevrons (Figs. 30B,C; *starosadka*: Valiashko, 1951b; Wardlaw and Schwerdtner, 1966). Where chevron halite has been developed, it significantly attests to the continued effectiveness of the diurnal cycle, despite the overlying cover of brine. The very important conclusion is that such a texture indicates brine shallow enough that the variation of night to day conditions are not suppressed (A. E. Kliske, pers. comm.,

Table 10. Some Important Textures and Structures in Halite Rocks

Cloudy halite	*Clear halite*
Hopper crystals	Clear cement
Chevron overgrowths	Coarse recrystallized halite
Cubic crystals	
	Mineral banding

A

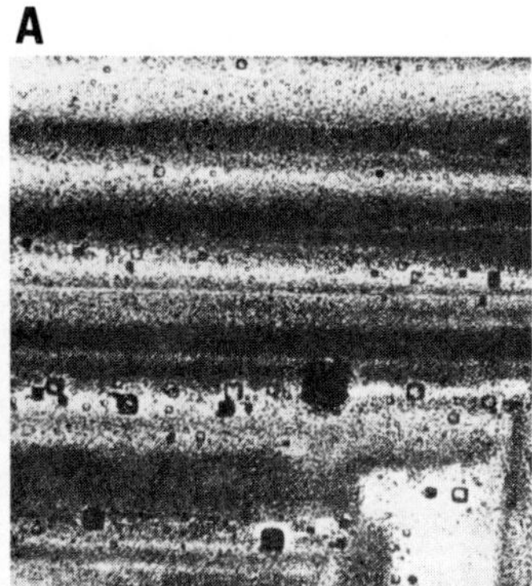

B

C

D

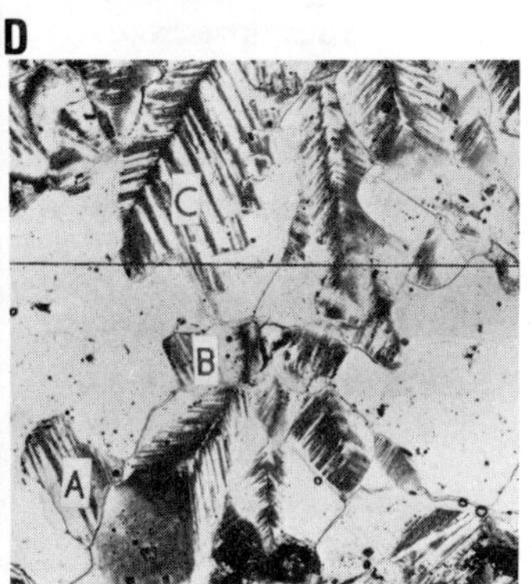

Figure 30. Textures of halite rock. (A) Zones of fluid-inclusion-rich (dark) and -poor (clear) halite in chevron halite rock, 30X. (B) Chevron texture, section perpendicular to bedding, 4.5X. (C) Chevron texture, parallel to bedding, 4.5X. (D) Chevron halite, apparently replaced by clear halite, 3X. It is suggested that A, B, and C are parts of a once continuous chevron grain. All from Devonian Prairie Evaporite Formation, at 3275 feet in Winsal Osler 3-28-39-W3, Saskatchewan, Canada. From Wardlaw and Schwerdtner (1966), courtesy of the authors and Geological Society of America.

1967; Smith, 1973). However, the controlling physical parameter (e.g., temperature, humidity, wind, light, gas evolution in the brine, density inversion) has not been demonstrated, so *how* shallow a brine is required for chevron growth is not yet clear. A complicating factor is the occasional presence of fainter ultra-fine lamellae, whose cause is unknown.

Some halite rock, and some modern salt pans, also contain whole cubes of halite in which the crystal form and the internal zoning are both sharply parallel to (100) in all directions (*granatka*: Valiashko, 1951b). They have been attributed to recrystallization, but it seems more likely (Shearman, 1978) that they represent a post-depositional overgrowth on sedimented hopper crystals that preceded upward extension as chevron texture. In other instances halite evidently grew diagenetically as isolated cubes in muddy sediments.

Clear halite (*kornevoi galit*: Valiashko, 1951b), without fluid inclusions, is commonly associated with cloudy halite of any of the above three subtypes. In many cases the clear halite appears to fill original open space among the aggregate of halite hopper crystals as a cement (Shearman, 1970); in other cases the clear halite transects laminae and crystal boundaries of the cloudy halite and may be considered a replacement, recrystallization, or "clarification" of the cloudy halite (Dubinina, 1951; Fuller and Porter, 1969). In many thick beds of clear halite rock only a few wisps of cloudy halite remain. Recrystallized halite may be coarsened in grain size, up to several centimeters. Commonly such recrystallized halite has a mosaic texture like marble or quartzite, with smoothly curving grain boundaries of random crystallographic direction. This texture is almost universal in halite rock of salt domes that has suffered the heat and deformation of diapirism. Some clear halite may be completely without fluid inclusions. But particularly where the clarity is a product of recrystallization rather than cementation, the halite may appear clear but at least some fluid inclusions are still present, in either of two forms: (a) large fluid inclusions, of several millimeters or even centimeters in size, randomly situated in the crystal; or (b) sheets of fluid inclusions concentrated along the curving boundaries of the mosaic crystals. The latter are particularly evident in salt dome halite. Bromide and other geochemistry of both clear halite crystal and its fluid inclusions indicates that the process of grain growth and fluid aggregation has proceeded in an essentially closed system with little change in chemistry.

Mineral banding is a common structure of halite rock. It is analogous to lamellar anhydrite, but the bands are centimeters thick compared to millimeters in the anhydrite (Table 8). A typical couplet is an alternation of clear to transluscent halite, and dark gray to black halite. The dark color is caused by a dispersion of anhydrite crystals that diffuse the light. In some basins the anhydrite is in spherulites (Vakhramayeva, 1956); elsewhere in an equally common type, the anhydrite is concentrated in one or a few dead-white lamellae (Fig. 28) rather than being dispersed. Authigenic silicates, such as talc, are not uncommon. Other banded halite rock shows the influence of clastic inflow to the basin, with the dark unit of the couplet being rich in clay minerals or quartz silt. The clear and cloudy textures of halite, previously described, also take part in the banding. Although Dellwig (1955) found no regular sequence of clear and cloudy layers in the Silurian Salina Salt of Michigan, a sequence silt-cloudy chevron-clear is normal in the Permian Upper Kama Deposits of the U.S.S.R. (Fiveg, 1948; Vakhramayeva, 1956). Banded halite may have the same variety of origins as lamellar anhydrite.

The smooth and regular bands like those of Figure 28 may generally be attributed to varve sedimentation. They are continuous and can be correlated for several kilometers to tens of kilometers (Vakhramayeva, 1956; Kunaz, 1970). Clastic influences of shallow turbulent water may introduce cross bedding, clasts, and rounding of grains of anhydrite and carbonate (Dellwig and Evans, 1969; Treesh and Friedman, 1974). Even diagenetic growth of halite in algal stromatolites has been described (Shearman and Fuller, 1969; Treesh and Friedman, 1974).

An additional striking feature of banded halite is truncation of growth textures such as chevrons, due to the inflow of undersaturated waters before deposition of the following band (Fig. 31; Wardlaw and Schwerdtner, 1966). Penecontemporaneous solution of halite bands may be pervasive, with the residual anhydrite, carbonate or silt bands in some cases being let down so gently as to be virtually undeformed (and the former presence of salt unsuspected) (Schreiber, 1978). Salt solution is more usual and more evident in subrosion related to the present land surface and ground water circulation, and is evident by brecciation of the immediately overlying evaporite section (Anderson *et al.*, 1972).

Figure 31. Banded halite rock, with a clay interlayer truncating underlying chevron halite. 20X. From Devonian Prairie Evaporite Formation at 3312 feet in Saskatoon 12-19-37-4-W3, Saskatchewan, Canada. From Wardlaw and Schwerdtner (1966), courtesy of the authors and the Geological Society of America.

A well-documented description of a large variety of textures in halite rock (Permian of Kansas) has recently been published by Holdoway (1978).

Much less work has been done on the textures and structures of the potash-magnesia facies, and a general synthesis seems to be lacking. Figure 7 illustrates varve banding of halite-sylvite in the Oligocene of the Rhine Valley (Barr and Kühn, 1962). Because of the much greater temperature coefficient of sylvite than halite, the dark (red) sylvite bands are believed to represent precipitation during the high summer evaporation. The interested reader is referred to particular descriptions of the petrography of the potash in the Devonian of Western Canada by Schwerdtner (1964) and Wardlaw (1968), in the Permian Zechstein by Armstrong *et al.* (1951), Elert and Peter (1975), Gottesmann (1965), Smith (1973), and Stewart (1951a,b, 1956), in the Permian Salado Formation of New Mexico by Schaller and Henderson (1932) and Jones and Madsen (1968), in the Permian Upper Kama Deposits of the U.S.S.R. by Vakhramayeva (1956), in the Cretaceous of Brazil by Wardlaw (1972) and Szatmari *et al.* (1979), and the Oligocene of the Rhine Valley (Barr and Kühn, 1962).

MINERALOGY OF EVAPORITES: USEFUL GENERAL REFERENCES

Bersticker, A.C. *et al.* (1963-1974) *Symposium on Salt, 1st-4th*, Northern Ohio Geological Society, Cleveland, Ohio.

Borchert, H. and R.O. Muir (1964) *Salt Deposits: The Origin, Metamorphism and Deformation of Evaporites*. D. Van Nostrand Co., Ltd., London, 338 p.

Braitsch, O. (1970) *Salt Deposits, Their Origin and Composition*. Springer-Verlag, Berlin, 297 p.

Buzzalini, A.D., F.J. Adler and R.L. Jodry (1969) Evaporites and petroleum. *Am. Assoc. Petrol. Geol. Bull. 53 (4)*, 775-1011 [Am. Assoc. Petrol. Geol. Reprint Ser. *2* (1971)].

Dean, W.E. and B.C. Schreiber (1978) Marine evaporites: *Soc. Econ. Paleon. Mineral. Short Course Notes 4*, 185 p.

Kirkland, D.W. and R. Evans (1973) *Marine Evaporites: Origin, Diagenesis, and Geochemistry*. Dowden, Hutchinson and Ross, Stroudsburg, PA.

Mattox, R.B. *et al.* (ed.) (1968) Saline Deposits. *Geol. Soc. Am. Mem. 88*.

Richter-Bernburg, G.V. (1972) Geology of Saline Deposits. *Earth Science Ser.*, vol. 7, UNESCO, Paris, 316 p.

Stewart, F.H. (1963) Marine evaporites. *In*, M.F. Fleischer (ed.), Data of Geochemistry, 6th edition. *U. S. Geol. Surv. Prof. Pap. 440-Y*, 52 p.

Adams, J.E. (1932) Anhydrite and associated inclusions in the Permian limestones of West Texas. *J. Geol., 60,* 30-45.

_____, and M.L. Rhodes (1960) Dolomitization by seepage refluxion. *Amer. Assn. Petrol. Geol. Bull., 44,* 1912-1920.

Anderson, R.Y., W.E. Dean, Jr., D.W. Kirkland, and H.I. Snider (1972) Permian Castile varved evaporite sequence, West Texas and New Mexico. *Geol. Soc. Amer. Bull., 83,* 59-86.

_____, and D.W. Kirkland (1960) Origin, varves, and cycles of Jurassic Todilto Formation, New Mexico. *Amer. Assn. Petrol. Geol. Bull., 44,* 37-52.

_____, and _____ (1966) Intrabasin varve correlation. *Geol. Soc. Amer. Bull., 77,* 241-256.

Ans, Jean D', D. Bretschneider, H. Eick, and H.E. Freund (1955) Untersuchungen über die Calcisumsulfate. *Kali Steinsalz, 1,* 17-38.

Armstrong, G., K.C. Dunham, C.O. Harvey, P.A. Sabine, and W.F. Waters (1951) The paragenesis of sylvine, carnallite, polyhalite, and kieserite in Eskdale boring nos. 3, 4, and 6, north-east Yorkshire. *Mineral. Mag., 29,* 667-689.

Barcelona, M.J. and D.K. Atwood (1978) Gypsum-organic interactions in natural seawater: Effect of organics on precipitation kinetics and crystal morphology. *Marine Chem., 6,* 99-115.

Barr, A. and R. Kühn (1962) Der Werdegang der Kalisalzlagerstätten am Oberrhein. *Neues Jahrb. Mineral. Abhandl., 97,* 289-336.

Beets, D.J. and T.B. Roep (1978) Evidence for shallow water deposition of Messinian gypsum, below a standing body of water (10 to 100 m), Sorbas Basin, SE Spain. *10th Intern. Cong. Sediment. Abstr., 1,* 62-63.

Bloch, M.R. and L. Picard (1970) The Dead Sea--a sinkhole? *Z. Deutsch. Geol. Gesellsch. Sonderh. Hydrogeol. Hydrogeochem., 1970,* 119-128.

Blount, C.W., and F.W. Dickson (1969) The solubility of anhydrite ($CaSO_4$) in $NaCl-H_2O$ from 100 to 450°C and 1 to 1000 bars. *Geochim. Cosmochim. Acta, 33,* 227-245.

Borchert, H. and R.O. Muir (1964) *Salt Deposits: Their Origin, Metamorphism and Deformation of Evaporites.* London: D. Van Nostrand Co., Ltd., 338 p.

Braddock, W.A. and C.G. Bowles (1963) Calcitization of dolomite by calcium sulfate solutions in the Minnelusa Formation, Black Hills, South Dakota and Wyoming. *U. S. Geol. Surv. Prof. Pap., 475C,* 96-99.

Braitsch, O. (1963) The temperature of evaporate formation. In, A.E.M. Nairn (Ed.) *Problems of Paleoclimatology.* Interscience, London, 479-531.

_____ (1971) *Salt Deposits, Their Origin and Composition.* Berlin: Springer-Verlag, 297 p.

_____, and A.G. Herrmann (1963) Zur geochemie des Broms in salinaren Sedimenten. Teil I: Experimentelle Bestimmung der Br-Verteilung in verschiedenen natürlichen Salzsystemen. *Geochim. Cosmochim. Acta, 27,* 361-391.

Briggs, L.I. and H.N. Pollack (1967) Digital model of evaporite sedimentation. *Science, 155,* 453-455.

Brongersma-Sanders, M. and P. Groen (1970) Wind and water depth and their bearing on the circulation in evaporite basins. In J.L. Rau and L.F. Dellwig, Eds., *Third Symposium on Salt,* Northern Ohio Geol. Soc., Cleveland. V. 1, 120-125.

Butler, G.P. (1969) Modern evaporite deposition and geochemistry of coexisting brines, the sabkha, Trucial Coast, Arabian Gulf. *J. Sed. Petrol. 39,* 70-89.

_____ (1970) Holocene gypsum and anhydrite of the Abu Dhabi sabkha, Trucial Coast: An alternative explanation of origin. In J.L. Rau and L.R. Dellwig (Eds.), *Third Symposium on Salt.* Northern Ohio Geol. Soc., Cleveland, V. 1, 120-125.

Chave, K.E. (1960) Evidence of history of sea water from chemistry of deeper subsurface waters of ancient basins. *Amer. Assn. Petrol. Geol. Bull., 44,* 357-370.

Clark, F.W. (1924) Data of geochemistry, 5th ed. *U. S. Geol. Surv. Bull., 770,* 841 p.

Cody, R.D. (1976) Growth and early diagenetic changes in artificial gypsum crystals grown within bentonite muds and gels. *Geol. Soc. Amer. Bull., 87,* 1163-1168.

Coleman, J.M. (1966) Recent coastal sedimentation: Central Louisiana Coast. *U. S. Clearinghouse Fed. Sci. Tech. Info. Doc. AD653092.*

Cruft, E.F. and P.C. Chao (1970) Nucleation kinetics of the gypsum-anhydrite system. In J.L. Rau and L.F. Dellwig, Eds., *Third Symposium on Salt.* Northern Ohio Geol. Soc., Cleveland, V. 1, 109-118.

Culkin, F. (1965) The major constituents of sea water. In J.P. Riley and G. Skirrow (Eds.), *Chemical Oceanography, V. 1.* London: Academic Press.

Dean, W.E. (1978) Theoretical versus observed successions from evaporation of sea water. *Soc. Expl. Paleontol. Mineral. Short Course Notes 4,* 74-85.

_____, and R.Y. Anderson (1978) Salinity cycles: evidence for subaqueous deposition of Castile Formation and Lower part of Salado Formation, Delaware Basin, Texas and New Mexico. *New Mexico Bur. Mines Mineral Res. Circ., 159,* 15-20.

_____, G.R. Davies, and R.Y. Anderson (1975) Sedimentological significance of nodular and laminated anhydrite. *Geology, 3,* 367-372.

Deffeyes, K.S., F.J. Lucia and P.K. Weyl (1965) Dolomitization of Recent and Plio-Pleistocene sediments by marine evaporite waters on Bonaire, Netherlands Antilles. *Soc. Econ. Paleo. Mineral. Spec. Publ. 13,* 71-88.

Dellwig, L.F. (1955) Origin of the Salina Salt of Michigan. *J. Sed. Petrol., 25,* 83-110.

Dubinina, V.N. (1951) Halite from the Verkhnedamsk deposit. (transl.) *Akad. Nauk SSSR Dokl., 79,* 859-862.

Dunham, K.C. (1948) A contribution to the petrology of the Permian evaporite deposits of Northeastern England. *Yorkshire Geol. Soc. Proc., 27,* 217-227.

Eardley, A.J. (1970) Salt economy of Great Salt Lake, Utah. In J.L. Rau and L.F. Dellwig, Eds., *Third Symposium on Salt.* Northern Ohio Geol. Soc., Cleveland, V. 1, 78-105.

Elert, K.-H. and D. Peter (1975) Elektronenmikroskopische Untersuchungen an Mineralen aus dem Stassfurt-Salinar (Zechstein 2). *Z. Geol. Wiss. Berlin, 3,* 197-209.

Fabuss, B.M., A. Korosi and A.K.M.S. Huq (1966) Densities of binary and ternary aqueous solutions of NaCl, $NaSO_4$, and $MgSO_4$, of sea waters, and of sea water concentrates. *J. Chem. Eng. Data, 11,* 325-331.

Fiveg, M.P. (1948) The annual cycle of sedimentation of rock salt from the Upper Kama deposit. (transl.) *Akad. Nauk SSSR Dokl., 61,* 1087-1090.

Friedman, G.M. (1971) Staining. In R.E. Carver (Ed.), *Procedures in Sedimentary Petrology.* New York: Wiley-Interscience, 511-530.

Fuller, J.G.C.M. and J.W. Porter (1969a) Evaporite formations with petroleum reservoirs in the Devonian and Mississippian of Alberta, Saskatchewan, and North Dakota. *Amer. Assn. Petrol. Geol. Bull., 53,* 909-926.

_____, and _____ (1969b) Evaporites and carbonates; two Devonian basins of western Canada. *Bull. Can. Petrol. Geol., 17,* 182-193.

Garrels, R.M., F.T. Mackenzie, and C. Hunt (1975) *Chemical Cycles and the Global Environment.* Los Altos, California: William Kaufmann, Inc., 206 p.

_____, and M.E. Thompson (1962) A chemical model for sea water at 25°C and one atmosphere total pressure. *Amer. J. Sci., 260,* 57-66.

Garret, D.E. (1966) Factors in the design of solar salt plants. Part II Optimum operation of solar ponds. In J.L. Rau (Ed.), *Second Symposium on Salt.* Northern Ohio Geol. Soc., Columbus, Ohio, V. 2, 176-187.

_____ (1970) The chemistry and origin of potash deposits. In J.L. Rau and L.F. Dellwig (Eds.), *Third Symposium on Salt.* Northern Ohio Geol. Soc., Cleveland, V. 1, 211-222.

Gill, D. (1977) Salina A-1 sabkha cycles and the Late Silurian paleogeography of the Michigan Basin. *J. Sed. Petrol., 47,* 979-1017.

Goldberg, E.D. (1965) Minor elements in sea water. In J.P. Riley and G. Sirrow (Eds.), *Chemical Oceanography, V. 1.* London: Academic Press.

Gottesmann, W. (1965) Petrographische Argumente zur Kalisalzgenese im Stassfurtflöz. *Geologie, 14,* 1215-1223.

Grimm, W.-D. (1962) Idiomorphic quartzes as guide minerals for saliferous facies. (transl.) *Erdöl Kohle Erdgas Petroch., 15,* 880-887.

Hardie, L.A. (1967) The gypsum-anhydrite equilibrium at one atmosphere pressure. *Amer. Mineral., 52,* 172-200.

_____, and H.P. Eugster (1971) The depositional environment of marine evaporites: A case for shallow, clastic accumulation. *Sedimentology, 16,* 187-220.

Heard, H.C. and W.W. Rubey (1966) Tectonic implications of gypsum dehydration. *Geol. Soc. Amer. Bull., 77,* 741-760.

Herrmann, A.G., D. Knake, J. Schneider, and H. Peters (1973) Geochemistry or modern seawater and brines from salt pans: Main components and bromine distribution. *Contrib. Mineral. Petrol., 40,* 1-24.

Higashi, K., K. Nakamura, and R. Hara (1931) The specific gravity and the vapor pressure of concentrated sea water at 0°-175°C. *Sci. Rep. Tohoku Univ. (Japan), Sendai, 10,* 433-

Hite, R.J. and T. Japakasetr (1970) Potash deposits of the Khorat Pleateau, Thailand and Laos. *Econ. Geol., 74,* 448-458.

Holdoway, K.A. (1978) Deposition of evaporites and red beds of the Nippewalla Group, Permian, Western Kansas. *Kans. Geol. Surv. Bull., 215,* 43 p.

Holland, H.D. (1972) The geologic history of sea water- an attempt to solve the problem. *Geochim. Cosmochim. Acta, 36,* 637-652.

Holser, W.T. (1963) Chemistry of brine inclusions in Permian salt from Hutchinson, Kansas. In A.C. Bersticker (Ed.), *Symposium on Salt* [*First*]. Northern Ohio Geol. Soc., Cleveland, 86-95.

_____, and I.R. Kaplan (1966) Isotope geochemistry of sedimentary sulfates. *Chem. Geol., 1,* 93-135 [reprinted in D.W. Kirkland and R. Evans (1973) Marine Evaporites: Origin, Diagenesis, and Geochemistry. Stroudsburg, Pa.: Dowden, Hutchinson, and Ross, 374-398].

_____, _____, H. Sakai, and I. Zak (1979) Isotope geochemistry of oxygen in the sedimentary sulfate cycle. *Chem. Geol., 25,* 1-17.

_____, N.C. Wardlaw and D.W. Watson (1972) Bromide in salt rocks: Extraordinarily low content in the Lower Elk Point salt, Canada. In *Geology of Saline Deposits, UNESCO Earth Sci. Ser. 7,* 69-75.

Ichikuni, M. and S. Musha (1978) Partition of strontium between gypsum and solution. *Chem. Geol., 21,* 359-363.

Jones, C.L. (1965) Petrography of evaporites from the Wellington Formation near Hutchinson, Kansas. *U. S. Geol. Surv. Bull., 1201A.*

_____, and B.M. Madsen (1968) Evaporite geology of Fifth ore zone, Carlsbad district, southeastern New Mexico. *U.S. Geol. Surv. Bull., 1251-B.*

Jung, W. (1958) Zur Feinstratigraphie der Werraanhydrit (Zechstein 1) im Bereich der Sangerhauser und Mansfelder Mulde. *Geologie Beih., 24.*

Kartsev, A.A., Z.A. Tabasaranskii, M.I. Subbota, and G.A. Moglevskii (1959) *Geochemical Methods of Prospecting and Exploration for Petroleum and Natural Gas* (transl.). Berkeley: Univ. California Press, 242-243.

Kastner, M. (1971) Authigenic feldspars in sedimentary rocks. *Amer. Mineral., 56,* 1403-1442.

Katz, A.E. Sass, A. Starinsky, and H.D. Holland (1972) Strontium behaviour in the aragonite-calcite transformation: An experimental study at 40-98°C. *Geochim. Cosmochim. Acta, 36,* 481-496.

Kellogg Company, M.W. (1965) *Saline Water Conversion Engineering Data Book*, Washington: U.S. Department of Interior.

Kinsman, D.J.J. (1966) Gypsum and anhydrite of Recent age, Trucial Coast, Persian Gulf. In J.L. Rau (Ed.), *Second Symposium on Salt*. Northern Ohio Geol. Soc., Cleveland, V. 1, 302-326.

_____ (1969) Interpretation of Sr^{2+} concentrations in carbonate minerals and rocks. *J. Sed. Petrol., 39,* 486-508.

_____, (1969) Modes of formation, sedimentary associations, and diagnostic features of shallow-water and supratidal evaporites. *Amer. Assn. Petrol. Geol. Bull. 53,* 830-840 [reprinted in D.W. Kirkland and Robert Evans (Eds.) (1973) *Marine Evaporites: Origin, Diagenesis, Geochemistry*. Dowden, Hutchinson, and Ross, Stroudsburg, Pa., 80-91].

_____ (1970) Trace cations in aragonite. *Geol. Soc. Amer. Abstr. Progr., 2,* 596-597.

_____ (1974a) Calcium sulphate minerals of evaporite deposits: Their primary mineralogy. In A.J. Coogan (Ed.) *Fourth Symposium on Salt*. Northern Ohio Geol. Soc., Cleveland, V. 1, 343-348.

_____ (1974b) New observations of the Pleistocene evaporites of Montallegra, Sicily and modern analog. *J. Sediment. Petrol., 45,* 469-479.

_____ (1976) Evaporites: I. Relative humidity control of primary facies. *J. Sed. Petrol., 46,* 273-279.

Kirchheimer, F. (1976) Blaues Steinsalz und sein Vorkommen im Neckar- und Oberrheingebiet. *Geol. Jarhb. D18,* 139 p.

Kohler, M.A. and L.H. Parmele (1967) Generalized estimates of free-water evaporation. *Water Resources Res., 3,* 997-1005.

Kolosov, A.S., A.M. Pustyl'nikov, I.A. Moshkina, and Z.M. Mel'nikova (1969) Talc in Cambrian salts of the Kantaseyevka Depression. *Akad. Nauk SSSR Dokl. 185,* 174-177 (transl. *Akad. Nauk SSSR Proc. Earth Sci. Sec., 185,* 127-129).

Kopnin, V.I. (1964) On volumetric correlations of the liquid and solid phases in the formation of salt deposits [in Russian]. *Geol. Geofiz., 1964, No. 5,* 32-37.

Kühn, R. (1968) Geochemistry of the German potash deposits. *Geol. Soc. Amer. Mem., 88,* 427-504.

_____ (1969) Vorkommen and Verteilung des Broms in Salzlagerstätten Mittel-und Westeruopas and Nordamerikas nebst einigen Auswertungen. *Kalisforschungs-Inst. Wiss. Mitt. 111* [unpublished report].

Kunasz, I.A. (1970) Significance of laminations in the Upper Silurian evaporite deposit of the Michigan Basin. In J.L. Rau and L.F. Dellwig (Eds.), *Symposium on Salt, Third*. Northern Ohio Geological Society, Cleveland, V. 1, 67-77.

Langbein, W.B. (1961) Salinity and hydrology of closed lakes. *U. S. Geol. Surv. Prof. Pap., 412,* 20 p.

Link, G.B. and J. Ottemann (1968) Zur Frange des Einbaues von Strontium im Gips. *Jahrb. Landesamt Banden-Württemberg, 10,* 175-178.

Livingston, D.A. (1963) Chemical composition of rivers and lakes. *U. S. Geol. Surv. Prof. Pap., 440G,* 64 p.

Lotze, Franz (1957) *Steinsalz und Kalisalz, V. 1, 2nd ed.* Gebrüder Borntraeger, Berlin.

Maiklem, W.T. (1971) Evaporative drawdown, a mechanism for water-level lowering and diagenesis in the Elk Point Basin. *Bull. Can. Petrol. Geol., 19,* 485-501.

_____, D.G. Bebout, and R.P. Glaister (1969) Classification of anhydrite--a practical approach. *Bull. Can. Petrol. Geol., 17,* 194-233.

McDonald, G.T.F. (1953) Anhydrite-gypsum equilibrium relations. *Amer. J. Sci., 251,* 884-898.

Meleshko, E.P. (1959) Physical properties of brines of Perekop salt lakes. In *Pitannya Geologii i Fizikokhimii Mineral'noi Sirovini Krimii; Zbernik Disertatsunikh Prats.* Akad. Nauk Ukr. RSR, Kiev, 70-76 [in Ukranian].

Morey, G.W., and J.M. Hesselgesser (1951) Solubility of some minerals in superheated steam at high pressures. *Econ. Geol., 46,* 821-835.

Murray, R.C. (1964) Origin and diagenesis of gypsum and anhydrite. *J. Sed. Petrol., 34,* 512-523.

Myers, D.M. and C.W. Bonython (1958) The theory of recovering salt from sea water by solar evaporation. *J. Appl. Chem. (Australia), 8,* 207-219.

Nurmi, R.D. and G.M. Friedman (1974) The Salina Group of the Michigan Basin: Shallow-water and sabkha deposition. *Geol. Soc. Amer. Abstr. Progr. 6, 1052.*

_____ (1978) Use of well logs in evaporite sequences. *Soc. Expl. Paleontol. Mineral. Short Course Notes, 4,* 144-176.

Ogniben, L. (1957) Petrografia della serie solfifera siciliana e considerzioni geolgiche relative. *Ital. Mem. Descrit. Carta Geol., 33.*

Owen, B.B., and S.R. Brinkley, Jr. (1941) Calculation of the effect of pressure upon ionic equilibria in pure water and in salt solutions. *Chem. Rev., 29,* 461-474.

Patterson, R.J. and D.J.J. Kinsman (1977) Determination of marine versus continental sources of subsurface brines in a Persian Gulf coastal sabkha, using Cl^-/Br^- and K^+/Br^-ratios. *Amer. Assn. Petrol. Geol. Studies Geol., 1977,* 381-397.

Pittman, J.S. and R.L. Folk (1970) Length-slow chalcedony: A new testament for vanished evaporite minerals. *Geol. Soc. Amer. Abstr. Progr. 2,* 654-655.

Raup, O.B. (1966) Bromine distribution in some halite rocks of the Paradox Memkea, Hermosa Formation, Utah. In J.L. Rau (Ed.), *Second Symposium on Salt.* Northern Ohio Geol. Soc., Cleveland, V. 1, 236-247.

_____ (1970) Brine mixing: an additional mechanism for formation of basin evaporites. *Amer. Assn. Petrol. Geol. Bull. 54,* 2246-2259.

Richter-Bernburg, G.V. (1955) Über salinare Sedimentation. *Zeitschr. Deutsch. Geol. Gesellsch., 105,* 593-645.

_____ (1959) Die Korrelierung isochroner Warven im Anhydrit des Zechstein 2. *Geol. Jahrb., 75,* 629-646.

_____ (1964) Solar cycle and other climatic periods in varvitic evaporites. In A.E.M. Nairn (Ed.), *Problems in Palaeoclimateology.* London: Interscience, 510-519.

Rothbaum, H.P. (1958) Vapor pressure of sea water concentrates. *J. Chem Eng. Data, 3,* 50-52.

Rowe, J.J., G.W. Morey, and C.S. Zen (1972) The quinary reciprocal salt system Na,K,Ca/Cl,SO_4--A review of the literature with new data. *U. S. Geol. Surv. Prof. Pap., 741,* 37 p.

Rubey, W.W. (1951) Geologic history of sea water. *Geol. Soc. Amer. Bull., 62,* 1111-1147.

Schaller, W.T. and E.P. Henderson (1932) Mineralogy of drill cores from the potash field of New Mexico and Texas. *U. S. Geol. Surv. Bull., 833,* 124 p.

Schettler, H. (1972) The stratigraphical significance of idiomorphic quartz crystals in the saline formations of the Weser-Ems area, north-western Germany. *UNESCO Earth Sci. Ser. 7,* 111-127.

Schlager, W. and H. Bolz (1977) Clastic accumulation of sulphate evaporites in deep water. *J. Sed. Petrol., 47,* 600-609.

Scholle, P.A. and D.J.J. Kinsman (1974) Aragonitic and high-Mg calcite caliche from the Persian Gulf--a modern analog for the Permian of Texas and New Mexico. *J. Sed. Petrol., 44,* 904-916.

Schreiber, B.C. (1978) Environments of subaqueous gypsum deposition. *Soc. Econ. Paleont. Mineral. Short Course, 4,* 43-73.

_____ and D.J.J. Kinsman (1975) New observations on the Pleistocene evaporites of Montallegro, Sicily and a modern analog. *J. Sed. Petrol., 45,* 469-479.

_____ and E. Schreiber (1977) The salt that was. *Geology, 5,* 527-528.

Schwerdtner, W.M. (1964) Genesis of potash rock in Middle Devonian Prairie Evaporite Formation of Saskatchewan. *Amer. Assn. Petrol. Geol. Bull., 48,* 1108-1115.

Shearman, D.J. (1966) Origin of marine evaporites by diagenesis. *Trans. Inst. Min. Metall., 75B,* 208-215.

_____ (1970) Recent halite rock, Baja California, Mexico. *Trans. Inst. Min. Metallurgy 79B,* 155-162 (reprinted in D.W. Kirkland and Robert Evans (Eds.) (1973) Marine Evaporites: Origin, Diagenesis, Geochemistry. Stroudsburg, Pa.: Dowden, Hutchinson, and Ross, 210-217].

_____ (1974) Preservation of organic matter in evaporites. (abstr.) In A.H. Coogan (Ed.), *Fourth Symposium on Salt.* Northern Ohio Geol. Soc., Cleveland, V. 1, 329.

_____ (1978) Evaporites of coastal sabkhas. *Soc. Expl. Paleontol. Mineral. Short Course Notes, 4,* 6-42.

_____, and J.G. Fuller (1969) Anhydrite diagenesis, calcitization, and organic laminites, Winnipegosis Formation, Middle Devonian, Saskatchewan. *Bull. Can. Petrol. Geol., 17,* 496-525.

Shoemaker, E.M. and W.M. Newman (1959) Moenkopi Formation (Triassic? and Triassic) in salt anticline region, Colorado and Utah. *Amer. Ass. Petrol. Geol. Bull., 43,* 1835-1851.

Smith, D.B. (1972) Foundered strata, collapse-breccias and subsidence features of the English Zechstein. *UNESCO Earth Sci. Ser. 7,* 255-269.

_____ (1973) The origin of the Permian Middle and Upper Potash deposits of Yorkshire: An alternative hypothesis. *Proc. Yorkshire Geol. Soc., 39,* 327-346.

Smith, S.M.B. (1973) Halite crystallization in supratidal salina, Ometepec Lagoon, Baja California, Mexico. (abstr.) *Amer. Ass. Petrol Geol. Bull. 57,* 805.

Tanaka, Y., K. Nakamura, and Rosaburo Hara (1931) On the calcium sulphate in sea water. I. Solubilities of dihydrate and anhydrite in the sea waters of various concentrations at 0°-200°C: *Kogyo Kwagku Kwai, Tokyo. Zasshi.* Suppl. binding 34, 284-287 [Tables reprinted by R. Hara, Y. Tanaka, and K. Nakamura (1934): *Tohoku Univ. Tech. Reports, 11,* 199-221].

Treesh, M.E. and G.M. Friedman (1974) Sabkha deposition of the Salina Group (Upper Silurian) of New York State.

Vakhrameyeva, V.A. (1956) The stratigraphy and tectonics of the Upper Kama deposits (transl.) *Trudy Vses. Nauch.-Issled. Inst. Galurgii, 22,* 277-313.

Valiashko, M.G. (1951a) Volume relations of liquid and solid phase in evaporating sea water. *Dokl. Akad. Nauk SSSR, 77,* 1055-1058 [transl. SLA Translation Center 61-13083].

_____ (1951b) Structural features of deposits of modern halite. (transl.) *Mineral. Sborn., 5,* 65-74.

_____ (1956) Geochemistry of deposits of potassium salts. *In Voprosy Geologii Agronomicheskikh Rud.* Akad. Nauk SSR, Moscow, 182-207 [in Russian].

_____ (1972) Scientific works in the field of geochemistry and the genesis of salt deposits in the USSR. *UNESCO Earth Sci. Ser., 7,* 289-316.

Vaughan, F., D. Eby and W.J. Meyers (1977) Calcitization of evaporites--do fabrics reflect diagenetic environment? *Geol. Soc. Amer. Abstr. Prog., 9,* 1209.

Vonder Haar, S.P. (1975) *Evaporites and Algal Mats at Laguna Mormona, Pacific Coast, Baja California, Mexico.* Ph.D. Dissert., Univ. Southern California.

_____ and D.S. Gorsline (1975) Flooding frequency of hypersaline coastal environments determined by orbital imagery: Geologic implications. *Science, 190,* 147-149.

Wardlaw, N.C. (1968) Carnallite-sylvite relationships in the Middle Devonian Prairie Evaporite Formation of Saskatchewan. *Geol. Soc. Amer. Bull., 79,* 1273-1294.

_____ (1972) Unusual marine evaporites with salts of calcium and magnesium chloride in Cretaceous basins of Sergipe, Brazil. *Econ. Geol., 67,* 156-168.

_____ and W.M. Schwerdtner (1966) Halite-anhydrite seasonal layers in Middle Devonian Prairie Evaporite Formation, Saskatchewan, Canada. *Geol. Soc. Amer. Bull., 77,* 331-342 [reprinted in D.W. Kirkland and Robert Evans, Eds., (1973) Marine Evaporites: Origin, Diagenesis, Geochemistry. Stroudsburg, Pa.: Dowden, Hutchinson, and Ross, 195-209].

Wedepohl, K.H. (1969-1978) *Handbook of Geochemistry, 5 vols.* Berlin: Springer-Verlag.

West, I.M. (1964) Evaporite diagenesis in the Lower Purbeck Beds of Dorset. *Proc. Yorkshire Geol. Soc., 34,* 315-330.

_____, Y.A. Ali, and M.E. Hilmy (1979) Primary gypsum nodules in a modern sabkha on the Mediterranean coast of Egypt. *Geology, 7,* 354-358.

White, D.E. (1965) Saline waters of sedimentary rocks. *Amer. Assn. Petrol. Geol. Mem., 4,* 342-366.

Wolery, T.J. and N.H. Sleep (1976) Hydrothermal circulation and geochemical flux at mid-ocean ridges. *J. Geol. 84,* 249-275.

Chapter 9

TRACE ELEMENTS and ISOTOPES in EVAPORITES

William T. Holser

INTRODUCTION

The ways in which minerals of the different evaporite facies undergo minor variations in composition while retaining their dominant mineralogy will be considered in this chapter. These variations may either be in trace element solid solutions, or simply in different isotopes of the same elements. The principles of these two substitutions are similar and familiar to most mineralogists. They will be reviewed in outline form here.

The distributions of trace elements and isotopes in evaporite rocks and in their related brines are potentially useful both in tracing the origin of the rocks and brines, and in delineating those specific conditions, such as temperature and concentration, under which they were deposited. Evaporites are particularly susceptible to this kind of analysis because they represent a chemical system. We have already seen how the *minerals* of the various species delineate the stage of the evaporation of a brine, as required for their precipitation. Trace elements and isotopes have the potential to discriminate on a finer scale for such parameters as stage of evaporation and temperature. In addition to the discrete facies changes represented by the mineral species, solid solution can give us a continuously variable measure. This depends on the fact that the distribution between brine and solution follows relatively constant laws.

TRACE ELEMENTS

Distribution Between Brine Solution and Solid Solution

Trace ions compete with regular (or "Carrier") ions for places in both liquid and solid solutions, so that although the ratio of foreign to regular ions in the crystal is not the same as that in the solution, the ratio of the two ratios is a constant (e.g., McIntire, 1963). To be precise, we could write an equation for the thermodynamic equilibrium constant

$$K = \frac{a_{Bc}/a_{Ac}}{a_{Bb}/a_{Ab}} \qquad [1]$$

where the activities a are denoted by subscripts for regular element A, trace element B, in crystal c or brine b.

This is a physical-chemical relation that remains constant throughout all variations in composition of not only the trace element but also the other constituents of the brine, at one temperature. In general K does vary more or less with temperature. Now activities are difficult to know, particularly in brines. But inasmuch as the ratio between activities and concentrations is relatively constant if the composition of the brine is not changed radically, we can derive a *distribution coefficient*, in terms of the corresponding concentrations,* that will be approximately constant:

$$D = \frac{c_{Bc}/c_{Ac}}{c_{Bb}/c_{Ab}} = \frac{r_c}{r_b} \qquad [2]$$

And where we are dealing with a particular kind of brine we can assume that saturation of the regular (main) element A, in actually growing the crystal, will have a constant ratio c_{Ac}/c_{Ab}, from which we derive a practically useful *apparent distribution coefficient* in terms of the trace element concentrations alone:

$$d = c_{Bc}/c_{Bb} \qquad [3]$$

It turns out to be most convenient to discuss the relations in terms of this apparent distribution coefficient, despite the fact that it may show some variation with composition of the brine, as well as with temperature.

Whether we consider the ideal thermodynamic or the apparent distribution coefficient, the important thing is that it tells us the way a trace element divides between a crystal solution and a brine solution. As we shall see, this not only gives information concerning the trace element composition of the particular brine from which a given crystal grew, but it also leads to progressive accumulation or depletion in a brine as crystallization proceeds. The resulting "profile" of trace element composition in the crystals growing from a sequence of brines may give some clues concerning the conditions under which the sequence accumulated. Finally, the

---

*Units of concentration should be consistent, and in equation 3 d should be in mass units (such as ppm by weight or moles/kg); for other units the values of the distribution coefficients D or d will be different than given in Table 1.

relations of trace elements in solid solution can help to distinguish "second-cycle" brines formed by the solution of marine salts, or the second-cycle salts crystallized from such second-cycle brines. Before discussing the various kinds of trace element profiles and their interpretation, however, some data on the distribution coefficients will be useful.

Distribution Coefficients

Distribution coefficients depend on the relative ease with which the trace element can replace the main element in the crystal and in the brine. Ionic size, charge, and bonding type, relative to the regular ion in the crystal structure, are the most evident fundamental characteristics that may govern this relation. However, distribution coefficients must generally be measured, either directly in the laboratory or indirectly by observing relations of rocks and brines, rather than by derivation from the fundamental parameters involved. The distribution coefficients known for evaporite minerals are listed for reference in Table 1.

Of these distribution coefficients, the one with the longest history of application to a wide variety of problems is for bromide in halite. The applications will be discussed in detail in a following section.

Careful laboratory studies of bromide distribution coefficients were made on artificial solutions that effectively approximate seawater (Braitsch and Herrmann, 1963) with the results indicated in Table 1. The expected bromide contents of halite have been well confirmed by measurements in modern artificial (Hermann, 1972; Hermann *et al.*, 1973) and natural (Holser, 1966a) marine salt ponds. Since Table 1 was compiled, the work of Valiashko *et al.* (1976) has been published; it is an extensive reinvestigation of synthetic systems similar to that of Braitsch and Herrmann (1963), generally confirming their results. Valiashko and Lavrova (1976) have provided further confirmation from an entirely different direction by comparing the composition of fluid inclusions with enclosing crystal, and showed that the ratios scatter around the above described experimental values, regardless of age, all the way back to the Cambrian. Other experimental determinations give distinctly lower values of the distribution coefficients, so while the well-documented higher values should be used, it should be with some reservation (see footnote to Table 1).

Table 1. Distribution Coefficients for Trace Elements in Evaporite Minerals, at 25°C (78°F) [± values in square brackets refer to last decimal place.]

Element	Mineral	Distribution coefficient, D (Eq. 2)	Apparent distribution coefficient (Eq. 3)	Reference and remarks
Br	Halite[a]	0.032[2]	0.13[1]	Sea water, 33-42°C: Hermann (1972)
		0.40-0.025	0.14-0.07[1]	Braitsch & Herrmann (1963) decreasing from pure NaCl solution as $MgCl_2$ added (Fig. 1)
			0.12[3][b]	Valiashko & Mandrykina (1952), from salt lakes of approximately marine composition, by various methods.
		0.016	0.066	Sea water: Bloch & Schnerb (1953)
		0.014	0.053	Pure NaCl solutions, 40°C: Puchelt *et al.* (1972)
	Sylvite	0.21	0.81[5]	Pure KCl solutions;
			0.73[4]	Carnallite solutions: Braitsch & Herrmann (1963)
		0.17	0.66	Pure KCl solutions, 40°C: Puchelt *et al.* (1972)
	Carnallite	0.32	0.55[b]	Puchelt *et al.* (1972)
		0.30	0.52[3]	Braitsch & Herrmann (1963); Kühn (1968)
	Other potash minerals			See Kühn (1968)

- - - - - - - - - - - - - - - - - -

[a] Correspondence with the authors involved has failed to resolve completely the discrepancy between the high values of Braitsch & Hermann, Valyashko & Mandrykina, and Kühn, *vs* the low values of Bloch & Schnerb, and Puchelt *et al.* Puchelt asserts that the high values obtained by others are due to fluid inclusions, which he has eliminated by very slow growth in a thermostat, but both Holser (1966) and Herrmann *et al.* (1973) have made water determinations from which it can be calculated that brine inclusions do not give significant corrections to their results. Herrmann supposes that Puchelt's experiments give values that are too low for application to the natural situation of evaporite rocks, because Puchelt's crystallization rate was very slow. Herrmann (letter of 12 Sept., 1972) says, "The relation of impurity distribution to growth rate in the crystallization of isomorphous systems from solution is a known point," but Wardlaw (1970) had previously shown that fast rate of crystallization gave an increase of only 20% in *d* for Br-sylvite. Bloch & Schnerb comment (letter of 10 Jan., 1971) on the differences of their experiments from those of Braitsch & Herrmann: (1) "Braitsch & Herrmann's results are based on evaporation trials with artificial solutions...We took for our evaporation experiments natural Mediterranean sea water...The $MgSO_4$ may have some influence...The NaCl content of our mother liquor was 21.5% and the experiments of Braitsch & Herrmann showed a NaCl content of 23.14 - 23.42%." -- but later (1972, 1973) Herrmann did repeat his results with Mediterranean sea water. (2) "We analyzed the *very first* precipitate of halite starting from 20 liters of Mediterranean sea water. Braitsch & Herrmann started from 1-3 liters and they get 10-20 gr "Bodenkorper" in the case of halite. We know that with increasing amounts of halite precipitates the Br^- content in this precipitate increases considerably..." -- but it is easy to calculate (*e.g.*, Fig. 5) that such integration of crystal crops will never double the average Br value. Herrmann comments (letter of 11 Dec., 1970), "Indeed, it is very difficult to explain...From my own experience I know that the crystallized NaCl must be formed under stability conditions. Only then is the reproduction of the bromine partition coefficients possible. Unfortunately Bloch & Schnerb don't write under which condition the sea water was evaporated." -- but in his 1973 paper Herrmann seems to be arguing the other way, that Puchelt *et al.* (and Bloch & Schnerb?) were low because they slowly crystallized, and presumably did have equilibrium conditions, whereas the crystallizations used by Herrmann were faster and corresponded to the natural conditions. Herrmann *et al.* (1973) conclude that "It seems that the growth rates of halite in natural conditions, and consequently the bromine partition coefficient for halite, may be variable. However, investigations of this phenomenon [*d vs* growth rate] are necessary in any case." I find it difficult to believe that a faster growth rate can account for a two-fold increase especially when so many different experimenters using different techniques and starting material have arrived at the same larger value. But an experiment could lay that theory to rest -- if it did we would be left with an unresolved discrepancy. See text for new data by Valiashko.

[b] Estimated here from a distribution coefficient published by the author in a different form.

Table 1, continued

Element	Mineral	Distribution coefficient, D (Eq. 2)	Apparent distribution coefficient (Eq. 3)	Reference and remarks
I	Halite	<0.003[b]	<0.01[b]	Schobert (1912)
K	Halite	0.0015	0.006	Puchelt (1972)[c]
		<0.0008	<0.003[b]	Reichert (1966)
	Aragonite	0.0005		Kinsman (1970)
Na	Sylvite	<0.0013	<0.005[b]	Reichert (1966)
	Aragonite	0.0004		Kinsman (1970)
Rb	Halite	0.002[1]	.0.007[4][b]	40°C: Schock & Puchelt (1971)
	Sylvite	0.0875[15]	0.34[6][b]	Pure KCl solutions, 27°C: McIntire (1968)
		0.099[16]	0.38[6][b]	Pure KCl solutions, 40°C: Reichert (1966)
		0.085[4]	0.33[2]	NaCl-containing solutions, 30°C: Kühn (1972)
		0.16[1][b]	0.83[4]	35°C, system KCl-K_2SO_4-$MgCl_2$-H_2O: Malikova (1967)
		0.09[3]	1.35[5]	30°C, with $MgCl_2$: Kühn (pers. comm., 1970)
	Carnallite	1.8[5]	36[1]	NaCl solutions, 30°C; increasing with K/Mg: Kühn (1972a,b)
		2.17[11]	19	Schock & Puchelt (1971)
		3.0[1]	125[5]	16-20°C: Malikova (1967)
		1.87[b]	22[1]	Braitsch (1966)
	Other potash minerals			See Schock & Puchelt (1971); Kühn (1972)
Cs	Halite	<0.0002	<0.0007[b]	40°C: Schock & Puchelt (1971)
	Sylvite	0.000172[12]		40°C: Schock (1966)
	Carnallite	0.96[12]		Schock & Puchelt (1971)
	Kainite	0.003[15]	0.006[3][b]	60°C: Schock & Puchelt (1971)
NH_4	Halite	<0.002	<0.007[b]	35°C: Reichert (1966)
	Sylvite	· 0.082[9]		35°C: Reichert (1966)
Mg	Aragonite	0.001		Kinsman (1970)
	Calcite	0.20		Kinsman & Holland (1969)
		0.33		Glover & Sippel (1967)
Ca	Halite	<0.16		40°C: Reichert (1966)
	Sylvite	<0.002		40°C: Reichert (1966)
Sr	Aragonite	1.12[4]		Kinsman & Holland (1969)
	Calcite	0.14[2]		Holland *et al.* (1964); see also Katz *et al.* (1972)
	Dolomite	~1.0		Calc. from crystal/brine data of Butler (1973)
	Gypsum	0.25[5]		30°C: D.J.J. Kinsman (pers. comm., 1970)
		0.18[2]		33°C: by calc. from sabkha crystals/brines: Butler (1973)
		0.21[1]		20-60°C: Ichikuni & Musha (1978)
	Anhydrite	~0.35		30°C: D.J.J. Kinsman (pers. comm., 1970)
		0.37[3]		30°C: by calc. from sabkha crystals/brines: Butler (1973)
	Sylvite	<0.002		40°C: Reichert (1966)
Ba	Sylvite	<0.0001		40°C: Reichert (1966)
B	Anhydrite		0.000x	Ham *et al.* (1961)
Mn	Calcite	16.2		40°C: Bodine *et al.* (1965)
Zn	Calcite	4.1[1]		35°C: Crockett & Winchester (1966)
		15[3]		50°C in Cl brines: Tsusue & Holland (1966)
Tl	Halite	0.0003		95°C: Brauer (1953)
	Sylvite	2.5		95°C: Brauer (1953)
	Carnallite	0.7[2]	33[1]	16-20°C: Malikova (1967)
Pb	Sylvite	4[5]		40°C: results highly dependent on Pb concentration. Reichert (1966)

---

[c] *D* for K in NaCl was given by Puchelt in his oral presentation but is not in the abstract cited and has not yet been published.

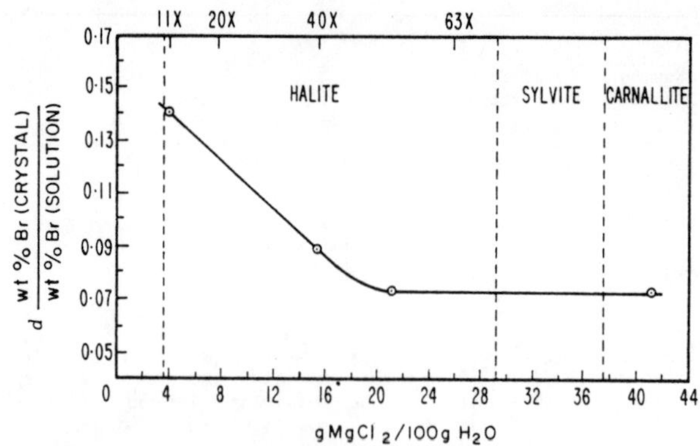

Figure 1. Apparent distribution coefficient *d*, for bromide in halite crystallizing from $MgCl_2$-bearing brines (Holser, 1966b, after Braitsch and Herrmann, 1963). The mineral names correspond to evaporite facies in seawater from which $MgSO_4$ has been removed.

The distribution coefficient decreases rather sharply with the increase of $MgCl_2$ concentrations in the brine, corresponding to the first half of the halite facies (Fig. 1). Kühn (1968) suggests that this is caused by the known complexing in $MgCl_2$-rich solutions. Experiments by Braitsch and Herrmann (1963) and Valiashko *et al.* (1976) established that other common seawater components such as K^+ and SO_4^{2+} have no measurable effect on the bromide distribution coefficient.

Effect of Temperature

As mentioned previously, distribution coefficients are generally affected by temperatures; however, bromide distribution coefficients only rise less than one percent with a rise in temperature of 5°C. The effect is nearly the same in the three minerals halite, sylvite, and carnallite. Within the range of temperatures encountered in crystallization of brines, the values of *d* (for *Br* only) given in Table 1 can therefore be considered constant. The pressure effects have not been measured, but theory indicates they would not be of any consequence even for the case of deep burial.

Braitsch (1963, p. 141) pointed out that this constant *d* when combined with the high-temperature coefficient of *solubility* of sylvite (Fig. 6 in Chapter 8) or carnallite, results in a strong increase with temperature of the bromide content of the first sylvite or carnallite to be deposited in a basin; he used this to study the variation of temperature with time and position in the Oligocene Rhine Valley deposits (Braitsch, 1966).

Surface and Kinetic Effects

All of the above discussion of distribution coefficients assumes thermodynamic equilibrium between the crystal solid solution and the liquid solution. This means that, at any particular stage of the crystallization, the concentration of trace ion in the bulk of the crystal is in equilibrium with the same kind of ion in the bulk of the solution. We are only talking here about some thin layer of crystal formed recently from substantially the same solution--we of course do not require that *all* previously formed crystal is of uniform trace element concentration and in equilibrium. But other things can upset the equilibrium situation. In the first place, the structure of the surface layer of the crystal and of the surface layer of solution may be sufficiently different from the bulk, that there on the surface where things are happening, the equilibrium between these thin layers may differ (in either direction) from bulk equilibrium. Furthermore, if the crystal is growing rapidly, it may grab up some extra trace element that does not represent equilibrium in any sense.

If crystallization is fast, the surface layers are more likely to be different from the bulk, and diffusion in the formed crystal is less likely to restore bulk equilibrium. So both the surface and kinetic effects can modify trace element distribution during fast crystallization in certain cases. Most experiemental studies of distribution coefficients seek to establish an equilibrium situation, and the extent of kinetic variations. Among the distribution coefficients in evaporites (Table 1), those for K^+ in sylvite (Wardlaw, 1970), Sr^{2+} in aragonite (Kinsman, 1969), and possibly Sr^{2+} and other elements in anhydrite (Dean and Anderson, 1974) have been shown to have kinetic effects.

Distribution of Trace Elements in Simple Evaporation

A direct consequence of the distribution coefficient is that the trace element is unevenly distributed between the growing crystals and the brine. In most cases such crystals do not react further and their removal from the system aggravates the segregation of the trace elements.

Thus, if a trace element has a distribution coefficient much less than one, such as bromide in halite, the trace ion is only slightly taken into the crystal and strongly segregated into the brine. Each batch of crystals removed from the brine further increases the bromide content of the brine. On the other hand, if a trace element has a distribution coefficient somewhat greater than one, like rubidium in carnallite (Table 1) the effect is opposite: The crystals soak up rubidium leaving less in the brine, and the more crystals that grow, the less rubidium in the remaining brine. Finally, if a distribution coefficient is near to one, as with the sulfur isotopes (see below), the crystals of gypsum or anhydrite will mirror nearly exactly the tracer composition of the brine, in whatever way it may be changed by inputs and outputs to the system. Figure 2 diagrams graphically the first case in which the distribution coefficient D is less than one, and the ratio r_c of concentrations of trace element B to main element A gradually builds up with time from an initial value of r_c° at the base, through successive layers of the crystals to values r_c'.

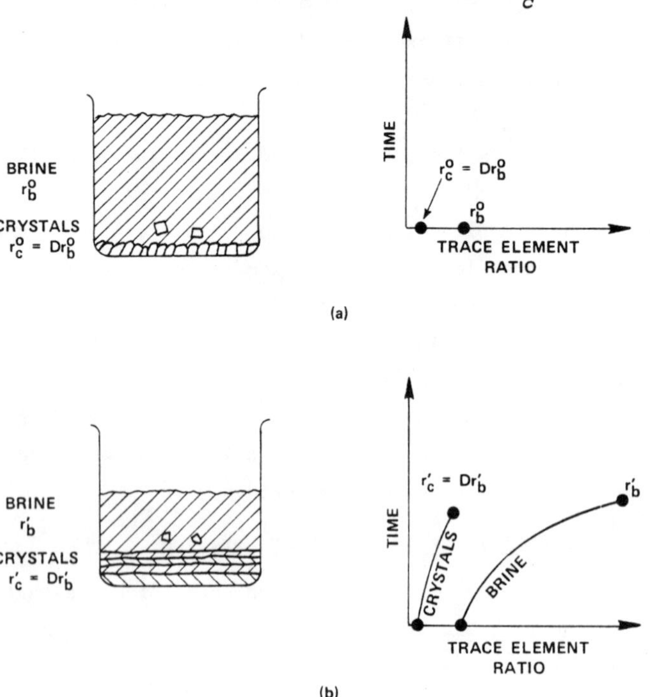

Figure 2. Model for trace element concentration $r = c_B/c_A$ in brine and crystals, for distribution coefficient $D < 1$ (reprinted with permission from the *Second Symposium on Salt*, Holser, 1966).

The *exact* way in which the trace element concentration changes, as larger and larger fraction of the main component is crystallized, is found by integrating the instantaneous distributions (equations 2, 3) according to Rayleigh's Law. If the amount (mass, volume, or thickness) of crystals deposited up to some level is m', and if the total amount that would deposit if evaporation were complete were m'', then the fraction yet to be crystallized is $(m''-m')/m''$. There is a logarithmic relation between this factor and the trace concentration ratio:

$$ln\left(\frac{r'_c}{r^\circ_c}\right) = (D-1)\ ln\left(\frac{m''-m'}{m''}\right) \qquad [4]$$

In words, as a larger and larger fraction of the brine is precipitated, the trace element concentration ratio follows a logarithmic curve, shown schematically in Figure 2.

A practical "trace element profile," which analogously plots trace element concentration c_{Bc} against thickness, is similarly logarithmic.

We have seen several lines of evidence indicating that simple evaporation is not a common mechanism for the crystallization of any consequential thickness of evaporites. However, that simple model is useful in demonstrating the general nature of trace element profiles. Consideration of the governing equation 4 suggests several deductions that can be made from the profile. First, the lowermost crystals are a sample of the original brine, undistorted by the effects of subsequent crystallization; consequently, if you know the distribution coefficient for the element and mineral, you can calculate the concentration of the trace element in the original brine, using equations 2 or 3. This trace element ratio in the original brine may be indicative of its source, for example, whether it is seawater or is derived from the solution of older evaporites. Second, the rapidity with which the trace element changes with thickness, that is, the slope of the profile, is a measure of the size of the body of brine. If the body of brine from which salts are crystallizing is large, then a trace element cannot accumulate or deplete very quickly, even if its distribution coefficient is much less or much greater than one, respectively. If the body of brine has a relatively small volume its trace element content can change rather rapidly, leading to a consequent change within a small thickness in the trace element profile of the salt. In other words, sharp changes of trace element content indicate shallow depths of brine and unchanging profiles indicate deep bodies of brine.

Third, any point on the trace element profile indicates how far along evaporation has taken the brine towards complete crystallization of the main salt. That is, the trace element content at any point in the profile is a measure of the stage of evaporation. All of these qualitative aspects will be considered further, with a more realistic model for the basin.

Distribution of Trace Elements in a Basin with Inflow and Reflux

The important evaporite basins that concern us must have had a considerable amount of inflow of new seawater and reflux of concentrated brine back to the sea. So we now consider how trace elements will be distributed in a basin in which the inflow, reflux, and evaporation constitute a regime that is constant for a period of time sufficient to build up a piece of a trace element profile. This regime may be characterized by a single mass-balance factor that describes how the composition of the reservoir is changing relative to crystallization (Holser, 1966; see also Kühn, 1968):

$$g = \frac{\text{Differential change of the amount of regular element } A \text{ in solution}}{\text{Amount of this change due to crystallization}}$$

$$= \frac{\text{Differential change of the amount of regular element } A \text{ in solution}}{-(\text{Change of the amount of regular element } A \text{ in crystals})} \qquad [5]$$

For example: Simple evaporation involves no inflow, all the change in the amount of A in the solution is due to crystallization, so for this simple case $g = 1$; for constant volume, there is no change in the amount of A in solution, so $g = 0$; if inflow is even more important, g is negative.

We also need to take into consideration the trace element concentration ratio of the net flow of solution into and out of the basin:

$$r_f = d_{Bf}/c_{Af} \qquad [6]$$

Now the distribution equation will be of the same general form as equation 4 for simple evaporation, but will include the factors g and r_f in the following way:

$$ln\,\frac{r'_c(D-g) + (g-1)Dr_f}{r^\circ_c(D-g) + (g-1)Dr_f} = \frac{D-g}{g}\,ln\,(\frac{m''-m'}{m''}) \qquad [7]$$

For straight evaporation, g = 1.0, and the equation reduces to the Rayleigh Equation 4. For a constant volume of brine in the basin, which is probably a very useful and often approximated geological model, g is zero and a special equation applies:

$$ln \frac{r'_o - r_f}{r^\circ_c - r_f} = -D \frac{m'}{m''} \qquad [8]$$

It is not the purpose to discuss any intricacies of such theoretical equations, but only to indicate some general principles that can be derived from their consideration.

Figure 3 indicates the variety of curve shapes that can be attributed to changes of the evaporation regime g with increasing importance of inflow versus evaporation from right to left. The main reason for showing these theoretical curves is to emphasize that none of these have maxima, minima, or even inflection points. Therefore, when (as is commonly the case) such anomalies do occur in trace element profiles, it means that some factor in the evaporation regime must have changed; such shapes cannot be derived from any combination of continuing and uniform conditions in an open evaporite basin. Stated another way, smooth profiles (either constant or rising or falling, even sharply) of a trace element indicate uniform conditions of the amount and composition of inflow, reflux, and evaporation; irregular profiles indicate many changes of these conditions.

A second consequence of the general profile equation is that the trace element content at any point on the profile is still a measure of the overall stage of evaporation of *all the brine in the basin at that time*. That is, the trace element ratio still tells how far along the process of evaporation has proceeded, despite the inflow and outflow.

Another consequence of the unique relation between trace element composition of brine and such profiles is that different parts of a basin in which salt is crystallizing should have profiles of correlative shape. If the basin is well mixed, the profiles should also match in their level of trace element content. On the other hand, if precipitation is taking place from brines of higher stages of evaporation, in parts of the basin farther from the oceanic inflow, such profiles may show similar form, due to changes with time of the conditions of inflow and reflux, but at a higher general level indicative of their further stage of evaporation.

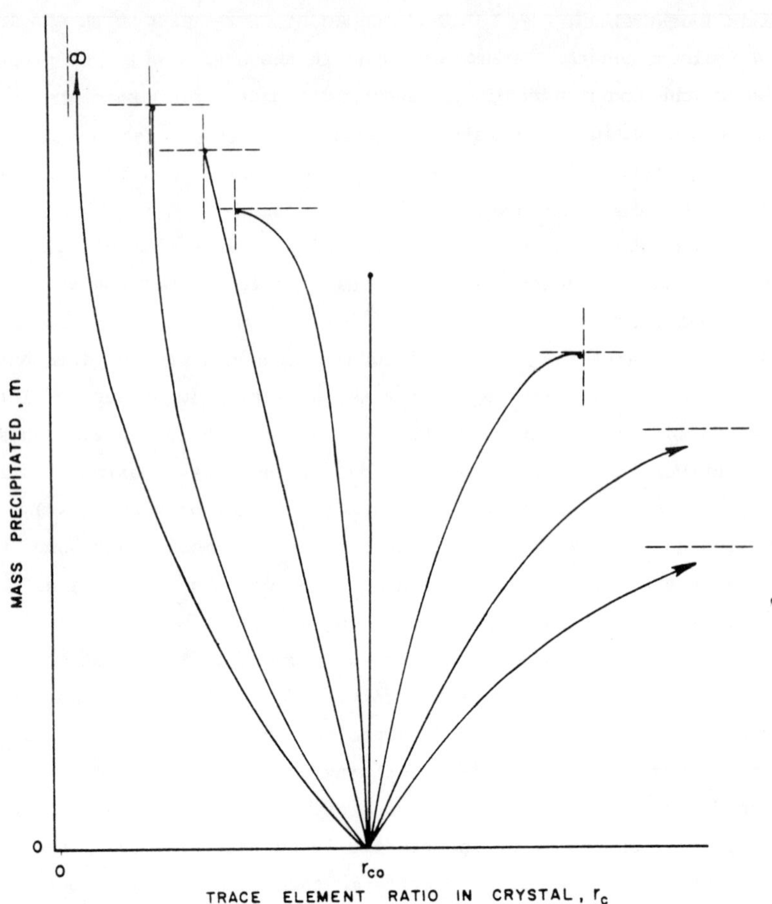

Figure 3. Trace element profiles in an open system in various evaporite regimes. The factor g (equation 5) increases from left to right, as evaporation increases relative to inflow. D is constant and less than one, and the trace element inflow concentration r_f is low, for all curves (reprinted from the *Second Symposium on Salt*, Holser, 1966).

Bromine Geochemistry

The geochemistry of bromine is particularly important in evaporite rocks, because (1) it can be studied in any of the several chloride-bearing minerals; (2) a great deal is already known about the distribution of bromide in such minerals, both by experiment and the analysis of actual evaporites; and (3) the geochemistry of bromine is on the whole strongly conditioned by its role in the evaporite cycle.

General distribution of bromine in rocks and waters is outlined in Figure 4 (Holser, 1966b). The keystone of bromine geochemistry is seawater, where the content of 65 ppm constitutes the major reservoir of that element. By the time the seawater has evaporated to saturation with halite, the bromide concentration has reached about 510 ppm. From the distribution coefficient, Table 1, one can then calculate that the first halite to crystallize should have 510X (0.14 ± 0.01) = 65-75 ppm Br. Then, as evaporation proceeds, either with or without inflow, bromide is segregated into the brine and rises in both brine and succeeding crops of crystals, as in the model curves of Figure 2. The theoretical curve is shown in

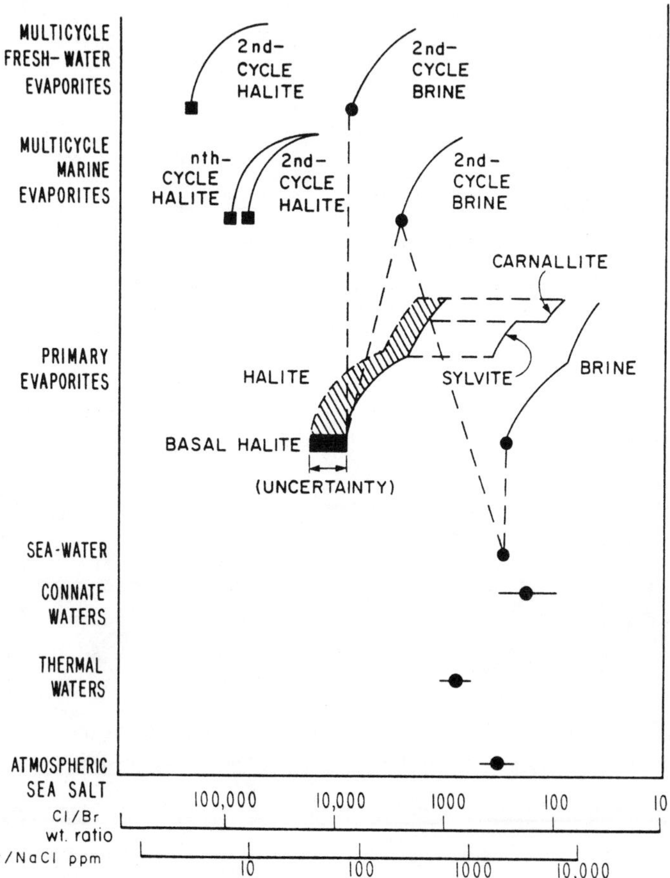

Figure 4. Distribution of bromine in rocks and waters (reprinted from the *Second Symposium on Salt*, Holser, 1966).

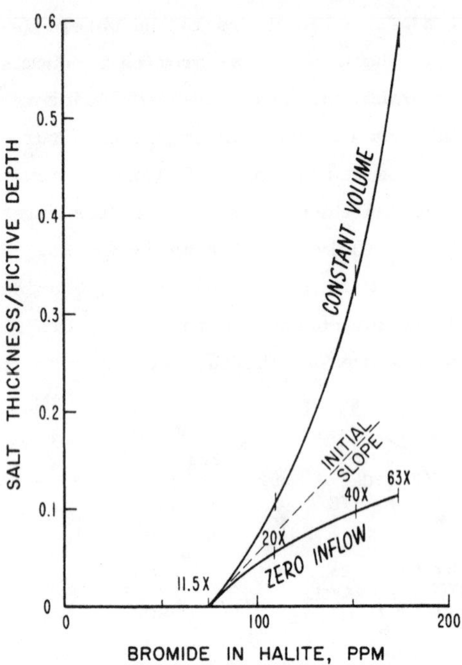

Figure 5. Ideal bromide profiles in halite rock, corrected for the variation in distribution coefficient with brine concentration (Fig. 1). The ordinate represents the fraction to which an original fictive depth of brine has been filled by solid salt. Note that bromide content is the same measure of evaporation index (see Table 1, Chapter 7, this volume), regardless of inflow regime. Reprinted with permission from the *Second Symposium on Salt*, Holser (1966).

Figure 5 for both zero inflow and constant volume, allowing for the variation in d of Figure 1.

In Figure 5 the mass crystallized, m' (Fig. 3), is replaced by a corresponding but more practical value of the *thickness* of the salt formation, relative to the fictional depth that the brine would have had if the basin were cylindrical. The bromide content of such an idealized curve rises to nearly 200 ppm, at which point minerals of the potash-magnesia facies begin to crystallize with even greater bromide contents (Table 1). The variation in bromide content of a succession of brines and crystals is also indicated diagramatically in the center of Figure 4.

The bromide content of halite is extensively used as a prospecting tool for potash deposits; as one example, a profile of the Salina salt in Michigan presented in 1965 (Holser, 1966b) led to the discovery of potash beds in that formation in 1966 (Anderson and Egleson, 1970). Re-solution of salt rock of course causes an equivalently high Cl/Br ratio in the resulting waters, and a consequent extremely low Br content in salt crystallizing from such waters in a "second-cycle" evaporite. As a consequence of these crystallization paths, bromide analyses fall into four categories: (1) seawater and derived formation waters, with Cl/Br near 300, trending down to 200; (2) marine salt rock and river or formation waters formed by their dissolution, with Cl/Br in the thousands; (3) second-cycle evaporite rocks (or waters dissolved from them) with Cl/Br in the tens of thousands; and (4) brines residual to the crystallization of salt rocks ("bitterns"), with Cl/Br often less than 200. Although the ideal cases here are very distinctive, real waters and salts tend to have a complex origin, so that many intermediate cases are actually found. Meteorites, volcanic gases, and some thermal waters lie in the general range Cl/Br = 500 to 1500. This is between the two main surface reservoirs of these halides, which would be expected if they represent any kind of sample of average primordial halides.

Bromide Profiles

Although the general outlines of the geochemistry of bromide can be understood in terms of the segregation between brines and salts as discussed above, the distribution in many evaporite rocks does not approximate the ideal profiles on Figure 5. Many such profiles are now available from salt sequences throughout the world and for ages back to the Late Proterozeoic. In general, these profiles reflect a wide variety and a general complexity of evaporative history. The subject has been reviewed by Holser (1966a), Kühn (1968), and Raup and Hite (1978).

Figure 6, from the second cycle of the Permian Zechstein 2 (Schulze, 1960, p. 181), is just about as ideal a profile as may be found anywhere. Figure 7 is a similar but slightly more irregular profile from Salt 3 of the Pennsylvanian Paradox Formation in Grand County, Utah (Raup, 1966). Although these are developed on scales that differ by an order of magnitude, both curves show a long steady rise through the central portion of the

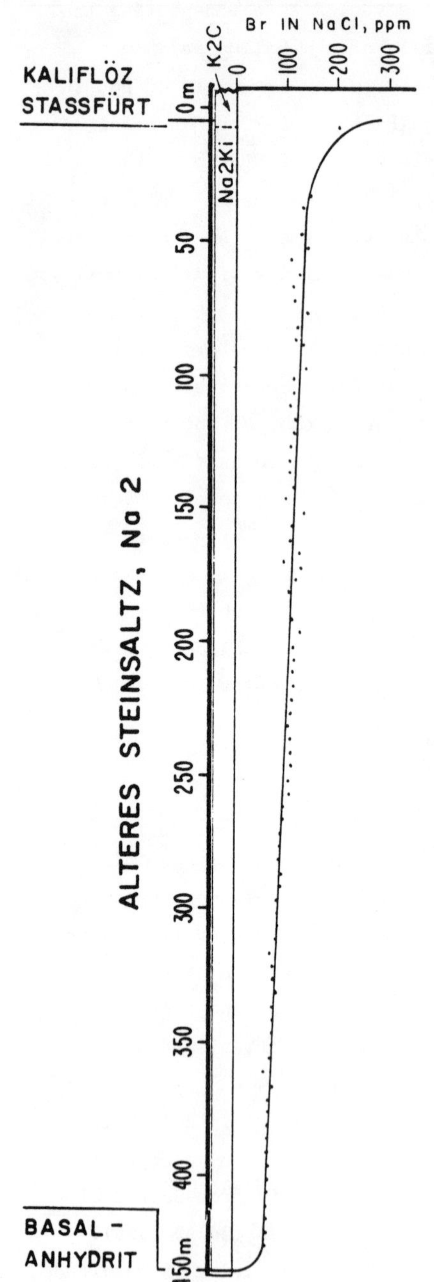

profile, terminated in a sharp rise at the top of the formation. In the case of the Zechstein 2, this rise coincides with the development of the world-famous Stassfurt potash bed; potash has not yet been found at this particular horizon in the Paradox Formation. In accordance with the principles outlined above, the rather sharp change near the top of these salt series attests to an important change of some kind in the regime of inflow, evaporation, and reflux; a probable interpretation is that inflow was virtually cut off, leading to a sharp rise in evaporation stage and eventual precipitation of potash minerals. The profiles differ at the lower end. Initial sharp rises such as that of Zechstein 2 are incompatible with the later slow rise, and have been attributed to some kind of post-depositional alteration that is not yet well understood. The negative slope at the base of Paradox 3 is probably due to hold-over of bromide-rich brine from a preceding evaporite cycle, as discussed below.

Bromide profiles are commonly much more irregular than the ideal example described above. A profile through the

Figure 6. Typical regular normal bromide profile, in the salt-potash sequence of the Permian Zechstein 2 (Stassfurt Series), Aschersleben, Germany (DDR) (Holser, 1966, after Schulze, 1960, p. 99). A short distance below the top of the profile the slow, regular climb of bromide content increases sharply, probably because of a change to simple evaporation leading to the potash bed at the top.

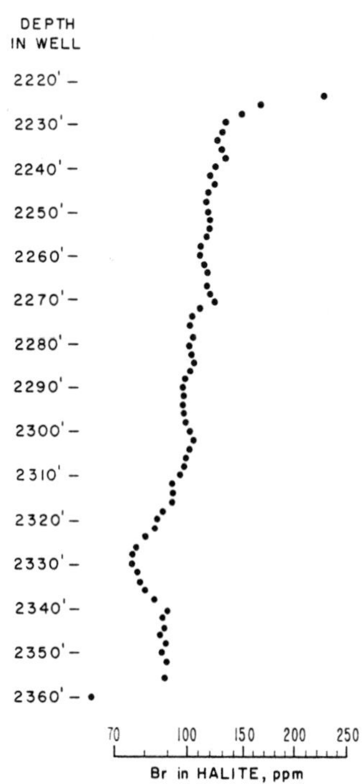

Figure 7. Regular bromide profile in Salt 3 of the Paradox Formation, Grand Co., UT, smoothed by 10-ft (five analyses) moving average. (Reprinted by permission from the *Second Symposium on Salt*, Raup, 1966.)

Devonian Prairie Evaporite in Central Alberta, shown in the upper part of Figure 8, is typical of such profiles. It was early recognized that such irregularity ". . . requires a modification of the conventional picture of an evaporite basin with uniform inflow and reflux. . . .It is difficult to see how the irregularities observed could be perpetuated in the presence of the flywheel effect of a body of brine continually overlying the depositing body of salt" (Holser, 1966b, p. 260). The years since that observation have seen a pervasive attribution of evaporites to the sabkha model, including the Prairie Evaporite (Maiklem, 1971; Fuller and Porter, 1969). Irregularity of bromide profiles is consistent with the intermittent aerial exposure and dessication of the sabhka regime, and this interpretation has been applied to the Miocene salt of the Mediterranean (Kühn and Hsü, 1975). The irregular profiles could also be generated in a brine shallow enough to permit quick buildups of bromide, but conversely, smooth profiles are good evidence for deep brine deposition. For many observed examples the limit is surely a few meters.

Many smooth bromide profiles, or even irregular profiles that have been smoothed by averaging bromide compositions over thicknesses of 10 to 50 feet, show a more or less regular rise and fall suggesting long-term evaporative cycles. Figure 9a gives an example from the Leine series, the Zechstein 3, overlying the Stassfurt Series of Figure 6. This profile shows, after an initial drop in bromide content at the bottom of the formation, two prominent peaks where the bromide content reaches about 200 ppm. Features such as this may be recognized throughout a large part of an evaporite basin, and serve to identify the particular evaporite series as well as possibly being used

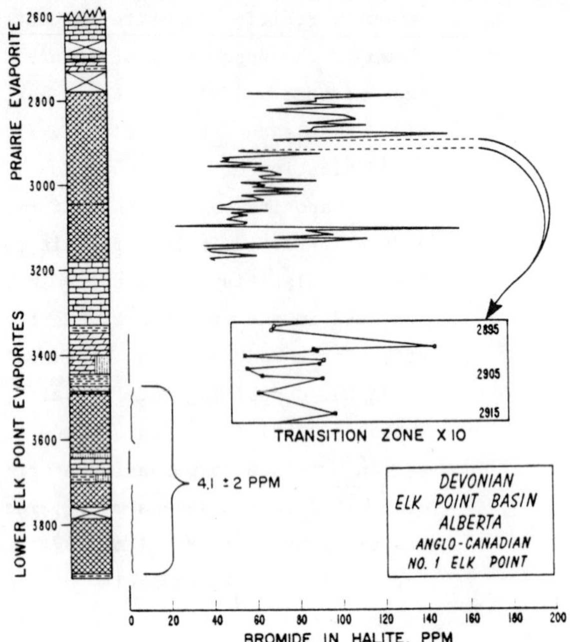

Figure 8. Bromide profiles through the Elk Point Group in the Alberta Basin, Canada. Detailed analysis for a 10-foot zone indicates the extreme variability. Solid symbols are gray, cloudy, possibly primary salt, and open symbols indicate clear, white, possibly recrystallized salt (see Chapter 8, section on Textures and Structures in Halite Rock, this volume). Reprinted with permission from the *Second Symposium on Salt*, Holser (1966).

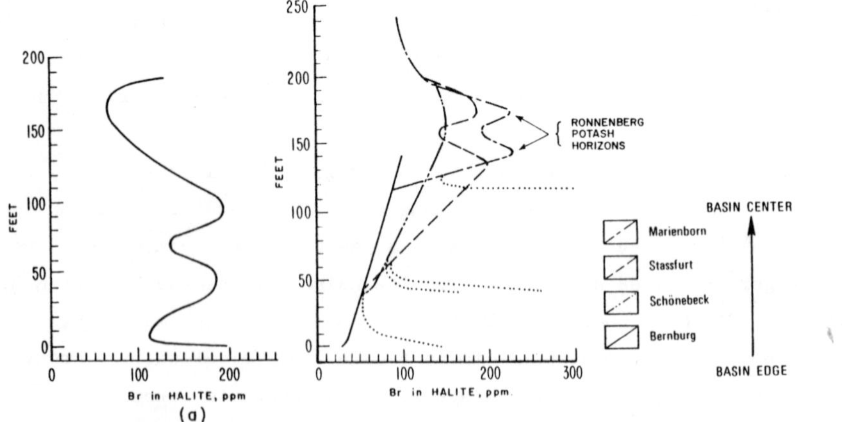

Figure 9. Bromide profiles in the lower part of the Leine Series, Zechstein 3: (a) profile at Stassfurt, Germany (DDR) showing two cycles with strong maxima; (b) comparison of generalized profiles at various sites from the edge towards the center of the basin. After Schulze (1960).

for time-correlation of particular parts of the series. An important feature of such profiles is that the bromide contents at the beginning of the sequence and at the peaks may both be greater in one locality than in another. This fact has been interpreted by Schulze (1960) as indicating a beginning of precipitation in the locality of low bromide content, which then does not commence at another point until the bromide content of the whole basin has risen. In the Zechstein basin (Fig. 9) this means beginning of salt crystallization at the edge of the basin on the south, before it began to fill in towards the center. On this model, however, it is difficult to see why the *peaks* should also be higher in the center of the basin. Although the interpretation of such profiles is not yet clear, it is evident that their characteristics are in some cases so distinctive that such comparisons will be very useful in working out a model for brine circulation and crystallization.

Raup *et al.* (1970) have described similar correlatable profiles from the Pennsylvanian Paradox salt of Utah-Colorado, and applied these to the detection of horizontal salinity gradients in the depositing basin.

In Zechstein 3 the rather steady decrease following the second peak continues (not shown) for an equal thickness to the top of this series; a similar decrease is evident in the upper part of the Prairie Evaporite (Fig. 8). In accordance with the general theory (Fig. 3), such decreases are a natural consequence if the evaporation regime is regressing back through the halite facies due to excessive inflow (as on the left side of Fig. 3). In other cases, as in Zechstein 2 (Fig. 6) and Paradox 3 (Fig. 7) the salt formation tops out at a high value; this high value may continue to be evident at the base of the next overlying salt bed, the Zechstein 3 (Fig. 9) and the Paradox 2 (Raup, 1966, p. 241), respectively. This had led to the suggestion that such basal high bromide contents represent a carryover of bromide-rich brine from the preceding cycle (Kühn, 1968, p. 455), perhaps expelled from intervening shale beds (Raup, 1966). In support of this idea, such initially high and rapidly dropping bromide profiles are not found in the earliest salt in a given basin, but only in later sequences.

Figure 10 shows an even steeper and greater decrease in bromide at the top of the Wellington Formation in the Permian of Kansas. The later Permian Nippewalla Group has an even lower bromide content, averaging less

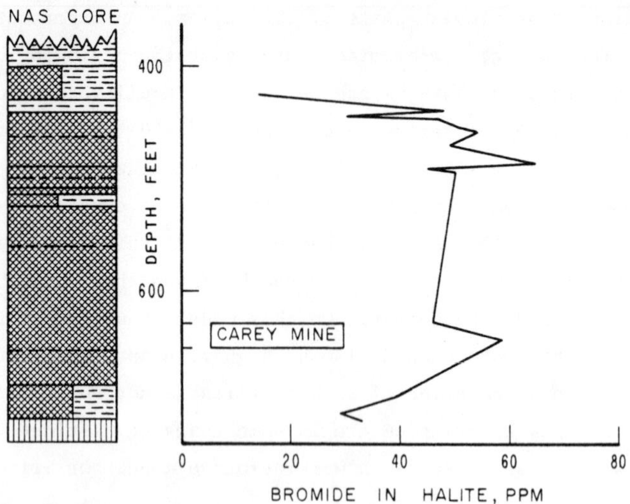

Figure 10. Bromide profile in the Permian Wellington Formation, Kansas. Crosshatching is mainly salt, vertical lines are anhydrite, and dashed lines are shale; for detailed section see Jones (1965). Note sharp drop in bromide at the top of the section, caused by inflow and/or recrystallization. Reprinted with permission from the *Second Symposium on Salt*, Holser (1966).

than 5 ppm (Holdoway, 1978). This bromide content is so low that it could not have been the result of a simple dilution with new seawater. Apparently, the uppermost parts of the Permian in Kansas are at least in part second-cycle salts (Fig. 4), which Kühn (1968, p. 455) has called "descendent salts." The occurrence of such second-cycle salts are naturally more likely at the top of a salt sequence, and they also require tilting or general uplift of a previously deposited salt in some part of the basin, in order to facilitate re-solution.

In some basins the bromide content is so extremely low, and continuously so, that it could not have been formed by even numerous cyclings of solution in seawater. Five hundred feet of salt core from the Rotliegendes Formation, in the Lower Permian underlying the Zechstein in the Heide Field, near Hamburg, Germany, all has bromide content in the range 2.5 to 4.5 ppm Br. This formation is surrounded by red beds and was already thought to be non-marine, and the source of the chloride may be in salt domes of the Devonian Pripiat and Donete basins (Holser, 1979). Several hundred feet of salt rock in the lower Elk Point Subgroup underlying

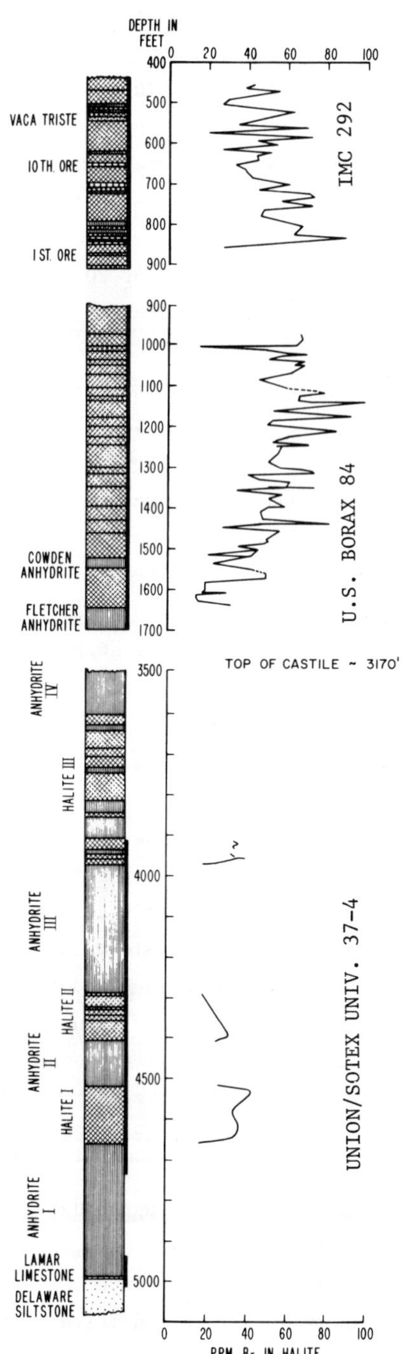

the Prairie Evaporite in Alberta, as shown in the lower part of Figure 8, is more surprising in its low average Br content of 4 ppm, minimum 1.8 ppm, as its non-marine origin had not previously been guessed (Holser *et al.*, 1972). Bromide contents in both of these salt sequences are analogous to those found in Tertiary to Recent salts of the interior Great Basin, USA, where the halides have apparently been derived by solution of marine salt rocks of the Paradox and other formations outcropping near the eastern edge of the basin (Holser, 1970; Eaton *et al.*, 1972).

Although not all non-marine salt rocks need necessarily have such low bromide contents, all those studied so far do have (except for very localized concentrations). So bromide contents below 20 ppm are generally indicative of re-solution, those below 10 ppm of re-solution in non-marine waters, and contents above 60 ppm Br are highly likely to be marine.

Bromide data from Permian Delaware Basin of Texas-New Mexico considerably strains our abiding faith in geochemistry. The profile of Figure 11 is representative

Figure 11. Bromide profile in the Permian Ochoan Series of the Delaware Basin, Texas-New Mexico. Castile section from detailed profile of one-foot samples that are nearly as smooth as this small-scale version. In contrast, the upper, Salado, salts are somewhat variable, perhaps indicating dessication as the basin filled. The whole profile is abnormally low in bromide.

of the extensive and well-documented data from many core and mine samples extending through several thousand feet of evaporites of both the Castile and Salado Formations. In general form the profiles are ordinary, starting from low values, building up slowly and steadily in several cycles, then more quickly and irregularly into the zones near the top where they move into the potash-magnesia facies (600-900 feet depth in Fig. 11). The anomaly is that the entire basin seems to be deficient in bromide by a factor of two or three. The profiles start at the low value of 20 ppm, are into the potash facies at less than 60-70 ppm Br, and even the sylvite and carnallite are correspondingly low in bromide. On the face of it, this would seem to indicate halite redissolved by at least seawater (Holser, 1966b, p. 262; Adams, 1969). Subsequently, we have made bromide analyses at one-foot intervals from the long core of Union Sotex University 37-4, 75 miles south and one-half mile below the previous main source of information in the Carlsbad potash district (described by Anderson *et al.*, 1972). The low values found there (smoothed data in Fig. 11) are not only completely consistent with the previous work in the Carlsbad district, but the profile in the regularly varved Castile Formation is unusually smooth, suggestive of uniform conditions in this deep basin. It seems difficult to derive such a large body of uniformly low-bromide salt by solution and redeposition of salt from the earlier Permian evaporites to the north, so the mystery is still unsolved.

An analogous method of study can be applied to Br analyses of brines, using the general concept laid out in Figure 4. Various Soviet authors, and more recently Rittenhouse (1967) and Collins (1967) in this country, have applied bromide analyses to deduce the origin of oil-field waters. The details of such relations, however, are best displayed in a modern evaporite, where the various changes can be caught in the act, or nearly so. We assisted D.J.J. Kinsman (Patterson and Kinsman, 1977) to demonstrate in the Persian Gulf area that the sabkha evaporites are marine on the seaward edge and non-marine on the landward edge. On the landward side, ground waters seeping from the desert area have undergone capillary concentration in the same way as the more commonly recognized supratidal marine sabkha deposits. A sequence of brines from these sediments, numbered approximately along a section from the lagoon (No. 1) to the interior (No. 15) is plotted in Figure 12. As discussed above, Cl/Br is sensitive to the

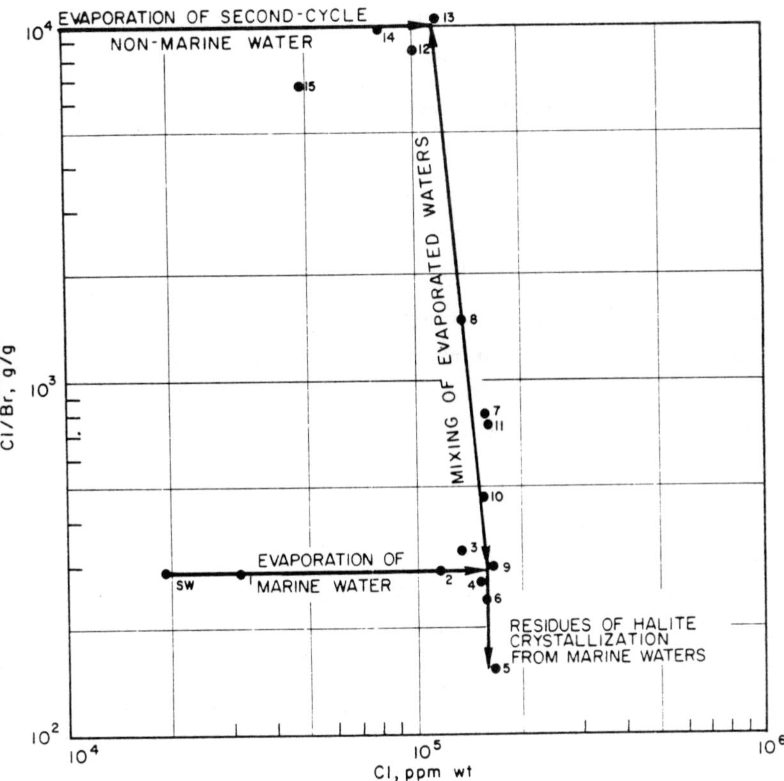

Figure 12. Use of bromide and chloride analyses to trace the history of evaporation and mixing in a modern sabkha, Persian Gulf. Sampling and chloride analysis from Patterson and Kinsman (1977).

origin of the water as marine (Cl/Br∿300) as second-cycle solution of marine salt (Cl/Br∿1000), and brine residues of halite crystallization (Cl/Br<250). Chloride concentration, plotted on the horizontal axis, is indicative of evaporation, up to halite saturation. Some of the conclusions that can be reached from this diagram are as follows:

1. Most of the brines can be explained as the result of evaporation and/or mixing of two sources of water.

2. The landward, non-marine water is second-cycle from the solution of marine halite, either old salt rocks or wind-blown halite from modern sabkhas, and *not* from wind-blown sea spray that is found in normal rain.

3. All waters seaward of number 7 are probably purely marine, and all those landward of number 11 are porbably purely nonmarine.

4. Four samples have a mixed marine-nonmarine origin.

5. All of these mixed waters are saturated or nearly so. This means either that mixing took place near the surface after evaporation, or that if mixing took place before evaporation, sampling was not sufficiently deep to obtain any mixed but not completely evaporated waters.

6. Two samples have low Cl/Br indicative of residues of halite crystallization. Although halite is not presently found as a permanent mineral of the sabkha, these numbers indicate that halite crystallization has been sufficiently important to segregate such brines, even though the halite may have been subsequently dissolved by fresher marine water.

The coordinates of Cl/Br and Cl in Figure 12 are similar to those used by Valiashko (1956). Chloride is preferable to the "total solids" used by Rittenhouse (1967) because it confines attention to geochemically related elements. Processes of evaporation, mixing, crystallization and various chemical reactions would of course be best confirmed by plots of additional components determined in the brines.

Rubidium Geochemistry

As an alkali element rubidium (also cesium) might be expected to substitute in evaporite minerals for another alkali element, potassium, which is nearest to it in ionic radius. As indicated in Table 1 its large distribution coefficient markedly concentrates Rb in the carnallite; but in sylvite the distribution coefficient is close to one. The distribution coefficient greater than one means that rubidium is grabbed up by carnallite in greater proportion to potassium than in the saturated solution in which the carnallite is growing. Consequently, rubidium is *impoverished* in succeeding brines and their carnallite products. This is dramatically demonstrated in Figure 13, a profile through the famous Stassfurt carnallite. (See also Kühn, 1972a,b.) In contrast, rubidium profiles in the Rhine Valley potash deposits are apparently strongly modified by solution and reprecipitation (Braitsch, 1966).

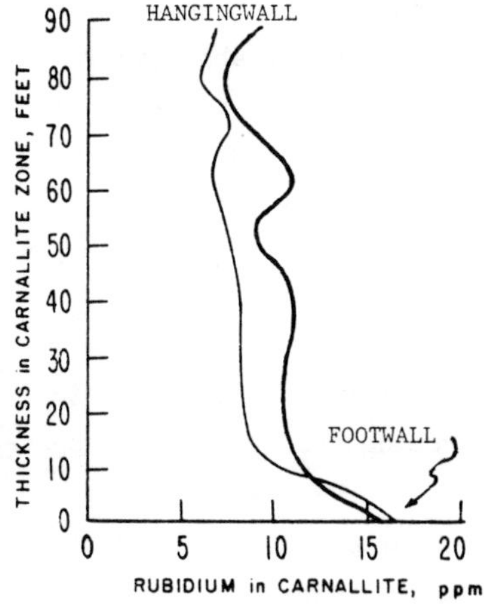

Figure 13. Rubidium profile through carnallite of the Stassfurt potash zone, at two levels of the Salzdetfruth mine, Germany (BRD). After Kühn (1963).

Table 2. Bromine and Rubidium Concentrations Characteristic of Various Histories of Marine Potash Salts (in ppm wt.) (after Kühn, 1968, p. 476)

Genetic Type (Kühn)	Description	Carnallite		Sylvite	
		Br	Rb	Br	Rb
Primary	Crystallization from sea water	3000-5000	250- 70	3000-4000	10- 50
Descendent	Crystallization from second-cycle brines	1700-3000	300-100	1500-3000	100-150
Secondary	Alteration at depth by brines and temperature	1000-2000	200-700	1000-2000	
Posthumous	Late alteration by ground water, e.g., cap-rock	1000-1700	190- 60		

Table 3. Characterization of Some Formation Waters from Germany by Bromide and Rubidium (after Kühn, 1963)

	Water Type	K/Rb, wt.	Cl/Br, wt.
(a)	Bitterns from carnallite precipitation	∞	65- 75
(b)	Brines from solution of carnallite rock	3.2- 420	25- 530
(c)	Brines from intensive alteration and potassium extraction ("Vertaubung") of potash beds	1300-5100	55- 160
(d)	Participation of fresh waters or nonsaline formation waters	>>1200	220-12,000
(e)	Transitions between (c) and (d); oil field waters	530-1100	75- 240

The contrasting distributions of bromine and rubidium can be used to characterize the origin of minerals in the potash facies, as shown in Table 2, taken from Kühn (1968). Of more direct consequence for us is the application of such analyses to determine the origin of formation waters that may be related to potash deposits. Kühn (1963) has applied this method to a list of formation waters with the results shown in Table 3 [p. 319].

Other Trace Elements

Although, as mentioned above, each element is distinctive, other elements have certain similarities in behavior to bromine and rubidium, respectively. Figure 14 summarizes the behavior of most of the trace elements that are readily detectable in brines evaporated from seawater. As would be expected from the above discussion, bromide is substantially accumulated in the brine because the distribution coefficient much less than one removes relatively little bromide; a noticeable flattening of the curve corresponds to significant removal of bromide during carnallite precipitation. In contrast to this, rubidium accumulates in the brine until carnallite begins to precipitate; thereafter, it decreases sharply as it is soaked up by the carnallite.

Lithium and boron are also accumulated in the brine, without even the flattening of the profile that was observed for bromide in the carnallite facies. Ham *et al.* (1961) give good geological and mineralogical evidence that perhaps 100 ppm of boron is taken up in anhydrite during its crystallization, but the amount is probably not sufficient to visibly affect the accumulation of boron in the brine. They find that the anhydrite → gypsum transformation results in the segregation of nodules of the calcium borate minerals proberite and ulexite, apparently because borate is less soluble in the gypsum form of calcium sulfate than it is in anhydrite. If boron accumulates in the manner of the experiment of Figure 14 and if primary anhydrite is being precipitated in both calcium sulfate and halite facies, boron should show a steady increase in anhydrites of these facies. In detailed studies of the Zechstein, Fabian *et al.* (1962) found no regular variations of boron through the anhydrite members. The final accumulation of boron should precipitate as magnesium borates in the final stages of the potash-magnesia facies (Valîashko, 1969). But various borate minerals are widely but sparsely distributed in all facies [including carbonate of

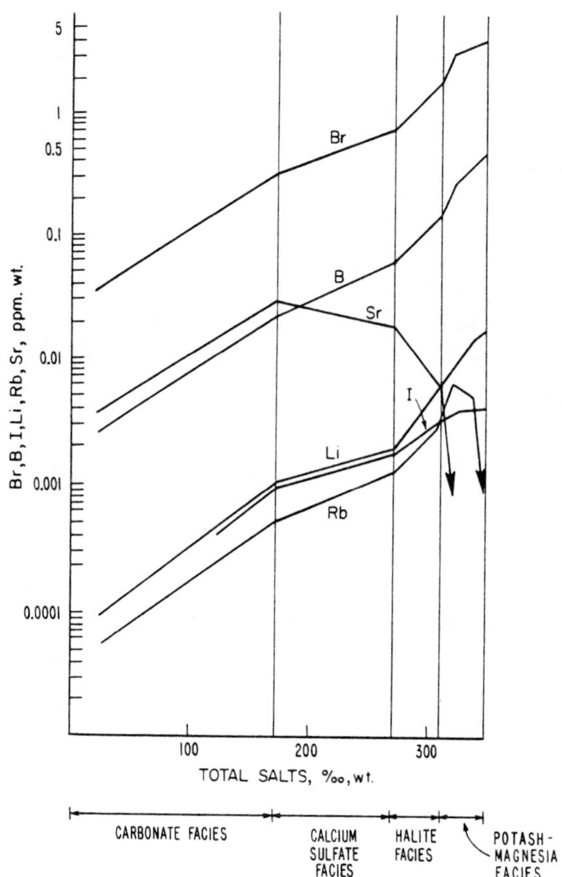

Figure 14. Behavior of trace elements in marine (Black Sea) water during evaporation. Data from Zherebtsova and Volkova (1966).*

evaporites in Germany (Braitsch, 1963; Kühn, 1968) and the Soviet Union (Yarzhemski, 1968)], and these minerals correspond to amounts of boron several orders of magnitude larger than would have been brought in by the amount of seawater needed to form the evaporites. The irregular distribution of boron may have been introduced by local volcanic rocks (Yarzhemskii, 1968), thermal waters (Kühn, 1968), or rivers fed by weathering of igneous rocks (Ozol, 1967). Within the range of variations dictated by such local sources, it may be that further study of boron geochemistry will furnish

*Figure 14 is plotted from Zherebtsova and Volkova's experimental data in their Table 1; their own plots, particularly their Figure 6, are inconsistent and apparently drawn erroneously in several respects.

information as to whether calcium sulfate was initially crystallized from the brine as gypsum or anhydrite. Boron geochemistry has been used extensively to determine the paleosalinity of brackish to marine waters in which shales were deposited. According to the most widely-held theory, the detrital illitic clays coming freshly from rivers soak up boron in proportion to the amount of borate-rich seawater that is mixed with the river water, but serious questions have been raised about this simple theory (Dewis *et al.*, 1972; Harder, 1974).

In evaporite basins, the amounts of such illitic clays, which should extract much of the boron (of whatever origin) from the brine, will be a further variable in the course of concentrating this element (Ozol, 1967). So far, nobody seems to have tried to use the boron content of illite in evaporites as a measure of salinity, and the complications that we have listed would make such studies difficult to interpret.

Figure 14 shows lithium rising regularly in brines, right through the potash-magnesia facies; the differences in slope at low concentrations, compared with bromine and boron, are certainly due to errors in determinations at extremely low concentrations. The distribution coefficient with all crystallizing salts is apparently so low that no lithium has been detected in either synthetic or natural marine salts. On the other hand, its regular accumulation in the brine may make it valuable in the study of formation waters. Its high energy of hydration and low energy of adsorption make it most likely to remain in solution during long distances of migration. As with boron, high contributions from thermal waters (White, 1957) are a complicating factor. Further study of lithium distribution in formation waters seems indicated.

Potassium, of course, accumulates in the brines up to the potash-magnesia facies. Hite (1974) has published a remarkable profile of potassium in halite (through 160 m of the Cretaceous Upper Khorat Series of Thailand) that shows a regular rise through the halite facies from 10 to 80 ppm K. Hite supposes that the K is included as a fairly regular amount of microscopic brine inclusions in the halite, but if we can believe the distribution coefficient of Puchelt (1972) (Table 1) this is about the amount to be expected in solid solution. Whether it is in solid solution or in a uniform content of fluid inclusions, it could be an additional useful indicator of the stage of evaporation of rocks within the halite facies. Distribution coefficients have been determined for K^+, Na^+, and

Mg^{2+} in aragonite (see Table 1), which have aided in determining the source of diagenetic solutions altering the aragonite to calcite (Kinsman, 1973), and in estimating paleosalinity (Laud and Hoops, 1973). The same elements have a regular distribution in extensive sections of rocks in the calcium sulfate (Dean and Anderson, 1974) and halite facies (Dean and Tung, 1974), and consequently are probably governed by solid solution distributions. Although a final interpretation of these data awaits determinations of the appropriate distribution coefficients, these trace elements apparently have potential for paleosalinity studies. However, some other results of Dean and Anderson (1974) suggest caution: They found that quickly deposited anhydrite (thick varve lamellae) had much higher trace element contents than slowly deposited anhydrite (thin lamellae). This indicates control by kinetic or adsorption effects rather than solid solution equilibrium, which would distort any salinity estimate.

Iodine also accumulates in brines, and, like lithium, its distribution coefficient into salt crystals (in this case as a substitute for chloride) is so low as to be hardly measurable. The flattening out of iodine concentration in brines of the potash-magnesium facies, as exemplified in Figure 14, is the result of another process. Iodine ion is much more easily oxidized (to the element) than bromide, particularly from the brines of high concentration. Consequently, the higher concentrations of iodine are removed by oxidization and evaporation into the atmosphere as iodine vapor. Iodine could be an important tracer in formation waters where even more than bromide, and for as vaguely defined reasons, it seems to be more concentrated in formation waters associated with oil fields than in those which do not have this association (Schoeller, 1955; Hitchon, 1974). However, this preference is only a statistical one, with such a high statistical dispersion that the use of iodine in formation waters to prospect for oil (Kartsev *et al.*, 1959) is not very definitive. Half of the world's supply of iodine comes from oil field waters, and concentrations greater than about 50 ppm are said to be of commercial value.

Thallium occurs in seawater at only about half the concentration of rubidium, and is partitioned into carnallite with a lower but still positive *D*. Consequently, it is impoverished less quickly than rubidium in the succeeding brines forming a carnallite deposit, as shown in Figure 15 (Malikova, 1967, p. 99). Thallium has not been analyzed in brines.

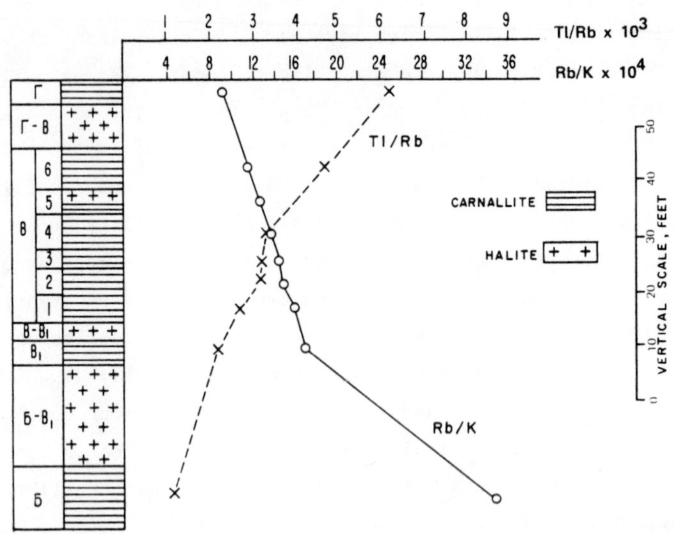

Figure 15. Variation of thallium and rubidium through the carnallite zone of the Verkhnekamsk potash deposits, Pre-Urals, USSR. After Malikova (1967, p. 99).

Most of the above trace elements are dominantly controlled by the fractionation as corresponding trace amounts in solid solution of the evaporite minerals that are crystallizing. Strontium undergoes similar distribution in both carbonate and sulfate minerals, as indicated by the distribution coefficients of Table 1. Dean (1978a) believes that most trace strontium in evaporite rocks is in the form of such solid solutions. However, a major complicating factor is the appearance of the mineral celestite, $SrSO_4$, with strontium as a major rather than trace element. When the solubility product of celestite has been exceeded, which is generally in the calcium sulfate facies,* from then on the concentration of Sr in the brine is controlled by the balance between celestite solubility and sulfate concentration. The latter is, in turn, controlled by a number of factors that may be of more or less importance in various evaporite sequences:

1. Evaporation may uniformly increase sulfate concentration.

---

*Unfortunately, even for the solubility of $SrSO_4$ in concentrated salt solutions, data in the literature are still in disagreement (Müller, 1960; Müller and Puchelt, 1961; Lucchesi and Whitney, 1962; Bount and Dickson, pers. comm., 1961).

2. Normal precipitation of gypsum and anhydrite will take up some sulfate but still allow it to increase in the brine, as shown by data from both artificial (Gravau, 1920; Herrmann *et al.*, 1973) and natural salt pans (Morris and Dickey, 1957; W.T. Holser, unpublished data from Laguna Ojo de Liebre, B.C., Mexico); consequently, Sr^{2+} decreases in the brine, as is the case in Figure 14.

3. Dolomitization by the magnesium-rich brines of (a presumably large volume of) previously precipitated aragonite or calcite, releases Ca^{2+}, which precipitates additional anhydrite (or gypsum), depresses SO_4^{2-} nearly to zero, and allows Sr^{2+} to rise in the brine (Kinsman, 1966, 1969b; Butler, 1970).

4. Biological reduction of sulfate to sulfur, or any other mechanism leading to the common sulfate deficit in evaporites, will have an effect parallel to dolomitization.

5. Organisms may segregate highly variable amounts of Sr^{2+} in shell or skeletal growths.

6. Strontium may be taken up by clays (Bausch, 1965). These Rube Goldberg mechanisms of variation in the brine will, in turn, be mirrored through the distribution coefficients by the strontium concentrations in minerals precipitated from each brine. Butler (1973) has done a masterful job of untangling these complexities in the environments of modern sabkha evaporites, by a detailed comparison of strontium analyses in gypsum and anhydrite of various forms and subfacies, with the brines found in association with them. The strontium in brine, and consequently in crystals, is controlled by different factors in each zone of seawater concentration, as outlined in Table 4. The sum of the concentrations of these minerals, plus the amount of celestite, gives the strontium analysis of the primary sediment.

The situation is further complicated by post-depositional redistribution of strontium and other elements, especially its association with mineral transformations: aragonite-calcite (Katz *et al.*, 1972), high-Mg calcite--low-Mg calcite, calcite-dolomite, gypsum-anhydrite, and anhydrite-dolomite. The first two reactions have been studied extensively, and in

Table 4. Strontium Geochemistry in a Modern Marine Sabkha Evaporite of the Calcium Sulfate Facies (data from Butler, 1973).

Subfacies	Evaporation Ratio	Sr^{2+}/Ca^{2+} in Brine (Mass)-Range	Sr^{2+} in Crystals, ppm	Controls
A. Lagoon to low upper intertidal	1.2→3.4	Constant 0.0185	Aragonite, 7400; calcite 1200	Aragonite precipitating, with D_{Sr}^{Ar}=1.0.
B. Intermediate upper intertidal	3.4→3.8	.0185→.022	Gypsum, 700→900	Gypsum beginning to precipitate, with D_{Sr}^{G}=0.18.
C. High upper intertidal	3.8→6.0	0.22→0.36	Gypsum, 900→1500	Gypsum precipitation dominant in brines superconcentrated by solution of evanescent halite, celestite precipitation suppressed, so Sr^{2+}/Ca^{2+} rises.
D. High upper intertidal	6.0→7.0	.036→.022	Gypsum, 1500→900; dolomite 660	High Mg^{2+} dolomitizes aragonite, released Ca^{2+} decreases Sr^{2+}/Ca^{2+} and precipitates gypsum as well as celestite.
E. Low supratidal	3.6→8.0	Constant .022	Gypsum 900; anhydrite 2400	Gypsum or anhydrite precipitation held down by lack of dolomitization, Sr^{2+}/Ca^{2+} balanced by celestite and discoid gypsum mush crystallization.
El. Low to intermediate supratidal	7.0→8.0	.022→.010	Gypsum 900→400; anhydrite 2400-1000	As in D, but usually with nodular anhydrite instead of gypsum.

general the changes in trace element content take an expected direction: For elements whose distribution coefficient (Table 1) in the product is less than one (e.g., Sr^{2+}, Mg^{2+}), the trace element content decreases, while for elements whose distribution coefficient is greater than one (e.g., Mn^{2+}, Zn^{2+}) the trace element content increases (Kinsman, 1969b, 1971; Veizer, 1974, 1977). Determination as to whether the system remained essentially open (Kinsman, 1971; Gavish and Friedman, 1969) or closed (Lohmann, 1978; Veizer, 1978) during alteration is critical to evaluation of the analyses. The analogous reactions involving sulfate minerals have been much less studied. In a modern sabkha, dolomitization removes the strontium from anhydrite as celestite (Kinsman, 1969a). Similarly, strontium is less soluble in gypsum than in anhydrite (Table 1), so that secondary celestite forms during surface alteration of anhydrite to gypsum (Ham, 1962). On this basis one should be able to distinguish primary anhydrite (high Sr) from anhydrite after primary gypsum (low Sr), but this hypothesis needs to be verified. In a contary example, an *increase* in trace elements (including Sr and B), was observed during gypsification of anhydrite; the postulated source was the altering ground water (Kropacher, 1960). Some regularities in the distribution of strontium have been observed in anhydrite rock sequences, both with and without celestite (Herrmann, 1961; Mokeyenko, 1965; Butler, 1973; Dean and Anderson, 1974); but many variations remain unexplained (Wazny, 1971).

ISOTOPES

General Considerations

The same general principles that have been described for the distribution of trace elements in solid solution also apply to isotopes. Both equilibrium and kinetic effects give slight preferences for lighter or heavier isotope species in particular crystals, waters, or the atmosphere, and in the processes of chemical reaction and diffusion that operate in these regimes. Logarithmic profiles like equation 3 and Figure 2 will result as such a process goes on. The final result is often complex, and this discussion will not go into detail, but will only indicate the ways that isotopes may be useful in the study of evaporites and brines, and some of the results so far obtained. A more general survey of isotope geochemistry may be found in the textbook by Faure (1977).

Compared with the trace element-regular ion pair, two isotopes of the same element are very nearly equal in all their properties. Consequently, the processes that separate isotopes in nature are very subtle, and the distribution coefficients for isotope pairs are very close to one. It is therefore more convenient to designate the effect by a fractional difference from some standard material, rather than by a ratio.

The fraction difference δ is usually expressed in parts per thousand, or "per mil." For example, for oxygen isotopes of masses 18 and 16:

$$\delta^{18}O = \frac{(^{18}O/^{16}O)_{sample} - (^{18}O/^{16}O)_{standard}}{(^{18}O/^{16}O)_{standard}} \times 1000 \qquad [9]$$

So positive values of δ mean that the isotopes in the sample are slightly heavier than the arbitrary standard,* negative values that the sample is slightly lighter.

Hydrogen and Oxygen in Water

One of the most obvious factors affecting the isotopic composition of both hydrogen and oxygen in water is that the major evaporation cycle leaves behind in the sea, water heavy in both isotopes, while atmospheric water vapor and the snow and rain that falls from them are variably light depending upon latitude, elevation, and history of the air mass. *Mixing* of such precipitated fresh waters with ocean waters gives the main line of variation shown on Figure 16. On the other hand, seawater left behind by strong *evaporation* gives the upper right curve of Figure 16, with much less increase in deuterium. Although D and ^{18}O are left behind during the initial slight evaporation, as might be expected on a simple model, at high salt concentrations they both begin to decrease again, and the plot of D *vs* ^{18}O hooks back (Fig. 16). This reversal is said (Gonfiantini, 1965) to be accounted for by the decrease in activity of water in highly saline brines as it affects the relative rates of evaporation and condensation at the surface, but I suggest that interaction between free water and the large amount of water in the hydration spheres of the cations in a brine may also play a part (Taube, 1954). The process has been followed

---

*The standards of reference usually used (including the following discussion) are: for carbon, "PDB" (Belemnitella from the PeeDee Formation, South Carolina); for oxygen and hydrogen, "SMOW" (Standard Mean Ocean Water); and for sulfur, "CD" (Troilite from the Canyon Diablo meteorite).

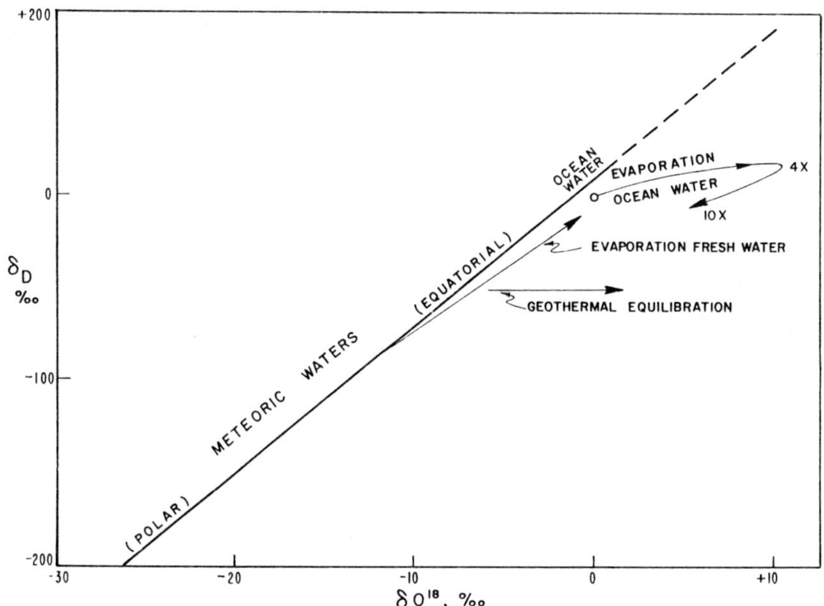

Figure 16. Variation of oxygen and hydrogen isotopes in waters. After Craig (1961) and Gonfiantini (1965). For variations on $\delta^{18}O$ for seawater evaporation see Lloyd (1966), and Fontes and Pierre (1978).

experimentally (as shown in Fig. 16) up to 10X seawater, and our own data from highly concentrated brines at Laguna Ojo de Liebre, Baja California (Fig. 17), demonstrate for hydrogen that the process continues up to the potash-magnesia facies. Analysis of water hydrogen from evaporite minerals by Borshchevskiy and Khristianov (1965) are also in the range δD +25 to -40°/oo, and they cite other published data in support of such changes.

Gonfiantini and Fontes (1963) have determined that oxygen in water of crystallization of gypsum is about 4°/oo heavier than the water from which the gypsum crystallized. The fractionation of hydrogen into water of crystallization has been determined for several nonmarine evaporite minerals: For gaylussite, $Na_2Ca(CO_3)_2 \cdot 5H_2O$, the crystalline hydrogen was slightly lighter, and for (total hydrogen) trona, $Na_3CO_3(HCO_3) \cdot 2H_2O$, considerably lighter; borax, $Na_2B_4O_7 \cdot 10H_2O$ did not appreciably fractionate (Matsuo *et al.*, 1972). This data base was applied to interpret the crystallization temperature of Searles Lake, California, evaporites (Smith *et al.*, 1970). Crystallization with light hydrogen, as found here, is in line with experimental work on a series of other, nonmineral hydrates (Barrer and Denny, 1964). In contrast, comparison of the very preliminary data from the two different sources shown in Figure 17 suggest that the crystal water in gypsum may have heavier hydrogen. Borschevskiy and Khristianov (1965) analyzed deuterium in water of hydration of a few evaporite minerals, but without comparison to either a standard of analysis or to the water from which they were crystallized. Any further interpretations of hydrogen isotopes in gypsum awaits a definitive laboratory experiment.

Sulfur and Oxygen in Sulfate and Sulfide

The general features and much of the data on sulfur isotopes in evaporites have been reviewed repeatedly (e.g., Thode and Monster, 1965;

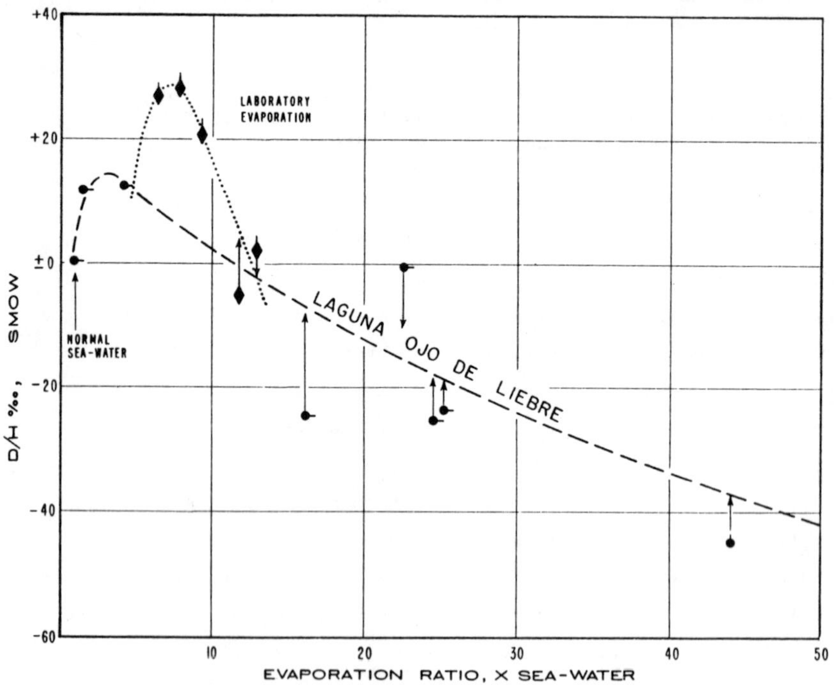

Figure 17. Variation of hydrogen isotopes in evaporite brines in the sabkha at Laguna Ojo de Liebre, Baja California, Mexico (dashed line), and in gypsum from a laboratory evaporation of seawater at constant volume. Isotope analyses by S. R. Silverman.

Holser and Kaplan, 1966; Nielsen, 1972; Dean, 1978b; Claypool *et al.*, 1979). Oxygen isotopes in the sulfate cycle were dealt with in detail by Holser *et al.* (1979). The principal geochemical feature that governs the distribution of sulfur isotopes is bacterial reduction on the sea-floor, which forms a light sulfide in the muds and leaves behind sulfur ^{34}S about +40 heavier in the sulfate of seawater. Evaporite sulfates, including not only gypsum and anhydrite but minerals of the potash-magnesia facies, crystallize from concentrated seawater brines with only a minor fractionation of $\delta^{34}S$ of $+1.7^{o}/oo$ $\delta^{18}O$ increases by +3.5% (Lloyd, 1968). Butler *et al.* (1973) and Olson and Schwarz (1979) have documented decreases of up to $4^{o}/oo$ in modern sabkhas. But in the absence of large-scale

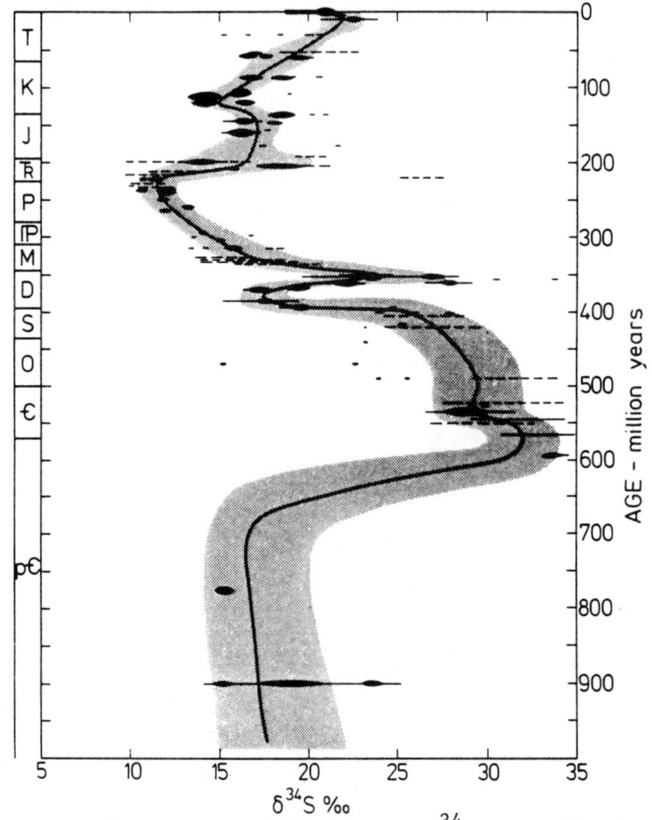

Figure 18. Isotope age curve for $\delta^{34}S$ in evaporite sulfate in equilibrium with the surface ocean. Note sharpness of rises in Late Precambrian, Middle Devonian, and Early Triassic, each of which would be even more striking if divergently high values were not discounted in drawing the mean curve. A compilation of published data and that to be published in Claypool *et al.* (1979).

biological fractionation in the evaporite basin, evaporite sulfates will generally have just about the same sulfur isotope composition as the brine in the basin, and for dominantly marine evaporites that means the same as in the world surface ocean. One might therefore have expected little variation in sulfur isotopes from evaporite minerals.

Surprisingly enough, wide variations of $\delta^{34}S$ were found with geological age of the evaporites, as shown by the "sulfur isotope age curve" of Figure 18. These variations apparently resulted from a shift with geologic time of the balance of sulfur stored in three main reservoirs: seawater, evaporites, and shales. The first two reservoirs are sulfate and heavy, and the third is sulfide and light. Holser and Kaplan (1966) proposed that during times in which a greater amount of sulfur is transferred to the sulfate of the sea by uplift, erosion, and oxidation of pyrite in shales than is transferred to the sulfide reservoir by the activity of sulfur-reducing bacteria on the seafloor, $\delta^{34}S$ in the world ocean should decrease, and vice-versa. Variations on the interpretation of Figure 18 have been published by Rees (1970) and Holland (1973), involving transfers into and out of the evaporite reservoir. Analyses of oxygen isotopes in many of the same anhydrites of Figure 18 also show excursions of $\delta^{18}O$ with age (Sakai, 1972; Claypool *et al.*, 1979). By playing the differences of these two curves against each other, it should in principal be possible to calculate shifts of sulfur among all three reservoirs (seawater, evaporites, and shales) with time (Claypool *et al.*, 1979). However, the rises of $\delta^{34}S$ (just below the Cambrian, in the Middle Devonian, in the Lower Triassic, and possibly in the Lower Cretaceous) are so sharp that they are difficult to explain by a sudden increase in bacterial reduction of oceanic sulfate to sulfide. This has led to speculation that the $\delta^{34}S$ of oceanic sulfate does not necessarily measure $\delta^{34}S$ of the entire ocean--that it may at these times have measured only the surface mixed zone of an ocean that was stratified by the accumulation of dense reflux brines in the deeps (Holser, 1977). The isotope age curves of sulfur and oxygen in evaporites have great potential as a record of the sediment-ocean-atmosphere cycle on a world-wide scale.

These interpretations are still controversial, but whatever the origin of the curve of $\delta^{34}S$ *vs* age, it can be useful in determining the age of sediments. An age is not given in a direct and unambiguous fashion as it would be by radiometric determinations, because the sulfur isotope age curve is both multi-valued and fuzzy. But in certain instances the age of an evaporite can be refined by $\delta^{34}S$. Some unpublished examples from my file: The Couva Marine anhydrite, Gulf of Paria, Trinidad, was shown to be pre-Upper Cretaceous, later confirmed by paleontology; the Ft. Dodge gypsum, in Iowa, was shown to be Jurassic-Traissic rather than Permian, later proved by palynology; evaporites in the Canning Basin, Australia, were shown to be Devonian. The sulfate ion is a very stable unit, and consequently, in the absence of any biological reduction, second-cycle sulfates reflect their first-cycle derivation from the ocean. Thus, anhydrite filling porosity in the San Andres Limestone in Texas was shown to be of Permian and not of later age. Further examples are given by Nielsen (1972).

This same stability of the sulfate ion also allows sulfur isotopes (probably oxygen also) to trace the origin of sulfate contents of formation waters (Muller *et al.*, 1966) and surface waters (Nielsen and Rambow, 1969; Hitchon and Kraus, 1972), and in particular to determine the relative importance of evaporites (even *which* evaporites) in their origin.

Very light $\delta^{34}S$ deposits of native sulfur in salt dome cap rock (Feely and Kulp, 1957) and in bedded evaporites (Davis and Kirkland, 1970) were formed by bacterial reduction of evaporite sulfate to H_2S and back-oxidation by oxygenated waters to native sulfur.

Carbon and Oxygen in Carbonates and Organic Carbon

The isotope geochemistry of carbonate has been studied more than any other. Among the isotopes of carbon, the greatest natural distinction occurs between the carbon of living matter (both plant and animal) and inorganic carbon. In general, ordinary limestones, including fossil shell material, have δC in the range $\delta C^{13} = -1$ to $+3^{o}/oo$ which is consistent with the bicarbonate of modern seawater, which is in turn in equilibrium with atmospheric CO_2 at $\delta^{13}C = -7^{o}/oo$. Organic material in sediments, including hydrocarbons, is in the range $\delta^{13}C = -20$ to $-30^{o}/oo$, with a mean of about $\delta^{13}C = -28^{o}/oo$. Some limestones, such as in salt dome cap rock (Feely and Kulp, 1957), are evidently formed from CO_2 that resulted from the oxidation of such organic matter, rather than from the ordinary carbonate of seawater that is in equilibrium with the atmosphere. Replacement of varve lamellar anhydrite by calcite was demonstrated on the same basis

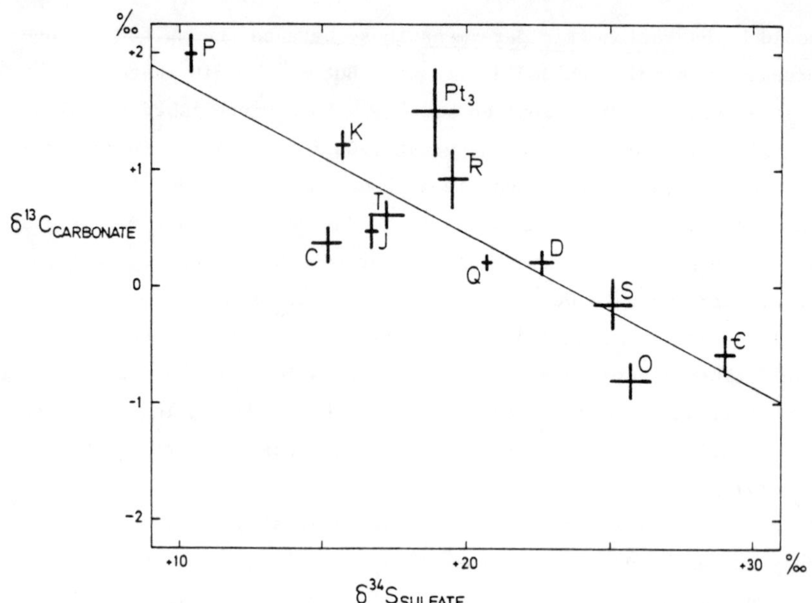

Figure 19. Scatter diagram of $\delta^{13}C_{carbonate}$ *vs* $\delta^{34}S_{sulfate}$ in marine rocks; all published data aggregated by geological period. Bars represent two standard deviations of the mean for a period. From Veizer *et al.* (1979).

by Kirkland and Evans (1976). Except for limestones of such unusual origin, most of those in the geological column, back to the earliest Precambrian, lie within the above-mentioned limits. But within this narrow range there is some consistency to variation of $\delta^{13}C$ with age. A statistical summary of all published carbon isotope analyses of limestones (Fig. 19) not only displays variations with age but also exhibits a negative correlation of $\delta^{13}C_{carbonate}$ with $\delta^{34}S_{sulfate}$ (Veizer *et al.*, 1979). The significance of this relation is that the oxidation-reduction processes of the carbon and sulfur cycles are tied together in some way that makes them mutually compensating. Consequently, the effect of the oxidation-reduction processes on the level of oxygen (plus carbon dioxide) in the atmosphere is minimized. Nonmarine limestones lie between ocean water and organic material with an average of perhaps $\delta^{13}C = -7$. However, $\delta^{13}C$ is not a salinity indicator, *per se*, and carbonates from evaporites, so far as have been studied, have essentially the same $\delta^{13}C$ as marine limestones.

The $\delta^{18}O$ of carbonates is often used to measure the temperature of crystallization, particularly of fossils, assuming normal marine water with

$\delta^{18}O_{H_2O}$ and $\delta^{18}O_{HCO_3^-}$ in equilibrium. But under evaporite conditions the temperature effect is overwhelmed by the effect of salinity on $\delta^{18}O_{H_2O}$ (Fig. 16). It has been generally asserted that most $\delta^{18}O_{CaCO_3}$ in limestones older than Certaceous were extremely altered by equilibration with later formation waters, but recent work has found a wide range of values in single samples that can be related to primary and diagenetic sequence of crystallization (Dickson and Coleman, 1978). Dean and Anderson (1978) published one profile, a series of varve lamellar anhydrite rocks of the Permian Castile Formation of Texas in which low salinity as evidenced by low $\delta^{18}O_{CaCO_3}$ was correlated with thin anhydrite lamellae. Other than this study, oxygen isotopes have not been much applied to evaporite carbonates. Carbon and oxygen isotope studies are important in determining the origin of dolomite, a perennial question that is pertinent here because dolomite is the most common carbonate in evaporites. Experimental studies indicate that primary dolomite should have $\delta^{18}O$ about 5 to $7^{o}/oo$ heavier, and $\delta^{13}C$ about $2.5^{o}/oo$ heavier, than calcite formed under the same conditions (e.g., Hoefs, 1973, p. 99). Protodolomite (dolomite in a disordered structure when first precipitated) gives $\delta^{18}O$ about 3 to $4^{o}/oo$ heavier than calcite (Fritz and Smith, 1970). Most dolomite/calcite pairs that have been studied have closely similar $\delta^{18}O$ and $\delta^{13}C$, suggesting that these dolomites formed by diagenetic alteration of a calcite (or aragonite) precursor; these include many dolomites of somewhat evaporitic association in both modern (Clayton *et al.*, 1968b) and ancient sediments (Degens and Epstein, 1964). However, in exceptional cases, as in the modern dolomite sediments at Deep Springs Valley, California (Clayton *et al.*, 1968a), at Mound Lake, Texas (Parry *et al.*, 1970), and in the Eocene evaporites of the Paris Basin (Fontes *et al.*, 1967, 1978) dolomites are found with the high $\delta^{18}O$ suggestive of direct precipitation. The conclusion, which must be tentative, is that most dolomites, even those of evaporitic associations, are diagenetic after $CaCO_3$, but that a few are probably precipitates. So far all examples of the precipitates are in dominantly continental evaporites, which is either a strange coincidence or a suggestion of factors not presently understood. For the present, these questions must remain open.

By combining data on clay minerals, trace elements, and oxygen and carbon (but not hydrogen) isotopes in both carbonates and gypsum, Fontes *et al.* (1967) have completed a definitive study of the upper Eocene evaporitic sediments of the central Paris basin. The data all point to crystallization of the evaporites from continental waters. Similar studies promise to be useful in other evaporite basins, but of course only in those cases where gypsum is primary. In most older evaporites, the study of isotope variations in outcrop gypsums will be pertinent only to considerations of the type of ground water that has generated the gypsum from buried anhydrite.

TRACE ELEMENTS AND ISOTOPES IN EVAPORITES: USEFUL GENERAL REFERENCES

Borisenkov, V.I., M.G. Valiashko, A.P. Vinogradov, I.I. Volkova, and I.K. Zherevtsova (eds.) (1976) *Bromide in Saline Deposits and Solutions as a Geochemical Indicator of Genesis, History, and Prospecting Indications*. Izd. Moskovskogo Universiteta, 453 p.

Dean, W.E. (1978) Some uses of stable isotopes of carbon, oxygen, and sulfur in solving problems related to evaporite deposits. *Soc. Expl. Paleontol. Mineral. Short Course 4*, 124-143.

Dean, W.E. and B.C. Schreiber (eds.) (1978) Marine Evaporites. *Soc. Expl. Paleontol. Mineral. Short Course Notes 4*, 193 p.

Faure, G. (1977) *Isotope Geology*. Wiley, New York, ch. 18-21.

Fredrickson, A.F. (1962) Partition coefficients--new tool for studying geological problems. *Am. Assoc. Petrol. Geol. Bull. 46*, 518-528.

Holser, W.T. (1966) Bromide geochemistry of salt rocks. *In*, J.L. Rau (ed.), *Second Symposium on Salt*, Northern Ohio Geological Society, Cleveland, v. 1, p. 248-275.

Kühn, R. (1968) Geochemistry of the German potash deposits. *Geol. Soc. Am. Spec. Pap. 88*, 427-504.

McIntire, W.L. (1963) Trace element partition coefficients--a review of the theory and application to geology. *Geochim. Cosmochim. Acta 27*, 1209-1264.

TRACE ELEMENTS AND ISOTOPES IN EVAPORITES: REFERENCES

Adams, S.S. (1969) Bromine in the Salado Formation, Carlsbad potash district, New Mexico. *N. Mex. Bur. Mines Mineral Res. Bull.*, *93*.

Anderson, R.J. and G.C. Egleson (1970) Discovery of potash in the A-1 Salina Salt in Michigan. *Mich. Geol. Surv. Miscellany, 1, 33p.*

Anderson, R.Y., W.E. Dean, Jr., D.W. Kirkland, and H.I. Snider (1972) Permian Castile varved evaporite sequence, West Texas and New Mexico. *Geol. Soc. Amer. Bull.*, *83*, 59-86.

Barrer, R.M. and A.F. Denny (1964) Water in hydrates. Part 1. Fractionation of hydrogen isotopes by crystallization of salt hydrates. *J. Chem. Soc. London*, *1964*, 4677-4684.

Bausch, W.M. (1965) Strontiumgehalte in Süddeutschen Malmakalken. *Geol. Rundschau*, *55*, 86-96.

Bloch, M.R. and J. Schnerb (1953) On the Cl^-/Br^- -ratio and the distribution of Br-ions in liquids and solids during evaporation of bromide-containing chloride solutions. *Bull. Res. Counc. Israel*, *1*, 151-158.

Bodine, M.W., H.D. Holland, and M. Borcsik (1965) Co-precipitation of manganese and strontium with calcite. In *Problems of Postmagmatic Ore Deposition*, Geological Survey of Czechoslovakia, Prague, V. 2, 401-406.

Borshchevskiy, Yu. A. and V.K. Khristianov (1965) Isotopic composition of water of crystallization in evaporate minerals. *Geokhimiya, 1965, 844-850* [transl. *Geochem. Internat. 1965*, 653-659].

Braitsch, O. (1963) The temperature of evaporate formation. In, A.E.M. Nairn (Ed.) *Problems of Palaeoclimatology*. Interscience, London, 479-531.

_____ (1966) Bromine and rubidium as indicators of environment during sylvite and carnallite deposition of the Upper Rhine Valley evaporites. In J.L. Rau (Ed.), *Second Symposium on Salt*. Northern Ohio Geol. Soc., Cleveland, V. 1, 293-301.

_____ and A.G. Herrmann (1963) Zur Geochemie des Broms in salinaren Sedimenen. Teil I: Experimentelle Bestimmung der Br-Verteilung in verschiedenen natürlichen Salzsystemen. *Geochim. Cosmochim. Acta*, *27*, 361-391.

Brauer, P. (1953) Thallium in alkali halides. *A. Naturforsch. 8A*, 273-274.

Butler, G.P. (1970) Holocene gypsum and anhydrite of the Abu Dhabi sabkha, Trucial Coast: An alternative explanation of origin. In J.L. Rau and L.R. Dellwig (Eds.), *Third Symposium on Salt*. Northern Ohio Geol. Soc., Cleveland, V. 1, 120-125.

_____ (1973) Strontium geochemistry of modern and ancient calcium sulphate minerals. In B.H. Purser (Ed.), *The Persian Gulf*. Springer-Verlag, New York, 423-452.

_____, R.H. Crouse, and R. Mitchell (1973) Sulphur-isotope geochemistry of an arid, supratidal evaporite environment, Trucial Coast. In B.H. Purser (Ed.), *The Persian Gulf*. Springer-Verlag, New York, 453-462.

Claypool, G.C., W.T. Holser, I.R. Kaplan, H. Sakai and I. Zak (1979) The age curves of sulfur and oxygen isotopes in marine sulfate and their mutual interpretation. *Chem. Geol.* (submitted).

Clayton, R.N., B.F. Jones and R.A. Berner (1968a) Isotope studies of dolomite formation under sedimentary conditions. *Geochim. Cosmochim. Acta, 32,* 415-432.

_____, H.C.W. Skinner, R.A. Berner and M. Rubinson (1968b) Isotopic composition of recent South Australian lagoonal carbonates. *Geochim. Cosmochim. Acta, 32,* 983-988.

Collins, A.G. (1967) Geochemistry of some Tertiary and Cretaceous age oil-bearing formation waters. *Environ. Sci. Technol., 1,* 725-730.

Craig, H. (1961) Isotopic variations in meteoric waters. *Science, 133,* 1702-1703.

Crockett, J.H. and J.W. Winchester (1966) Coprecipitation of zinc with calcium carbonate. *Geochim. Cosmochim. Acta, 30,* 1093-1109.

Davis, J.B. and D.W. Kirkland (1970) Native sulfur deposition in the Castile Formation, Culberson County, Texas. *Econ. Geol., 65,* 107-121.

Dean, W.E. (1978a) Trace and minor elements in evaporates. *Soc. Explor. Paleontol. Mineral. Short Course Notes, 4,* 86-104.

_____ (1978b) Some uses of stable isotopes of carbon, oxygen, and sulfur in solving problems related to evaporite deposits. *Soc. Explor. Paleontol. Mineral. Short Course Notes, 4,* 125-145.

_____ and R.Y. Anderson (1974) Trace and minor element variations in the Permian Castile Formation, Delaware Basin, Texas and New Mexico, revealed by varve calibration. In A.H. Coogan (Ed.), *Fourth Symposium on Salt.* Northern Ohio Geol. Soc., Cleveland, V. 1, 275-285.

_____ and _____ (1978) Salinity cycles: evidence for subaqueous deposition of Castile Formation and Lower part of Salado Formation, Delaware Basin, Texas and New Mexico. *New Mex. Bur. Mines Mineral. Res. Circ., 159,* 15-20.

_____ and A.L. Tung (1974) Trace and minor elements in anhydrite and halite, Supai Formation (Permian), East-central Arizona. In A.H. Coogan (Ed.), *Fourth Symposium on Salt.* Northern Ohio Geol. Soc., Cleveland, V. 1, 187-199.

Degens, E.T. and S. Epstein (1964) Oxygen and carbon isotope ratios in coexisting calcites and dolomites from Recent and ancient sediments. *Geochim. Cosmochim. Acta, 28,* 23-44.

Dewis, F.J., A.A. Levinson and P. Bayliss (1972) Hydrogeochemistry of the surface waters of the MacKenzie River drainage basin, Canada--IV. Boron-salinity-clay mineralogy relationships in modern deltas. *Geochim. Cosmochim. Acta, 36,* 1359-1375.

Dickson, J.A.D. and M. Coleman (1978) Changes in carbon and oxygen isotope values during the deagenesis of Lower Carboniferous grainstones. *Tenth Intern. Cong. Sediment. Abstr., 1,* 178-179.

Eaton, G.P., D.L. Peterson, and H.H. Schumann (1972) Geophysical, geohydrological, and geochemical reconnaissance of the Luke salt body, central Arizona. *U. S. Geol. Surv. Prof. Pap., 753,* 28 p.

Fabian, H.J. and G. Klenert (1962) Der Bor-Gehalt der Zechstein-Anhydrite. *Erdöl Kohle Erdgas Petrochem., 15,* 603-606.

Faure, G. (1977) *Isotope Geology.* Wiley, New York, 464 p.

Feely, H.W. and J.L. Kulp (1957) Origin of Gulf coast salt-dome sulphur deposits. *Amer. Assn. Petroleum Geol. Bull., 41,* 1802-

Fontes, J.C., P. Fritz, J. Gauthier, and G. Kulbicki (1967) Minéraux argileaux, éléments-traces et compositions isotopiques ($^{18}O/^{16}O$ et $^{13}C/^{12}C$) dans les formations gypsiféres de l'Éocène Supérieur et de l'Oligocéne de Cormeillesen-Parisis. *Bull. Centre Rech. Pau SNPA, 1,* 315-366.

_____, and C. Pierre (1978) Oxygen 18 changes in dissolved sulphate during sea water evaporation in saline ponds. *Tenth Intern. Cong. Sediment. Abstr., 1,* 215-216.

Fritz, P. and D.G.W. Smith (1970) The isotopic composition of secondary dolomites. *Geochim. Cosmochim. Acta, 34,* 1161-1173.

Fuller, J.G.C.M. and J.W. Porter (1969) Evaporites and carbonates; two Devonian basins of western Canada. *Bull. Can. Petrol. Geol., 17,* 182-193.

Gavish, E. and G.M. Friedman (1969) Progressive diagenesis in Quaternary to Late Tertiary carbonate sediments: Sequence and time scale. *J. Sed. Petrol., 39,* 980-1006.

Glover, E.D. and R.F. Sippel (1967) Synthesis of magnesian calcites. *Geochim. Cosmochim. Acta, 31,* 603-613.

Gonfiantini, R. (1965) Effetti isotopi nell'evaporazione di acque salate. *Atti Soc. Tosc. Sci. Nat. Ser. A, 72,* 1-22.

_____ and J.C. Fontes (1963) Oxygen isotopic fractionation in the water of crystallization of gypsum. *Nature 200,* 644-646.

Grabau, A.W. (1920) *Geology of the Nonmetallic Mineral Deposits Other Than Silicates, V. l, Principles of Salt Deposition.* New York: McGraw-Hill Book Co., Inc.

Ham, W.E. (1962) Economic geology and petrology of gypsum and anhydrite in Blaine County. *Okla. Geol. Surv. Bull., 89,* 100-151.

_____, C.J. Mankin, and J.A. Schleicher (1961) Borate Minerals in Permian gypsum of westpcentral Oklahoma. *Okla. Geol. Surv. Bull. 92.*

Harder, H. (1974) Boron. In K.H. Wedepohl (Ed.), *Handbook of Geochemistry, V. II/1, Sec. 5.* Springer-Verlag, Berlin.

Herrmann, A.G. (1961) Zur Geochemie des Strontiums in den salinaren Zechsteinablagerungen der Stassfurt-Serie des Südharbezirkes. *Chemie Erde, 21,* 138-194.

_____ (1972) Bromine distribution coefficients for halite precipitated from modern sea water under natural conditions. *Contrib. Mineral. Petrol., 37,* 249-252.

_____, D. Knake, J. Schneider, and H. Peters (1973) Geochemistry of modern seawater and brines from salt pans: Main components and bromine distribution. *Contrib. Mineral. Petrol.*, *40*, 1-24.

Hitchon, B. (1974) Application of geochemistry to the search for crude oil and natural gas. In A.A. Levinson (Ed.) *Introduction to Exploration Geochemistry*. Applied Publishing Co., Calgary, 509-545.

_____ and H.R. Krouse (1972) Hydrogeochemistry of the surface waters of the Mackenzie River drainage Basin, Canada--III. Stable isotopes of oxygen, carbon and sulphur. *Geochim. Cosmochim. Acta*, *36*, 1337-1357.

Hite, R.J. (1974) Evaporite deposits of the Khorat Plateau, Northeastern Thailand. In A.H. Coogan (Ed.), *Fourth Symposium on Salt*, Northern Ohio Geol. Soc., Cleveland, V. 1, 135-146.

Hoefs, J. (1973) *Stable Isotope Geochemistry*. Springer-Verlag, Heidelberg, 140 p.

Holdoway, K. (1978) Deposition of evaporites and red beds of the Nippewalla Group, Permian, Western Kansas. *Kans. Geol. Surv. Bull.*, *215*, 43 p.

Holland, H.D. (1973) Systematics of the isotopic composition of sulfur in the oceans during the Phanerozoic and its implications for atmospheric oxygen. *Geochim. Cosmochim. Acta*, *37*, 2605-2626.

_____, H.J. Holland, and J.L. Munoz (1964) The coprecipitation of cations with $CaCO_3$--II. The coprecipitation of Sr^{2+} with calcite between 90 and 100°C. *Geochim. Cosmochim. Acta*, *28*, 1287-1302.

Holser, W.T. (1966a) Diagenetic polyhalite in Recent salt from Baja California. *Amer. Mineral.*, *51*, 99-109 [reprinted in D.W. Kirkland and R. Evans (Eds.) (1973) *Marine Evaporites: Origin, Diagenesis, and Geochemistry*. Dowden, Hutchinson, and Ross, Stroudsburg, Pa., 69-79].

_____ (1966b) Bromide geochemistry of salt rocks. In J.L. Rau (Ed.), *Second Symposium on Salt*. Northern Ohio Geol. Soc., Cleveland, Ohio, V. 1, 248-275 [reprinted in D.W. Kirkland and R. Evans (Eds.) (1973) *Marine Evaporites: Origin, Diagenesis, and Geochemistry*. Dowden, Hutchinson, and Ross, Stroudsburg, Pa., 333-360].

_____ (1970) Bromide geochemistry of some non-marine salt deposits in the southern Great Basin. *Mineral Soc. Amer. Spec. Pap.*, *3*, 307-319.

_____ (1977) Catastrophic chemical events in the history of the ocean. *Nature*, *267*, 399-403.

_____ (1979) Rotliegend evaporites, Lower Permian of northwestern Europe. *Erdöl Kohle Erdgas Petrochem.*, *32*, 159-161.

_____ (1966) Diagenetic polyhalite in Recent salt from Baja California. *Amer. Mineral.*, *51*, 99-109 [Reprinted in D.W. Kirkland and R. Evans, (Eds.) (1973) *Marine Evaporites: Origin, Diagenesis, and Geochemistry*. Dowden, Hutchinson, and Ross, Stroudsberg, Pa., 69-79].

_____ (1977) Catastrophic chemical events in the history of the ocean. *Nature*, *267*, 399-403.

_____ (1979) Rotliegend evaporites, Lower Permian of Northwestern Europe: Geochemical confirmation of the non-marine origin. *Erdöl Kohle Erdgas Petroch.*, *32*, 159-161.

_____, and I.R. Kaplan (1966) Isotope geochemistry of sedimentary sulfates. *Chem. Geol.*, *1*, 93-135 [Reprinted in D.W. Kirkland and R. Evans (Eds.) (1973) *Marine Evaporites: Origin, Diagenesis, and Geochemistry*. Dowden, Hutchinson, and Ross, Stroudsburg, Pa., 374-398].

Hudec, P.P. and P. Sonnenfeld (1974) Hot brines on La Roques, Venezuela. *Science*, *185*, 440-442.

Hudson, J.D. (1964) Sedimentation rates in relation to the Phanerozoic time scale. *Quart. J. Geol. Soc. London*, *120S*, 37-42.

International Union of Geodesy and Geophysics (1966) Proceedings of the Second Intern. Symp. on Recent Crustal Movements, Aulanko, Finland, August 1965. *Ann. Acad. Scient. Fennicae, Ser. 3A, 90*, 1-498.

Jones, C.L. (1965) Petrography of evaporites from the Wellington Formation near Hutchinson, Kansas. *U.S. Geol. Surv. Bull.*, *1201A*.

_____, and B.M. Madsen (1968) Evaporite geology of Fifth ore zone, Carlsbad district, southeastern New Mexico. *U.S. Geol. Surv. Bull.*, *1251-B*.

Jung, W. (1958) Zur Feinstratigraphie der Werraanhydrit (Zechstein 1) im Bereich der Sangerhauser und Mansfelder Mulde. *Geologie Beih.*, *24*.

Kartsev, A.A., Z.A. Tabasaranskii, M.I. Subbota, and G.A. Moglevskii (1959) *Geochemical Methods of Prospecting and Exploration for Petroleum and Natural Gas* (transl.). Berkeley: Univ. California Press, 242-243.

Kastner, M. (1971) Authigenic feldspars in sedimentary rocks. *Amer. Mineral.*, *56*, 1403-1442.

Katz, A., E. Sass, A. Starinsky, and H.D. Holland (1972) Strontium behaviour in the aragonite-calcite transformation: An experimental study at 40-98°C. *Geochim. Cosmochim. Acta*, *36*, 481-496.

Kay, M. (1955) Sediments and subsidence through time. *Geol. Soc. Amer. Spec. Pap.*, *62*, 665-684.

Kellog Company, M.W. (1965) *Saline Water Conversion Engineering Data Book*, Washington: U.S. Department of Interior.

Kinsman, D.J.J. (1969) Modes of formation, sedimentary associations, and diagnostic features of shallow-water and supratidal evaporites. *Amer. Assn. Petrol. Geol. Bull.*, *53*, 830-840 [reprinted in D.W. Kirkland and Robert Evans (Eds.) (1973) *Marine Evaporites: Origin, Diagenesis, Geochemistry*. Dowden, Hutchinson, and Ross, Stroudsburg, PA, 80-91].

_____ (1971) Diagenetic history of limestones determined from Sr^{+2} distribution. *In* O.P. Bricker (Ed.) *Carbonate Cements*. Johns Hopkins Press, Baltimore, 259-263.

Kinsman, D.J.J. (1973) Trace cation concentrations and the origin and diagenetic history of carbonate rocks. *Symposium "The Chemistry of Sedimentary and Diagenetic Processes."* London Mineral. Soc., April 1973 [unpublished].

_____ (1974a) Calcium sulphate minerals of evaporite deposits: Their primary mineralogy. *In* A.J. Coogan (Ed.) *Fourth Symposium on Salt*. Northern Ohio Geol. Soc., Cleveland, V. 1, 343-348.

_____ (1974b) New observations of the Pleistocene evaporites of Montallegra, Sicily and a modern analog. *J. Sediment. Petrol., 45*, 469-479.

_____ (1976) Evaporites: I. Relative humidity control of primary facies. *J. Sediment. Petrol., 46*, 273-279.

_____ and H.D. Holland (1969) The co-precipitation of cations with $CaCO_3$-- IV. The co-precipitation of Sr^{2+} with aragonite between 16 and 96°C. *Geochim. Cosmochim. Acta, 33*, 1-18.

Kirchheimer, F. (1976) Blaues Steinsalz und sein Vorkommen im Meckar- und Oberrheingebiet. *Geol. Jahrb. D18*, 139 p.

Kirkland, D.W. and R. Evans (1976) Origin of Limestones Buttes, Gypsum Plain, Culberson County, Texas. *Amer. Assn. Petrol. Geol. Bull., 60*, 2005-2018.

Kohler, M.A. and L.H. Parmele (1967) Generalized estimates of free-water evaporation. *Water Resources Res., 3*, 997-1005.

Kolosov, A.S., A.M. Pustyl'nikov, I.A. Moshkina, and Z.M. Mel'nikova (1969) Talc in Cambrian salts of the Kantaseyevka Depression. *Akad. Nauk SSSR Dokl., 185*, 174-177 (transl. *Akad. Naul SSSR Proc. Earth Sci. Sec., 185*, 127-129).

Kropacher, A.M. (1960) Minor elements in anhydrites and epithermal gypsums of the Permian of the Pre-Urals. (transl.) *Zap. Vses. Mineral. Obshch., 89*, 598-602.

Kühn, R. (1963) Rubidium als geochemisches Leitelement bei der lagerstättenkundlichen Charackterisierung von Carnalliten und natürlichen Salzlösungen. *Neues Jahrb. Mineral. Monatsh., 1963*, 107-115.

_____ (1968) Geochemistry of the German potash deposits. *Geol. Soc. Amer. Mem., 88*, 427-504.

_____ (1972a) Combined evaluation of Br- and Rb-contents for the genetic characterization of carnallites and sylvite rocks. In *Geology of Saline Deposits, UNESCO Earth Sci. Ser., 7*, 77-89.

_____ (1972b) Zur Kenntnis der Rubidiumgehalte von Kalisalze Ozeanischer Salzlagerstätten nebst einigen lagerstättenkundelichen Ausdeutungen. *Geol. Jahrb., 90*, 127-220.

_____ and K.J. Hsü (1975) Bromine content of Mediterranean halite. *Geology, 2*, 213-216.

Land, L.S. and G.K. Hoops (1973) Sodium in carbonate sediments and rocks: a possible index to the salinity of diagenetic solutions. *J. Sed. Petrol., 43*, 614-617.

Lloyd, R.M. (1966) Oxygen isotope enrichment of sea water by evaporation. *Geochim. Cosmochim. Acta, 30,* 801-814.

_____ (1968) Oxygen isotope behavior in the sulfate-water system. *J. Geophys. Res., 73,* 6099-6110.

Lohmann, K.C. (1978) Closed system diagenesis of high magnesium calcite and argonite cement. *Geol. Soc. Amer. Abstr. Progr. 10,* 446.

Lucchesi, P.J. and E.D. Whitney (1962) Solubility of strontium sulphate in water and aqueous solutions of hydrogen chloride, sodium chloride, sulphuric acid and sodium sulphate by the radiotracer method. *J. Appl. Chem., 12,* 277-279.

Maiklem, W.R., D.G. Bebout, and R.P. Glaister (1969) Classification of anhydrite--a practical approach. *Bull. Can. Petrol. Geol. 17,* 194-233.

Malikova, I.M. (1967) *Distribution of Rubidium, Thallium and Bromine in Potassium Salt Deposits.* Akad. Sci. USSR Sibir. Otd. Inst. Geol. Geofiz. Novosibirsk (in Russian), 149 p.

Matsuo, S., I. Friedman, and G.I. Smith (1972) Studies of Quaternary saline lakes--I. Hydrogen isotope fractionation in saline minerals. *Geochim. Cosmochim. Acta, 36,* 427-435.

McIntire, W.L. (1968) Effect of temperature on the partition of rubidium between sylvite crystals and aqueous solutions. *Geol. Soc. Amer. Spec. Pap., 88,* 505-524.

Mokiyenko, V.V. (1965) Some features of the geochemistry of strontium in the Lower Permian of the Volgograd Oblast. *Dokl. Akad. Nauk SSSR 162,* 189-191 [transl. *Dokl. Earth Sci. Sec. 162* (1965), 189-191].

Morris, R.C. and P.A. Dickey (1957) Modern evaporite deposition in Peru. *Amer. Assn. Petrol. Geol. Bull., 41,* 2467-2474 [reprinted in D.W. Kirkland and R. Evans (Eds.) (1973) *Marine Evaporites: Origin, Diagenesis, and Geochemistry.* Dowden, Hutchinson and Ross, Stroudsburg, Pa., 21-28].

Müller, G. (1960) Die Löslichkeit von Coelestin ($SrSO_4$) in wasserigen NaCl-und KCl-Lösungen. *Neues Jahrb. Mineral. Monatsh., 1960,* 237-239.

_____, H. Nielsen, and W. Ricke (1966) Schwefelisotopen-verhältnisse in Formationswässern und Evaporiten Nord- und Süddeutschlands. *Chem. Geol., 1,* 211-220.

_____ and H. Puchelt (1961) Die Bildung von Coelestin aus Meerwasser. *Naturwiss., 48,* 301-302.

Nielsen, H. (1972) Sulphur isotopes and the formation of evaporite deposits. In *Geology of Saline Deposits. UNESCO Earth Sci., Ser. 7,* 91-102.

_____, and D. Rambow (1969) S-Isotopenuntersuchungen an Sulfaten hessischer Mineralwässer. *Notizbl. Hess. L.-Amt. Bodenforsch., 97,* 352-366.

Olson, E.R. and H.P. Schwarz (1979) Sulfur and oxygen isotope geochemistry of the Abu Dhabi sabkha. *Amer. Assn. Petroleum Geol. Bull., 6,* p. 505.

Ozol, A.A. (1967) Geochemistry of supergene boron and the formation of halogensedimentary boron deposits. *Lithol. Poleznye Iskop., 1967, no. 5*, 143-155 [transl. *Lithol. Mineral Resources, 1967*, 645-653].

Parry, W.T., C.C. Reeves, Jr., and J.W. Leach (1970) Oxygen and carbon isotopic composition of West Texas lake carbonates. *Geochim. Cosmochim. Acta, 34*, 825-830.

Patterson, R.J. and D.J.J. Kinsman (1971) Determination of marine versus continental sources of subsurface brines in a Persian Gulf coastal sabkha, using Cl^-/Br^- ratios. *Amer. Assn. Petrol. Geol., Studies Geol., 1977*, 381-397.

Puchelt, H. (1972) Distribution of bromide, rubidium and cesium in salt minerals. *Intern. Geol. Cong., 24th, Montreal, Abstr.*, 319.

_____, F. Lutz, and H.H. Schock (1972) Verteilung von Bromid zwischen Lösongen und chloridischen Salzmineralen. *Naturwiss. 59*, 34-35.

Raup, O.B. (1966) Bromine distribution in some halite rocks of the Paradox Member, Hermosa Formation, Utah. In J.L. Rau (Ed.), *Second Symposium on Salt.* Northern Ohio Geol. Soc., Cleveland, V. 1, 236-247.

_____, R.J. Hite and H.L. Groves, Jr. (1970) Bromine distribution and paleosalinities from well cuttings, Paradox Basin, Utah and Colorado. In J.L. Rau and L.F. Dellwig (Eds.), *Third Symposium on Salt.* Northern Ohio Geol. Soc., Cleveland, V. 1, 40-47.

_____ and R.J. Hite (1978) Bromine distribution in marine halite rocks. *Soc. Expl. Paleontol. Mineral. Short Course Notes, 4*, 105-123.

Rees, C.E. (1970) The sulphur isotope balance of the ocean: an improved model. *Earth Planet. Sci. Lett., 7*, 336-370.

Reichert, J. (1966) Verteilung anorganischer Fremdionen bei der Kristallen von Alkalichloriden. *Contrib. Mineral. Petrolog., 13*, 134-160.

Rittenhouse, G. (1967) Bromine in oil-field waters and its use in determining possibilities of origin of these waters. *Amer. Assn. Petrol. Geol., 51*, 2430-2440.

Sakai, H. (1972) Oxygen isotope ratios of some evaporites from Precambrian to Recent ages. *Earth Planet. Sci. Lett., 15*, 201-205.

Schobert, E. (1912) *Uber die Kristallisation von Chlornatrium, Bromnatrium, und Jodnatrium aus Schmelzen und wässerigen Lösungen. Diss. Univ. Leipzig.*

Schoeller, H. (1955) Geochemie des eaux souterraines. *Rev. Inst. Fr. Petrol., 10*, 181-213, 219-246, 507-552.

Schock, H.H. (1966) Bestimmung sehr kleiner Verteilungskoefficienten von Cs, Na and Ba zwischen Lösung and KCl-Einkristallen mit Hilfe radioaktiver Isotope. *Contrib. Mineral. Petrolog., 13*, 161-180.

_____, and H. Puchelt (1971) Rubidium and cesium distribution in salt minerals--I. Experimental investigations. *Geochim. Cosmochim. Acta, 36*, 307-317.

Schulze, G. (1960) Stratigraphische und genetische Deutung der Bromverteilung in den mitteldeutschen Steinsalzlagern des Zechsteins. *Freiberger Forschungshefte, C83*, 114 p.

Smith, G.I., I. Friedman, and S. Matsuo (1970) Salt crystallization temperatures in Searles Lake, California. *Mineral Soc. Amer. Spec. Publ.*, *3*, 257-259.

Taube, H. (1954) Use of oxygen isotope effects in the study of hydration ions. *J. Phys. Chem. 58*, 523-528.

Thode, H.G.O. and J. Monster (1965) Sulfur-isotope geochemistry of petroleum, evaporites, and ancient seas. *Amer. Assn. Petrol. Geol. Mem.*, *4*, 367-377.

Tsusue, A. and H.D. Holland (1966) The coprecipitation of cations with $CaCO_3$--III. The coprecipitation of Zn^{2+} with calcite between 50 and 250°C. *Geochim. Cosmochim. Acta*, *30*, 439-453.

Valiashko, M.G. (1956) Geochemistry of deposits of potassium salts. In, *Voprosy Geologii Agronomicheskikh Rud.* (in Russian). Akademiia Nauk SSR, Moscow, 182-207.

_____ (1969) Genesis and exploration of borate deposits related to marine salt deposits. *Sovet. Geol. 1969, no. 6*, 88-100 [transl. *Internat. Geol. Rev. 12*, 711-719 (1970)].

_____ and L.N. Lavrova (1976) On some new possibilities of the treatment of bromide/chloride relations in the investigation of the conditions of formation of salt deposits (in Russian). In V.I. Borisenkov, M.G. Valiashko, A.P. Vinogradov, N.N. Volkova, and I.K. Zherbtsova (Eds.) *Bromine in Salt Deposits and Solutions as a Geochemical Indicator of their Genesis, History, and Prospecting Indications.* Izd. Moskovskogo Universiteta, 343-353.

_____ and T.V. Mandrikina (1952) Bromine in salt deposits as a genetic and prospecting indicator (in Russian). *Trudy Vses. Nauchno-Islled. Inst. Galurgii 23*, 54-92.

_____, I.K. Zherebetsova, A.N. Lavrova, and U-Bi Khao (1976) On the distribution of bromide between crystalline salt and solutions of various compositions and concentrations (in Russian). In V.I. Borisenkov, M.G. Valiashko, A.P. Vinogradov, N.N. Volkova, and I.K. Zherebtsova (Eds.) *Bromine in Salt Deposits and Solutions as a Geochemical Indicator of their Genesis, History, and Prospecting Indications.* Izd. Moskovskoga Universiteta, 381-407.

Veizer, J. (1974) Chemical diagenesis of Belemnite shells and possible consequences for paleotemperature determinations. *Neues Jahrb. Geol. Paläontol. Abhandl.*, *147*, 91-111.

_____ (1977) Diagenesis of pre-Quaternary carbonates as indicated by tracer studies. *J. Sed. Petrol.*, *47*, 565-581.

_____ (1978) Simulation of limestone diagenesis--a model based on strontium depletion: Discussion. *Can. J. Earth Sci.*, *15*, 1683-1685.

_____, W.T. Holser, and C.K. Wilgus (1979) Correlation of $^{13}C/^{12}C$ and $^{34}S/^{32}S$ secular variations. *Geochim. Cosmochim. Acta* (submitted).

Wazny, H. (1971) Strontium and diagenetic processes in the carbonate Zechstein deposits of Poland. *Internat. Geochem. Cong.*, *Moscow, USSR, 202-25 July, 1971, Abstr.*, 758-759.

Wardlaw, N.C. (1970) Effects of fusion, rates of crystallization and leaching on bromide and rubidium solid solutions in halite, sylvite, and carnallite. In J.L. Rau and L.F. Dellwig (Eds.), *Third Symposium on Salt*. Northern Ohio Geol. Soc., Cleveland, V. 1, 223-231.

Wedepohl, K.H. (1969-1978) *Handbook of Geochemistry, 5 vols.* Springer-Verlag, Berlin.

White, D.E. (1957) Thermal waters of volcanic origin. *Geol. Soc. Amer. Bull., 68,* 1637-1658.

Wolery, T.J. and N.H. Sleep (1976) Hydrothermal circulation and geochemical flux at mid-ocean ridges. *J. Geol. 84,* 249-275.

Yarzhemskiy, Ya. Ya. (1968) Possibility of sedimentation of borates from eutonic brine in salinogenic basins of marine type. *Sovet. Geol., 1968, no. 2,* 15-24 [transl. *Internat. Geol. Rev., 10,* 1096-1102 (1968)].

Zherebtsova, I.I. and N.N. Volkova (1966) Experimental study of behavior of trace elements in the process of natural solar evaporation of Black Sea water and Sasyk-Sivash brine. *Geokhimiya, 1966,* 832-845 [transl. *Geochem. Internat., 3,* 656-670 (1966)].

Chapter 10

MARINE PLACER MINERALS

Virginia Mee Burns

INTRODUCTION

Placers are mineral deposits formed on the earth's surface by mechanical, and sometimes chemical, concentration of mineral particles from weathered debris by alluvial, marine, lacustrine, glacial and occasionally aeolian processes. Deposits occurring on land are commonly divided into alluvial (stream), eluvial (slope), residual (authochthonous), fossil (deep-lead), and marine (off-shore) (Hails, 1976). Economic placer minerals have been divided by Emery and Noakes (1968) into three groups: heavy heavy minerals including gold, platinum, and tin oxides (specific gravity 6.8-21); light heavy minerals including ilmenite, rutile, zircon, monazite, and magnetite (specific gravity 4.2-5.3); and gems including ruby, sapphire and diamond (relatively low specific gravities of 2.9-4.1, but extreme hardness). Emery and Noakes have also related the mineral groups to different depositional environments found in land placer deposits. The most common marine placer minerals, which are among the most dense and chemically or mechanically resistant of minerals, are listed in Table 1.

The marine placer mineral deposits which have been studied occur on the continental shelf generally within five miles of the coast. Submarine placer deposits in the near shelf region are predominantely late Tertiary to Recent in age. During the Pleistocene glacial ages, the sea level was lowered considerably as a result of increased ice formation in higher latitudes. The decrease is believed to have been up to 150 m below the present sea level, which approaches the present shelf edge (Hails, 1976). During the low stands of sea level in the Pleistocene, alluvial, eluvial, authochthonous and beach placers were formed on the exposed shelf areas. When the sea level rose again at the end of the Pleistocene, these placers were covered by the rising ocean and were partly reworked, destroyed, and/or partly covered by Recent sediments. The deep leads of the shelf region are sub-Recent or Glacial Age buried

Table 1. Marine placer minerals.

Mineral	Composition	Specific Gravity	Hardness
Cassiterite	SnO_2	6.8-7.1	6.0-7.0
Chromite	$(Mg,Fe)Cr_2O_4$	4.1-4.7	5.5-6.5
Diamond	C	3.5	10
Fergusonite	$(Y,Ce,Fe)(Nb,Ta,Ti)O_4$	5.4	5.5-6.5
Gold	Au	19.3	2.5-3.0
Ilmenite	$FeTiO_3$	4.5-5.5	5.0-6.0
Magnetite	Fe_3O_4	5.2	5.5-6.5
Monazite	$(Ce,La,Y,Th)PO_4$	4.6-5.4	5.0-5.5
Platinum	Pt	14-19	4.0-4.5
Rutile	TiO_2	4.25	6.0-6.5
Wolframite	$(Fe,Mn)WO_4$	7.1-7.5	4.0-4.5
Xenotime	YPO_4	4.4-5.1	4-5
Zircon	$ZrSiO_4$	4.6-4.7	7.5

river channels which have been preserved under younger sediments. Eluvial occurrences of marine placer deposits are found where weathered source rocks of heavy minerals outcrop on the shelf.

While numerous marine placer deposits are known worldwide, only a few illustrative examples will be described in this chapter. Published descriptions of these and other marine placers are cited in the bibliography at the end of the chapter.

TIN DEPOSITS

The Southeast Asian Tin Belt

The most widespread and economically important of the offshore mineral placers are those of the Southeast Asian Tin Belt shown in Figure 1. The Tin Belt stretches more than 3000 km from Burma and northern Thailand down Peninsular Thailand and western Malaysia to the Tin Islands of Indonesia (Singkep, Bangka and Belitung, formerly Billiton) in the western, most highly mineralized belt. The less valuable eastern belt

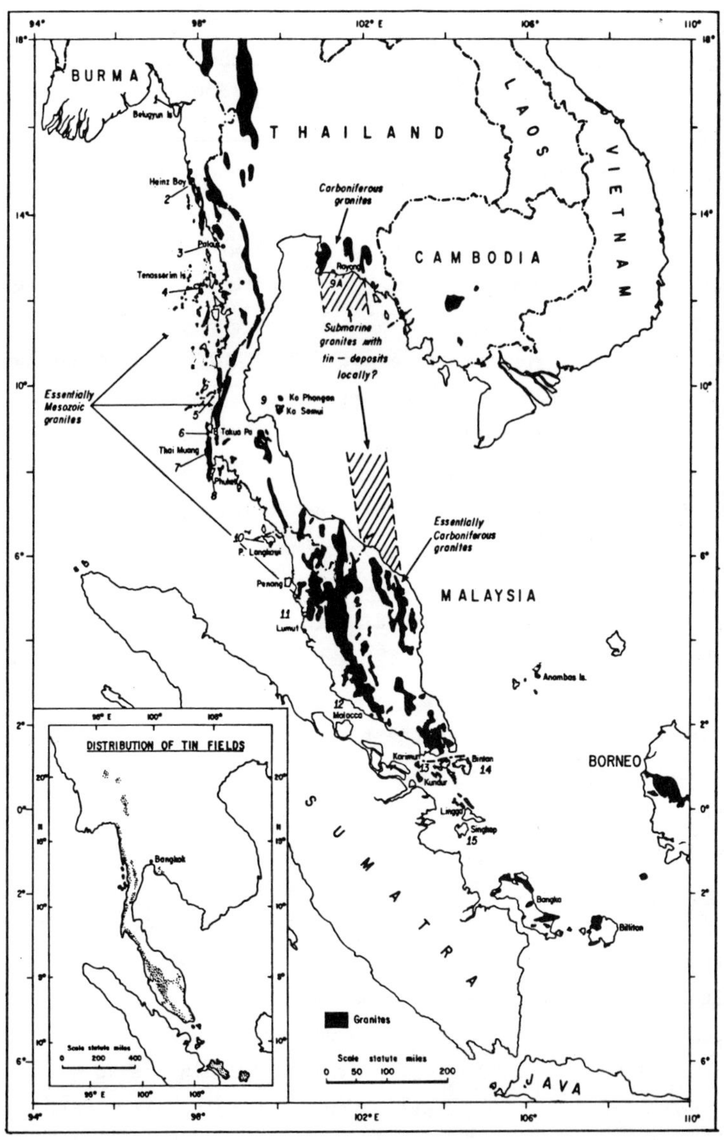

Figure 1. Location of producing and potential offshore tin areas of southeast Asia (from Hosking, 1971). (1) Belugyun Island, (2) Heinze Basin, (3) Spider Island, (4) Tenasserim Delta and Lampa and neighboring islands, (5) Ranong and coast to South, (6) Takua Pa, (7) Thai Muang, (8) Phuket, (9) Ko Phangan and Ko Samui, (9a) Rayong, (10) Langkawi Islands area, (11) Lumut-Dindings, (12) Malacca, (13) Karimun and Kundur, (14) Bintan, (15) The Tin Islands.

begins near the Cambodian border, disappears beneath the sediment-filled Thailand Basin and ends west of Borneo. The source rocks of the onshore and offshore tin placers, which together supply more than half the free world's tin, are biotite or biotite-muscovite granite intrusives of Mesozoic (mostly late Jurassic or late Cretaceous age), which are localized along tectonic belts.

The arc-trench system of the Indonesia Islands and Western Thai Peninsula is an example of a convergent juncture in the plate tectonic model. According to Katili (1974), the arc was formed by subduction of oceanic crust under continental crust. The continental crust is relatively thick in this area comprising volcanic arcs of Permian, Cretaceous, and Tertiary ages. The magmatic rocks formed above the Benioff zone here are mostly sialic and intermediate in character. According to Katili (1974) and Mitchell and Garson (1972), the tin-bearing biotite or biotite-muscovite granites of the Malaysia-Indonesian arc trend system were emplaced during intermittent igneous episodes above the Mesozoic paleo-Benioff zone. In late Tertiary times the subduction zone shifted towards the Indian Ocean and volcanism, which is still active in Indonesia, again accompanied the subduction (Katili, 1974).

The relationship between the tin-bearing granitic belts and the bordering marginal basins and island arcs has been postulated by Mitchell and Garson (1972) as follows. Fluorine, present in large residual amounts in the sedimentary and igneous rocks caught up in the subduction zone, helps to concentrate the tin in the biotite. The lithospheric slab descending beneath the Indonesian arc system resulted in upward transfer of hot volatiles and possibly magma. This would explain the quartz-fluorite bodies occurring on the continental side of the folded belt. The upward stream of volatiles would now partly underlie the Continental margins, remelting the alkali-rich fraction within the acid igneous belt, and an upward movement of volatile fluorides of alkali metals, silicon, tin, tungsten and other metals would take place. Mitchell and Garson suggested that alkali hydroxystannates or thio-stannates were formed by reaction with hydrous phases. Changing conditions of pH and falling temperature resulted in deposits of tin and tungsten minerals. Hydrofluoric acid reacted with available lime to form fluorite and with other elements to form lepidolite and topaz. While the biotite granite of the

Tin Belt contains about 20-50 ppm Sn, most of the tin is contained in the biotite. Bursukov (1957) has reported that 80% of the tin in tin-rich granites is incorporated into the biotite structure, replacing iron. In the later muscovite, tin, tungsten and niobium (and presumably tantalum) are trapped between sheets and in crystal defect sites. These ions, in contrast to those in biotite, are easily leached out to form secondary concentrations of ore (Mitchell and Garson, 1972). Perceptible amounts of tin also occur in the ilmenite and zircon from the granites.

Other geologic factors which contributed to the unusually high tin value of this belt were deep and rapid tropical weathering, which released large quantities of primary tin from source lodes; formation of alluvial and marine placers, many of which are now seaward of the present coast line; and preservation of onshore placers by later comparatively low terrain drained by low-velocity streams (Hails, 1976).

The Indonesian Tin Islands. The Indonesian Tin Islands of Bangka, Singkep, Tujuh Archipelago and Belitung (Billiton) shown in Figure 2 are elevated central parts of the submerged Sunda Platform. From the early

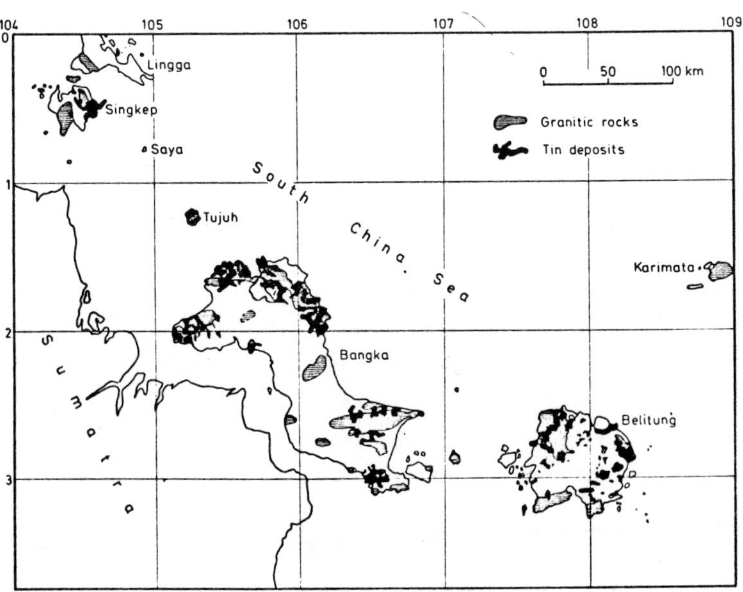

Figure 2. Indonesian Tin Belt from Singkep to Karimata (after Bon, 1979).

Teritary the orogenic zone containing the Southeast Asian tin deposits was a mountainous belt that constituted a part of the continental mass known as Sunda Land. Prolonged denudation up to Pleistocene times resulted in peneplanation to base level of marginal parts of the northern portions of the Tin Belt and virtually complete peneplanation of that part of it which is now occupied by the Tin Islands (Hosking, 1969). It is not known when the primary tin deposits were first uncovered in this area, but it is believed not to have occurred until very late Tertiary or early Pleistocene (Osberger, 1967). Major fluctuations in sea level during Pleistocene times, due to removal and return of sea water when ice formed or remelted, played a major part in the formation of the Indonesian placers by inducing rejuvenation of the rivers and rapid valley erosion, followed by transgression of land by the sea which resulted in reworking of placers previously developed on land. Local tilting on the Sunda Shelf region during the Quaternary also facilitated these processes.

The geology of the largely sea-covered Sunda Shelf within the Indonesian Tin Belt was originally inferred from studies on the Tin Islands (Aleva, 1973; Aleva *et al.*, 1973). On these islands the country rock consists of isoclinally folded sedimentary formations of alternating argillites and sandstones with occasional limestone beds ranging in age from Lower Permian to Triassic. Intrusive rocks range in composition from hypersthene-norite, gabbro, tonalite, granites and alkali granites. In many places a clear connection between cassiterite mineralization and the biotite-granite intrusives can be seen. The hornblende granites and more basic intrusives are not connected with the cassiterite deposits. On Singkep and Bangka all previous and existing cassiterite reserves occur within the granite boundaries in greisen veins or masses which are composed of quartz and yellowish-grey mica with cassiterite, altered feldspar, and kaolin. The Greisen zones vary in thickness from millimeters to tens of meters. The granites and country rocks are mostly chemically weathered to a considerable depth, increasing from the divide areas (2-5 m) to the valleys (over 30 m). The hills present are monadnocks, often with rather steep slopes composed of weathering-resistant rocks, such as quartzites, quartz veins, and granites.

On Belitung, the primary cassiterite mineralization is found in the sedimentary country rocks in zones several kilometers wide and tens of kilometers long. Within the zones, the cassiterite occurs in veins 0.1 to 10 cm wide in association with one or more of the minerals, columnar quartz, feldspar, tourmaline and topaz. Veins up to 75 cm wide composed of pure, fine-grained acicular cassiterite crystals also occur and occasionally cassiterite-bearing sulphide and magnetite veins are found (Aleva, 1973). The rivers on Belitung often follow the contacts between granites and country rocks. On the valley floors a layer of coarse-grained quartz particles and decomposed quartz veins is cemented together by iron oxyhydroxides forming "kaksa," as it it known in Indonesia. In the river valleys of these peneplained humid tropical areas the cassiterite diminishes in quantity and grain size within several hundred meters (Alvea, 1973). It is in these valleys that most of the billitonite (tektite) has been found.

The concentration process for cassiterite is enhanced by the presence of traps where it is protected from further transport and mechanical abrasion. While cassiterite is chemically resistant, it is brittle and easily fractured by mechanical action, and its limit of travel from a source area is believed to be less than 10 km (Emery and Noakes, 1968). The kaksa furnishes an effective trap. Other effective traps include the sinkholes formed in limestone bedrock in Malaya, Bangka and offshore Belitung.

No essential change is observed in cassiterite placers occurring in submerged valley floors a few kilometers offshore of the Tin Islands. The valley floors are somewhat deeper, locally to 50 m below mean sea level. A blanket of Recent to sub-Recent marine mud covers most of the offshore area. The submarine area near shore is a continuation of the mature morphology known from the islands. It is generally accepted that the shallow sea surrounding the islands represent a peneplain submerged by the last post-glacial eustatic rise in sea level. An aureole of submerged river valleys containing valuable tin ores surround the islands of Singkep, Bangka, Belitung and the Trujuh Archipelago. Offshore tin mining in the sheltered areas around the Tin Islands started before World War II, and by 1970 ten dredges were in operation. More recent effort has been concentrated around the Tujuh Islands, a small archipelago located between the islands of Singkep and Gangka (Fig. 2). Extensive seismic and drilling by a subsidiary of Billiton N/V on this remote unprotected area found a complex system of filled-in submarine gullies. Correlation of seismic profiles with drill data showed that two sedimentary episodes occurred, one when sand gravels were deposited followed by one characterized by humic and peat leyars. Cassiterite was found to be marginally restricted to the latter (Bon, 1979).

The small granite dome outcropping on Pulau Tujuh Island is considered to be the source rock of most of the cassiterite, although submerged granites may be the source of some of the tin. Below the seafloor the granite in most places is altered into a clayey mass several tens of meters deep. The enclosing country rock appears to be a mica schist, and the contact has been mapped over a considerable distance. The submarine granite-schist contact is surrounded by non-consolidated sediments consisting of alternating layers of clays and occasional coarse pebbly beds known locally as the "older sedimentary cover." The submarine basement surface and the older sedimentary cover are incised by a number of gullies that radiate outward from the center of the granite. In many places a layer of shelly marine mud, the "younger sedimentary cover," occurs on the abrasion level which truncates the weathered basement at a fairly consistent depth of 20-25 m below present mean sea level.

A small quantity of cassiterite is found in the basal pebble beds of the older sedimentary cover. It is most abundant in the gravel gullies. As well as cassiterite, the heavy mineral concentrates found in the submerged gullies contain abundant marcasite-pyrite, and minor amounts of ilmenite, zircon and monazite (Aleva, 1973).

Thailand Tin Deposits. Another country with a substantial offshore production of cassiterite is Thailand. In this case tantalum is an important by-product which is recovered from the tin smelting operation there. Offshore tin placers have been mined on the west coast of peninsular Thailand since 1908 when production began at Tongah Harbor, Phuket Island.

The Thailand tin deposits are closely associated with Late Cretaceous biotite-muscovite granites which occur as a series of northerly trending belts of irregular to elongate batholiths and stocks that form the backbone of the main mountain system of the Thailand-Malaya peninsula. In places in the offshore area of western Thailand, from Phuket Island in the south to Thai Muang and Takua Pa in the north, a distance of over 120 km, the granite-country rock contact extends seaward from the present coast line and localizes offshore placers that are now being mined. Most primary tin deposits in Thailand occur within 200-300 m of the granite contact and consist of pegmatites, pegmatitic veins, small irregular quartz veins, contact metasomatic deposits developed in altered crystalline limestone, and disseminated cassiterite in locally altered or greisenized outer margins of the granites (Straczek, 1968). On Phuket Island Sainsbury (1969) describes the lode deposits as consisting of cassiterite accompanied by tourmaline, topaz, white mica, and lesser amounts of zircon, ilmenite, monzonite and garnet in quartz veins and pegmatite dikes (Sainsbury, 1969). Few of the lode deposits are economically valuable, but weathering of the lode deposits forms numerous small placers. In the offshore deposits eluvial deposits can be distinguished from alluvial deposits by lack of evidence of stream transport such as grading, shape, and grain size of the cassiterite and associated heavy mineral components. The eluvial placers grade outwards to alluvial deposits of gravel and sand trains in former stream channels. The thickness, coarse size of the minerals, and distribution attest to the greater carrying capacity of the streams in this area during the Pleistocene. The localization of offshore and onshore placer deposits, which all occur within a few kilometers of the granite contact zone, is a function of the mechanical and chemical breakdown of cassiterite.

During the Pleistocene the extensive areas now covered by water to depths of 100 m or more of the west and east coasts of the Thai-Malay Peninsula were dry land, and the development of eluvial and alluvial deposits must have been similar to the existing placer deposits on land. The sea encroached during the interglacial stages, and the deposits were probably reworked by wave and current action before being covered by more recent marine sediments of soft mud which now overlie the more compact coarser tin-bearing sediments. Off of Phuket Island and peninsular Thailand, economic concentrations of tin placers in water depths up to 22 m extend from shore to 1200 m from shore and consist of tin-bearing sands and gravels, with interbedded clay lenses. The tin gravels range in thickness from 1-15 m and average 5 m. The deposits straddle the northerly trending granite contact which is an extension of the onshore contact zone along with small alluvial and eluvial placers occur. The country rocks are Paleozoic slate argillites. The coarseness of the cassiterites, the angularity of grains, and diminuition of grain size and grade away from the granites support the cassiterite source occurring within the narrow contact zone limits. Thus, the placers are largely eluvial with short-range alluvial extensions and were formed by subaerial weathering processes (Straczek, 1968).

The associated minerals of the offshore placers include ilmenite, monazite and xenotime plus lesser amounts of magnetite, zircon, topaz, rutile, garnet, sphene, spinel, epidote, hornblende, hematite, goethite, pyrite or marcasite, feldspar, siderite, tourmaline, anatase, barite, fluorite, epidote, hornblende, muscovite, biotite, and chlorite. Sporadically found minerals include silver, gold, fergussonite and diamonds (Hoffman, 1966; Bateson and Stephens, 1967). Monazite and xenotime are recovered as by-products. Diamonds, which are occasionally found on the jigging tables, are apparently of gem quality but of unknown origin. There are no known occurrences of kimberlite in the area, and there is a lack of associated kimberlite minerals. It has been suggested that the diamonds are derived from secondary sedimentary host rocks, possibly a pebbly greywacke of the Pelaozoic Phuket Series (Bateson and Stephens, 1967), or from contientnal clastic sediments of Mesozoic age (Straczek, pers. comm., 1979).

While tantalum recovered from the tin slags of the tin smelting operation on Phuket Island is an important by-product of the offshore placer tin operations, minerals of the tantalite-columbite series, rarely, if ever, occur offshore, although they have been mined locally from onshore placers (Hockin, 1957). Electron microprobe studies of the offshore cassiterite ores from here and another offshore Thailand tin deposit located about 120 km due north, Takuapa, have shown that up to 17.5% Ta_2O_5 is contained in the cassiterite and rutile phases in solid solution (Hosking, 1970; Hockin, 1957; Straczek, pers. comm., 1979). The ionic radii of tin ($Sn^{4+} = 0.73$ A), tantalum ($Ta^{5+} = 0.68$ A), niobium ($Nb^{5+} = 0.70$ A), and titanium ($Ti^{4+} = 0.68$ A) are close enough to facilitate the formation of substitutional solid solutions among the oxide phases. In the Thailand Peninsula cassiterite the average content of Ta_2O_5 is about 1% and Nb_2O_5 is about 0.5%, but the amount of tantalum and niobium present ranges widely among individual deposits (Straczek, pers. comm., 1979).

Hosking (1970) has reported that the strong red-green pleochroism observed in some cassiterites from Thailand and Malayan deposits is probably due to the presence of appreciable percentages of tantalum and niobium in the structure. J. Straczek (pers. comm., 1979) reports that in Thailand the cassiterite is commonly enriched in tantalum while in Malaya it is relatively enriched in niobium. Some of the cassiterite in primary lode deposits in Thailand is magnetic. Grub and Hannford (1966) have postulated that the ferromagnetic cassiterite which occurs associated with iron minerals is due to the presence of hydrated ferrous stannates and ilmenite-type $FeSnO_3$. According to Bradford (1961), paramagnetic cassiterite, which occurs associated with columbite and tantalite, may be due to the presence of tapiolite solid solutions. It has not been reported whether any of the cassiterites offshore are magnetic. The coarser grains of cassiterite in the offshore concentrates are sometimes embedded in quartz and/or feldspar.

The mineral malayaite, $CaSnSiO_5$, an analogue of sphene which is found in small amounts in skarn deposits of the Thai-Malay Peninsula, has not been reported from the offshore placers. However, it is possible that some of the sphene mentioned as an accessory in the Takua Pa deposit (Hoffman, 1966) is acutally malayaite. Most of the so-called pyrite is probably marcasite. It appears to be of secondary origin and forms branching and botryoidal shapes, although a few pyrite octahedrons have been reported. Ilmenite and magnetite, which are major associated minerals, are altered. The siderite which occurs as brownish spherical and "bowtie" aggregates is of secondary oceanic origin. In some locations lamandine-andradite garnets have been reported (Hoffman, 1966).

Cornwall Offshore Tin Deposits

Placer tin deposits much smaller in extent are found off the coast of Cornwall England, shown in Figure 3. Here cassiterite eroded from numerous tin lodes in granites intruded during the Armorican orogeny have been transported to the sea since Tertiary times (Dunlop and Meyer, 1973). In addition, several bays in Cornwall have been receiving tin-bearing material from mining operations. The St. Ives Bay, in particular,

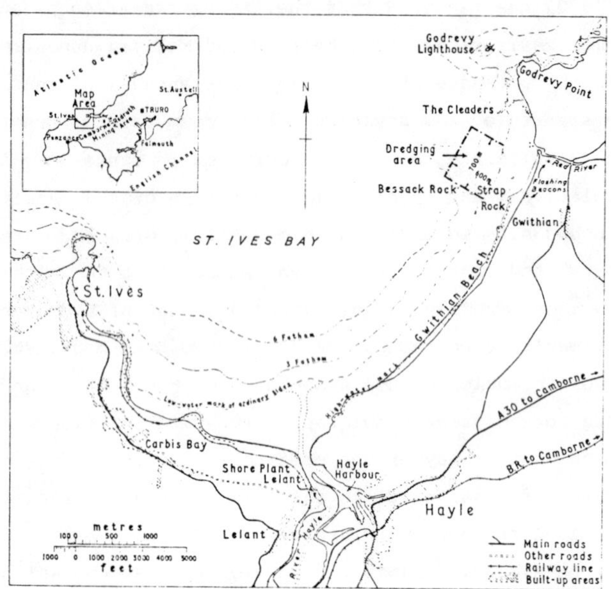

Figure 3. Locality map of the Cornwall offshore tin deposits (after Taylor, 1969).

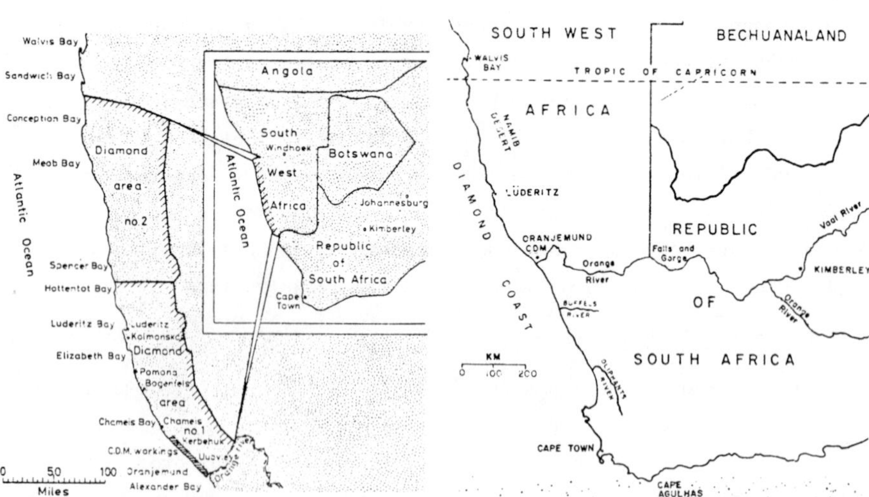

Figure 4. Locality map of the south-west Africa offshore diamond deposits (after Borchere *et al.*, 1969).

Figure 5. The Orange River, South Africa (after Bascom, 1964).

has been receiving tailings from mining operations which have been conducted continuously over the past 2000 years in the Camborne-Redruth area. Upwards of 100,000 tons of tin metal has been estimated to have been lost in the operations of a hundred or more mines in this area. These tailings, upon reaching the sea, although subjected to natural concentration processes, constitute what could be regarded as an artificially-produced offshore placer deposit (McGuiness and Hamilton, 1972).

The sand deposits of St. Ives Bay are widely distributed with their greatest depth at low water mark where they vary from 30 to 40 feet. The top 5 or 6 feet of the sand is generally dark in color although composed of about 56% quartz and 25% calcite. Other minerals include several percent muscovite, blue tourmaline, chlorite, about 1% fluorite and trace amounts of siderite, brown tourmaline, biotite, and cassiterite. The cassiterite content rarely exceeds 0.2%. The highest tin concentrations are found immediately off the mouth of the Red River from low water mark out to sea and in a strip about 100 feet wide along high water mark on the fore shore. In both places the tin is concentrated in the top two feet of sand. Except in a few placers, concentration is not found above the bedrock which is slate (Taylor, 1969). Much of the cassiterite is extremely fine grained, 5-25 microns, and is tightly locked in particles of quartz associated with chlorite, tourmaline, and iron minerals. A significant percentage of the cassiterite is so fine grained and/or tightly bound in gangue it cannot be recovered by beneficiation.

In the St. Ives Bay a similar layer exists at water depths of 30-60 feet. Again, the highest concentrations of tin occur in the upper few feet. By the mid-1970's virtually the entire area around Cornwall had been investigated for offshore-bearing tin sands. Beside the St. Ives Bay area, substantial reserves have also been reported in the St. Agnes Head-Cligga Head area. Again, much of the cassiterite is fine grained and difficult to beneficiate. In 1975 plans were announced to work the Cornwall stanniferous sands in the areas farthest offshore and away from the holiday beaches during the summer and the near-shore and beach zones during the winter (McGuiness and Hamilton, 1972).

DIAMOND

While rubies, sapphires, and diamonds are mined from onshore placers, the only precious gem minerals mined from offshore deposits are diamonds which have been recovered off the Namibian (southwest Africa) coast in the area north of the Orange River, illustrated in Figure 4.

Bedrock in this area consists of ancient Precambrian schists and quartzites of the Gariep System. These rocks are overlain by generally unconsolidated marine sediments which are Pleistocene to Recent in age. The marine sediments, which contain diamonds along with other heavy minerals, consist of gravel ranging from small pebbles to boulders and commonly directly overlie bedrock (Murray, 1969).

The source of the diamonds is known to occur within the Orange River drainage basin, because kimberlite accessory minerals such as pyrope have been found in the lower valley as sediments, but the exact source is not known. No kimberlite pipes have been reported from the lower reaches of the river. While the Orange River and its tributary, the Vaal River, shown in Figure 5, drain the area of Cretaceous kimberlite pipes of the Kimberly field, it is more than 800 km inland. According to Bascom (1964), the Kimberly area could not have supplied the diamonds, not only because of the distance factor, but also because diamond cutters can distinguish between the two types of diamonds produced at a glance. However, Bascom does not specify the criteria they use. Sochneva and Sukhodol'skaya (1974) have reported that peculiar mechanical abrasion surfaces found on diamonds in modern placers from the northeastern Siberian Platform distinguish them readily from non-abraded onshore placer diamonds. Also, the diamonds eroded from the Kimberly area during the Pleistocene might be somewhat different in character from those mined today.

During the Pleistocene, when the rise or fall in the sea level was temporarily stable for a while, cutting of a beach platform occurred. At least four such shelves occur above present day sea level and several below sea level. The most prominent below present sea level occurs at -20 meters and has been traced over a distance of 550 km. A typical profile is shown in Figure 6. It is believed that the -20 meter platform was cut during a stand in sea level as recently as 9000 years ago, judging from the presence of macro- and microfauna of Recent age (Murray *et al.*, 1970).

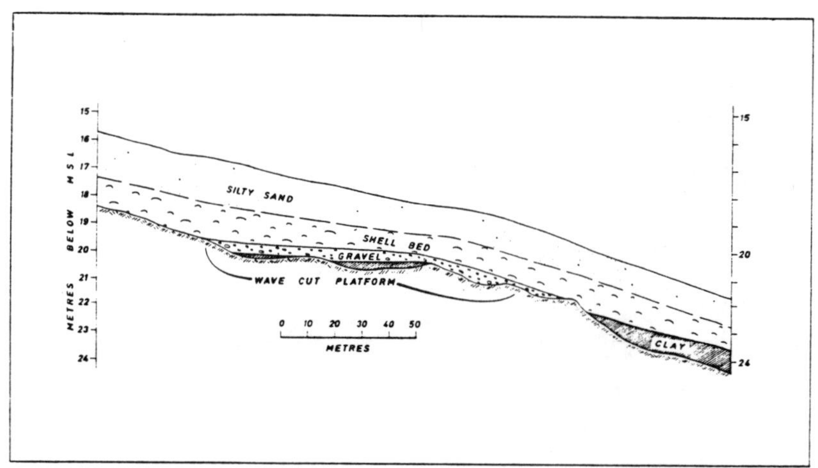

Figure 6. Section showing typical simplified stratigraphic succession of sediments on -20 m wave-cut platforms. Vertical exaggeration 10 x (after Murray *et al.*, 1971).

Another extensive wave-cut platform has been found at -75 meters and is thought to be late Pleistocene.

The marine deposits, where fully developed, comprise a series of beaches ranging in elevation from about 115 m above sea level to several additional beach platforms now below present sea level. Detrital material was moved along the platform by strong northward longshore currents and concentration of heavy minerals took place in gullies or pockets, or were trapped by ridges of hard quartzite. Such submarine topographic features are illustrated in Figure 7. The important transporting agents were surf and currents, and these caused heavy minerals to be selectively deposited and the lighter minerals were carried away. Waves would have had a "jigging" motion resulting in the minerals of higher density settling to the bottom (Murray *et al.*, 1970). Thus, heavy minerals are distributed in a relatively thin and discontinuous blanket on bedrock which acted to form traps and in channeling directions of movement. The marine deposits consist of a succession of interbedded marine sands and gravels, usually commencing with a well-developed basal gravel resting on a series of wave-cut rock platforms descending seaward in steps.

It is believed that the diamondiferous basal gravels were deposited contemporaneously with or shortly after the cutting of the underlying

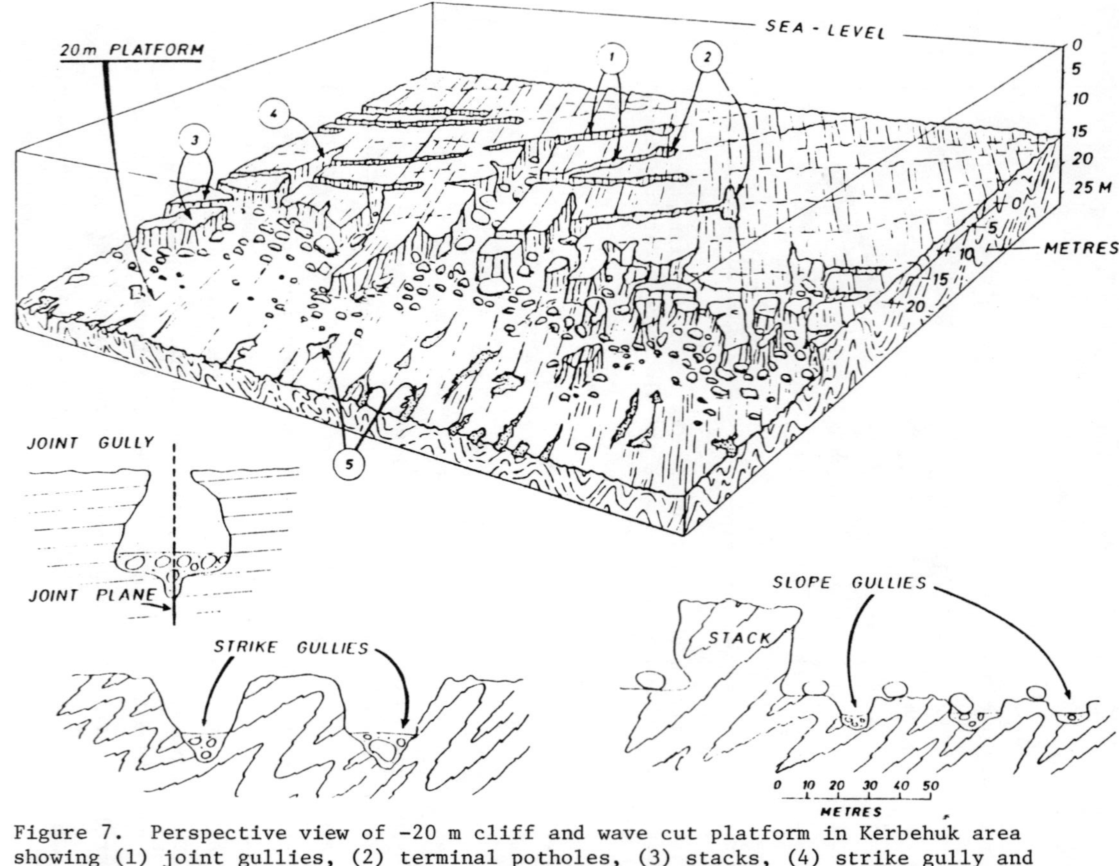

Figure 7. Perspective view of -20 m cliff and wave cut platform in Kerbehuk area showing (1) joint gullies, (2) terminal potholes, (3) stacks, (4) strike gully and (5) slope gullies in -20 m platform, and detailed cross-sections of joint, strike, and slope gullies. (After Murray *et al.*, 1971.)

rock shelves and were subsequently covered by barren quartzo-feldspathic sands and grits which are found on the present beach (Borchers *et al.*, 1969). The overburden varies from 20 m to the south to 3 m in the north on the present day beach and consists of a coarse beach sand with considerable quantities of magnetite and garnet. In the beach deposits the larger stones are generally found near the mouth of the Orange River where the median size in 1964 was 1.5 carats; 50 km north the median size was 0.8 carats, and 150 km north 0.2 carats (Bascom, 1964). Most of the stones are gem quality.

The wave-cut platforms offshore were mined by dredging in the 1960's and more recently by the construction of sea walls up to 200 m beyond low water (Stevens, 1975). (See Figure 8.) The overlying offshore sediments also consist of quartzo-feldspathic sand mixed with shell grit and contains minor fine-grained heavy minerals. The most common heavy mineral constituents in the diamond-bearing sand gravels are epidote, ilmenite, garnet, magnetite and pyrite. Chalcedony and jasper have also been recovered during dredging. Diamond in the mining areas seldom exceeds concentrations of one part diamond to 50 million parts sediment by weight, suggesting that recovery must approach 100% to be economically feasible. The diamonds are predominantly similar in character to the gemstones

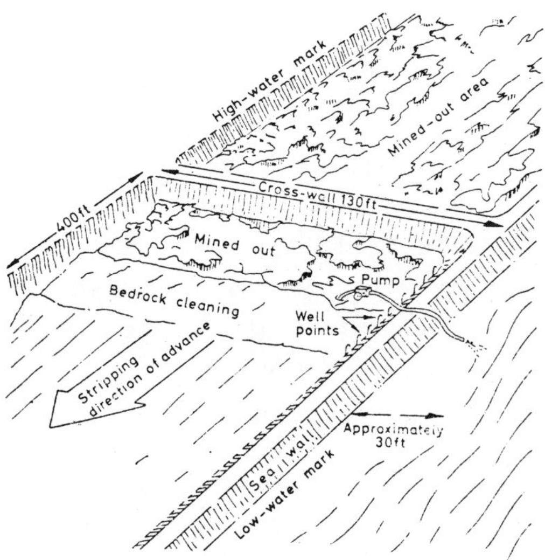

Figure 8. Diagram of sea walls built to mine diamond-bearing beach sands in southwest Africa.

recovered from adjacent onshore beach deposits, but were averaging slightly less than half a carat in 1964, which was somewhat below the average size obtained from the beach deposits. The offshore diamonds are typically colorless, with few inclusions. Colored stones are pale yellowish or smoky and constitute about one quarter of the total. The predominant crystal form is a combined rhombic-dodecahedron and octahedron (Murray *et al.*, 1970).

AMBER

While diamond is the only precious gem recovered from offshore deposits, another valuable material, amber, sometimes included as a gemstone, comes from offshore marine deposits. Amber is not a mineral *per se*, but is a yellow-brown translucent or transparent fossil resin derived from coniferous trees. Amber has variable C:H:O ratios and contains succinite (yielding succinic acid). On land it is vulnerable to weathering and alteration, but under water it remains relatively fresh and unfractured. The material occurs in large quantities along the Baltic coast principally in Prussia, but it is also found on the Danish, Swedish and Russian coasts. It has been recovered for over 3000 years from beaches where it was washed up after storms removed the material from the seafloor. Amber collectors also obtained it from the foreshore by following the outgoing tide on foot or horseback and collecting as much as possible before the rapidly returning tides trapped them. It was also collected from the bottom of shallow waters by boats with people using long-handled nets. By the 19th century amber lying in deeper water and/or embedded on the seafloor was being collected by divers and dredges. Dredging which commenced in 1860 on the Kurisches Haff, south of Memel on the Baltic coast where the floor of the lagoon which contained a bed of alluvium very rich in amber was excavated to depths of 7-11 meters. This operation employed about 1000 people but was mined out by 1890 (Bauer, 1968).

GOLD

Laptev Sea Marine Gold Placers

There has been a considerable amount of prospecting for marine gold placers, especially in Russia, Alaska and Nova Scotia. In the USSR there has been a substained effort during the last several five-year programs to study the metal potential of the USSR's continental shelves, particularly the Arctic and far eastern shelf areas (Igrevsky *et al.*, 1972). At South Primor'ye in the Laptev Sea of the Arctic Ocean, shallow marine gold placers have been found which reportedly have been mined. The placer gold is restricted to thin (0.5-2.0 m) Holocene coarse-grained marine deposits occurring on the submarine slope of the bay, where longshore drift has concentrated an area of relative concentration (Kogan *et al.*, 1974).

The coastal marine gold palcer is confined to the deep side of a triangular-shaped bay screened on the seaward side by an island. A number of narrow mountain streams flow into the bay. About 10 km from the bay, on the upper reaches of the middle of the three main streams draining the area, an outcrop lode of gold ore occurs in thin quartz veins with small quantities of sulfides. The vein is associated with a Late Cretaceous biotite-hornblende granodiorite. Several other occurrences of gold-containing mineralizations are known in the area, all associated with this intrusive (Kashcheyev *et al.*, 1972).

The Holocene marine deposits are fairly widely developed on the submarine slope, in the beach zone and around the mouths of the streams flowing into the bay. The marine deposit consists of coarse-grained sands and gravels, clayey and silty oozes and sands. Coarse-grained sands and gravels are found on the submarine slope, on the oozes, and in the submerged watershed area on bedrock, and form a three-meter marine terrace. The drilling program found gold in almost all the sediments making up the bay slope and floor, but the highest concentrations were confined to the Holocene marine deposits and the alluvium of the central stream submerged valley bottom. In the marine deposits the gold occurs in the near-surface layer of sediment which in places is overlain by a thin layer of muds. In the central part of the bay the gold occurs in both the alluvium and coastal portions of the overlying marine sediments. The placer width is about one meter, and the unconsolidated deposits have an average thickness of one meter. In the western area a submarine abrasional terrace occurs carved on middle Paleozoic granites covered partly by Holocene marine sediments about 0.5 meters thick. The gold is concentrated toward the basal section, and the grade rises sharply in individual depressions (Kascheyev *et al.*, 1972). The richest area of the deposit occurs in the former, now submerged, western stream valley where the gold is confined to a 200 meter wide sand belt of a regressive series of Holocene marine beds and is concentratated mainly in the upper reaches of the section. The thickness of the producing deposits reaches four meters and averages about two meters. In addition to the submarine placers the deposit also has a beach placer.

The mineral composition of the producing deposits are the typical heavy mineral assemblage of granodiorite source material and include magnetite, ilmenite, zircon, rutile and monazite. None of these minerals occur in sufficient amounts to be of economic interest. Traces of bismutite, barite, cinnabar and cassiterite also are found. The concentration of fine zircon grains characteristically rises with increasing gold concentration. The gold has been grouped into three grain size classes: -0.5 + 0.25 mm, -0.25 + 0.10 mm, and -0.1 mm. The average size is 0.2 mm. The gold particles in the -0.5 + 0.25 mm size range are tabular with elongate or irregular outlines and strongly corroded surfaces. In the -0.25 + 0.1 mm class thin abraded flakes and lumps occur. The rounding of the gold particles ranges from good to complete and the surfaces of the grains are corroded. The gold particles in the -0.1 class have spongy, lumpy and irregular shapes with poor to medium rounding and a surface which is slightly corroded to corroded. The spongy lumpy characteristics of the samll grain fraction might be regarded as giving some support to the idea that some of the gold found in placers is deposited from solution (Wopfner and Schwartzbach, 1976).

The main source of the placer gold is believed to be from alluvial deposits of the central stream valley which formed when the sea level was lower than at present and was eroded during the Holocene marine transgression. Concentration of gold in the marine deposits of the submarine slope was facilitated by currents and wave action. Strong currents are thought to have stirred up the sediments, picking up the gold particles from the sea floor and transporting them along the shore in a generally westward direction. The former valley bottom of the western stream served as a large sink in the path of the flow of sediments. To the west of this valley only minor amounts of gold occur (Kashcheyev *et al.*, 1972). The overlying of the deposit by a layer of silty and clayey oozes up to several tens of centimeters thick on the surface of the bay over the whole placer area secured the stability and preservation of the placer. The beach placer was formed as a result of the supply and partial redeposition of gold-bearing bottom material in the zone of tide and surf action with additional source of erosion from a higher Pleistocene shore terrace.

Alaskan Marine Gold Placers

The Seward Peninsula in Alaska has been an important source of placer gold production since the beginning of the 20th century. In the 1960's and 1970's there was considerable interest in the possibility of offshore gold placers in the surrounding Norton Sound and Bering Sea (Tagg and Green, 1973), shown in Figure 9.

The Nome coastal plain and adjacent offshore areas are underlain by Pliocene and Pleistocene marine and glacial sands and gravels. The glacial drift and marine sediments of the Nome coastal plain are, in

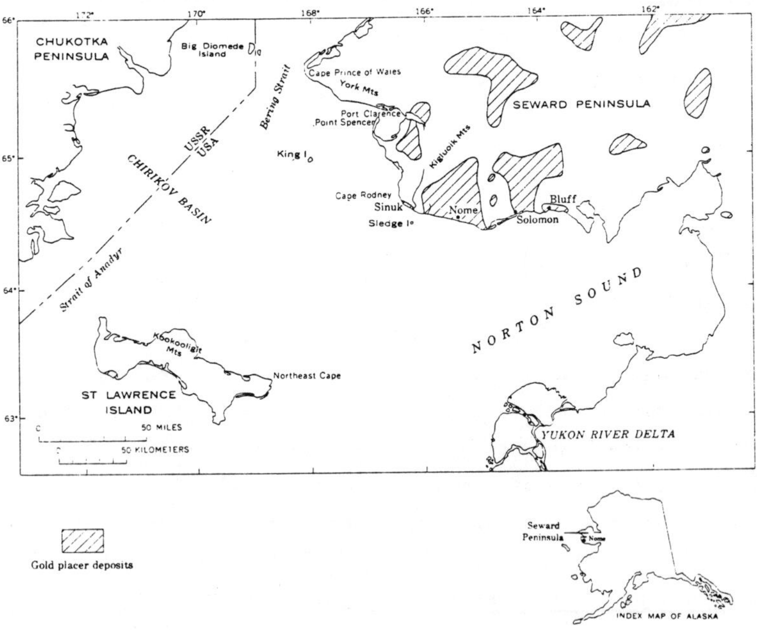

Figure 9. Map of the area in Alaska where marine placer gold deposits are found. Also shown are onshore placer gold deposits (after Nelson and Hopkins, 1972).

turn, overlain by alluvium, silt, and peat of Wisconsin and Holocene age (Tagg and Green, 1973).

The Bering Sea, an epicontinental sea, was most likely emergent until late Tertiary time. Crustal warping then created the present marine basins. Since late Cenozoic time the sea has transgressed beyond the present shore line several times, producing the presently emergent buried beaches which have been important gold producers onshore. Between some of these transgressions the sea level receded below the present level, producing relict offshore beaches now submerged. At least six separate stands of sea level during the Pleistocene-Recent age have created the onshore and offshore beach deposits. At least two of the recessions of sea level were accompanied by glacial invasions beyond the present shore line which may have carried placer gold offshore from existing lodes or onshore placers (Nelson and Hopkins, 1972).

The bedrock in most areas is a very hard, flatbedded, dark-colored schist, ranging from graphitic to calcareous, occasionally pyritic with quartz veins in places. The depth of surfical weathering into bedrock seldom exceeds 1.5 cm. The overlying sediments vary widely in character, and include peat, soft silt, tough or compact clays, silty sands, sand pebble and cobble gravels, boulders and extremely hard cemented gravels (Dailey, 1969).

Nelson and Hopkins (1972) in a study of the Bering Sea sediments off Nome found the richest concentrations and coarsest particles of gold 1 mm or larger occurring in seafloor relict gravels that mantle glacial drift. These bodies of relict gravel formed during transgression and regression of the shoreline during Pleistocene eustatic changes in sea level.

Gold is commonly scarce or absent on bedrock. Fine-grained marine deposits rarely contain visible gold. The glacial till, however, consistently contains small amounts as it does onshore. Outwash and alluvial gravel at the base of stream channels incised into glacial drift commonly contain some gold. The highest gold concentrations found in the study were in gravels in the first few feet of drill cores. The surface sediment with the highest gold content and coarsest particles in gold is the relict lag gravel that veneers glacial drift. The largest gold particles are several millimeters in diameter. The relict gravel is usually less than 15 cm but can occur up to 60 cm in thickness.

The gold in the relict gravel was apparently derived from the underlying drift when lighter particles were winnowed out during sedimentary processes of shoreline transgression or regression. The gold content in the submerged beach deposits is generally highest in places where the submerged beaches cross the glacial drift (Nelson and Hopkins, 1972). The heavy mineral suite associated with the gold is the characteristic metamorphic mineral assemblage found on the Seward Peninsula, with red garnet predominating. Other minerals present include epidote, staurolite, chloritoid and sphene (McManus *et al.*, 1977; Nelson, pers. comm., 1979).

Marine Gold Placers in Nova Scotia

Placer gold occurrences in Nova Scotia from Shelbourn to New Harbour were drilled in the late 1960's, and one location in Munenburg Bay mined for a short time. The survey party looked for seaward extensions of ancient and recent stream channels which might contain gold placers, seaward extensions of onshore gold-bearing strata, and scour channels and basins in gold-bearing areas. About three miles offshore of Isaacs Harbour, the drilling program found an intersection of a major local fault and an anticlinal crest, which is a typical environment of local gold deposits. Bathymetric contours showed it to be semicircular and domeshaped in outline. Seismic traverses indicated steeply-dripping beds beneath a fairly thick layer of eluvium. The site is in water 25-40 meters deep in an area of strong currents. Nearly all the drill holes yielded some coarse gold which was thought to come from a nearby alluvial or eluvial source. The highest grade deposits were found in Lunenburg Bay where ancient drowned and buried stream channels were found to be gold bearing and were buried beneath thick deposits of recent sediment, mainly silt. Most of the gold occurred in coarse gravels but some occurred in medium- and fine-grained sands.

OTHER PLACER MINERALS

Australian Rutile and Ilmenite

Heavy minerals such as ilmenite, rutile, zircon and monazite are capable of being transported long distances without being destroyed.

Much of the world's production of these minerals come from beach sands. For many years the major world producer of rutile plus associated phases ilmenite, zircon and monazite has been the the east coast of Australia where heavy mineral deposits were deposited on present-day and Pleistocene-Recent beaches by current and wave action, particularly during storms. The area shown in Figure 10, which extends approximately 1000 km between the Hawkesbury River in New South Wales to Fraser Island in Queensland, is fringed with sand beaches and by sand plains (McKeller, 1975). The sand mass is composed largely of quartz with little or no feldspars, micas or other light minerals. Many of the onshore beaches have now been depleted in heavy minerals by extensive mining, and most current productions comes from the offshore islands, mainly Fraser, North and South Stradbroke, and Moreton where parabolic beach sands up to several hundred feet high are found.

The sources of the heavy minerals are granites and schists subjected to deep weathering since the early Tertiary. The combination of semi-tropical environment, wave energy and littoral drift has removed unstable minerals such as feldspar, pyroxene, amphibole and mica. The area as a whole is noted for the high proportion of rutile and zircon in the sands compared to beach deposits elsewhere in the world. From south to north along the coast there is a decrease generally in the proportion of rutile and zircon and relative increase in the ilmenite and monazite fractions. Much of the ilmenite is chrome bearing, decreasing its value commercially. Other heavy minerals such as magnetite, garnet, tourmaline, cassiterite and very occasionally gold are sporadic and generally constitute only a few percent of the heavy mineral fraction (Gardner, 1955; McLeod, 1965). Anomalously high values of ilmenite and sometimes magnetite occur near source areas of Tertiary basalts.

Here, as elsewhere in the world, repeated eustatic changes in sea level during the Pleistocene has given rise to a series of beach deposits, some onshore and some now submerged. During the Pleistocene the land mass in Queensland was higher than present, and the cutting power of the rivers greater. Large amounts of sediments were transported by these relatively fast-flowing streams. Once the rivers deposited their sediments the formation of heavy mineral concentrations depended on the configuration of the sea floor and coast line. Under the climatic extremes characterizing

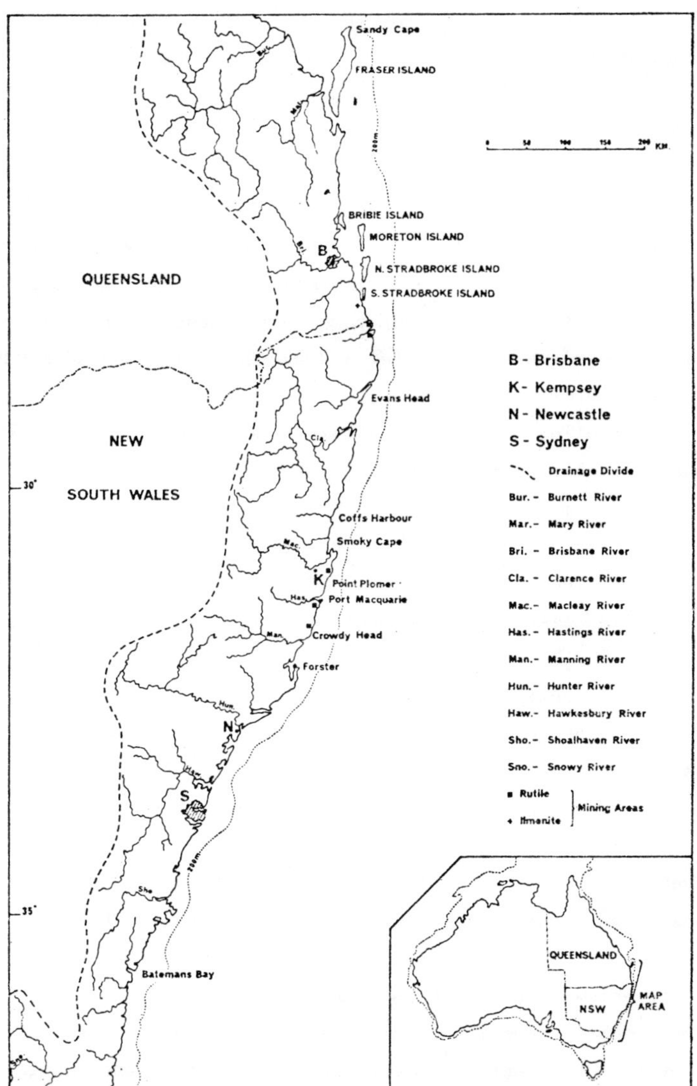

Figure 10. Locality map of New South Wales and Southeastern Queensland coast. Inset map shows continental shelf (after Hails, 1972).

the period, sands were swept inland forming dunes up to several hundred feet high. Remnants of this High Dune System, some of it now offshore forming Fraser, Moreton and North Stradbrooke Islands, are sufficiently numerous to suggest the system formed a more or less continuous coastal belt (McKellar, 1975).

Prospecting in the offshore areas around these islands as well as southward along the eastern coastlines of Queensland and New South Wales has found substantial amounts of heavy minerals but of lower grade than the onshore deposits. The drilling program found a fossil strand line occurring in about 30-35 meters of water and evidence of a second strand line in about 70-80 meters. Two more fossil beaches may exist at greater depths (Brown and McCulloch, 1970). The main occurrences of heavy mineral offshore are as a blanket or sand layer 1.5-5 meters thick below the sea-floor overlying barren material or, more commonly, as seams averaging 200-300 meters in width and 3-5 meters in thickness in elongate fossil beaches beneath 3-5 meters of low-grade or barren sediments. In places the two types intersect giving rise to a rich zone of mineralization (Brown and McCulloch, 1970).

In the offshore deposits variable amounts of heavy minerals are found. In one area a seam of 50 weight percent heavy minerals occurred, but the high concentration was due to a high proportion of ilmenite and magnetite, the contamination being due to close proximity of the Lismore Basalt near the Tweed River. The average heavy mineral content has been estimated at about 0.2%, and the rutile and zircon proportions average around 30%.

The sediments associated with the heavy mineral seams are fine-grained, well-sorted, rounded gray sands. The fine gray sand usually overlies an orange-brown sand which is usually void of heavy minerals. Two types of rutile have been reported from the offshore deposits. One type has the normal rutile optics; the other type has the same chemical composition but is opaque with an appearance similar to that of ilmenite. On heating, this opaque phase develops the normal optical characteristics of rutile (Brown and McConnell, 1970). The opaque phase may, in fact, be a late weathering product of ilmenite or leucoxene, which has a titanium content equivalent to rutile and results from preferential weathering of iron from the ilmenite structure, leaving the more chemically stable

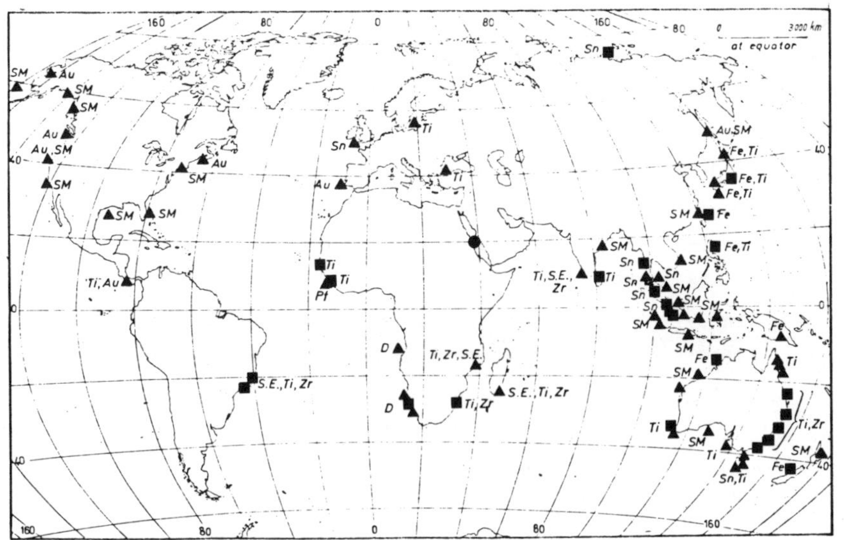

Figure 11. Offshore placer deposits and hydrothermal metallic deposits of the oceans. Au: gold; Pt: platinum; Fe: iron ore; Sn: tin; Ti: titanium minerals; Zr: zircon; SE: rare-earth minerals among which monazite; SM: heavy minerals, not further distinguished. (From Scott, 1970b, with supplements.) ■ = exploitation on beach or nearshore; ▲ = important exploration activity; ● = exploration of hydrothermal metallic deposits (after Scott, 1976).

titanium behind. Ilmenite, as mined in placer deposits, commonly contains much more TiO_2 than the chemical formula for ilmenite allows and actually consists partly of fine-grained alteration minerals rimming the ilmenite which converts entirely to leucoxene when the weathering is intense (Force, 1976). Leucoxene is the general term for fine-grained, opaque, whitish alteration products of ilmenite commonly consisting mostly of rutile and partly of anatase of sphene (Gary *et al.*, 1972). Rutile is among the minerals most resistant to weathering and does not normally form alteration products. However, according to Force (1976), where weathering is intense, rutile which is high in niobium, tantalum or tin may be unstable and recrystallize to fine-grained aggregates of anatase.

Offshore Magnetite Sands in Western Pacific

Another type of offshore heavy mineral occurrence is the iron sands found in shallow water off Japan and the Philippines (Fig. 11). In Japan, at Ariake Bay and Kagoshima Bay off the southern part of Kyushu Island, several million tons of iron sands were recovered during the 1960's. At Ariake Bay about 600 meters off the coast in water about 50 meters deep, the iron sands occurred as a layer about four meters in thickness and averaging about 3-5% titaniferous magnetite. The concentrates produced averaged 56% iron, 12% titanium dioxide, and 0.25% phosphours (Wilson, 1965). At Kagoshima Bay, the iron sands occurred at depths of 15 to 20 meters (Bouysse, 1972).

At Lingayen Gulf, of the northwest coast of Luzon Island in the Philippines, titaniferous magnetite iron sands were recovered during the 1970's in water 3-10 meters deep. The iron sand bed averaged about six meters in thickness and extended for 7-8 kilometers with a width of 300-500 meters. Reserves were estimated in 1972 to be about seven million tons of contained metal (Bouysse, 1972; Hidalgo, 1973, 1975).

Chromite Placers in Oregon

Possible offshore chromite placers may occur off the southwestern Oregon coast. The black beach sand deposits of southwest Oregon, which extend southward from Coos Bay into northwestern California, are found in unconsolidated sediments in present day and raised Pleistocene terraces. The high chromite conrent of the beach sands made the area of strategic interest during World War II and afterwards, when about 10,000 tons of chromite concentrates were obtained from the raised terrace deposits. Associated minerals include ilmenite, magnetite, garnet, zircon, gold and platinum metals (Bowman, 1972). The beach and upraised terrace deposits also have been worked intermittently on a small scale for gold and platinum minerals. According to Bowman (1972), an offshore drilling program during 1970 found some localized areas of heavy mineral enrichment in near-surface sediments. The areas of enrichment are believed to be associated with submerged terrace deposits which formed during the series of glacio-eustatic regressions of the ocean that occurred during the Pleistocene.

MARINE PLACER MINERALS:

REFERENCES AND BIBLIOGRAPHY

Aleva, G.J.J. (1973) Aspects of the historical and physical geology of the Sunda shelf essential to the exploration of submarine tin placers. Geol. Mijinbouw 52, 79-91.

Aleva, G.J.J., L.J. Fick, and G.L. Krol (1973a) Some remarks on the environmental influence on secondary tin deposits. Bur. Miner. Res., Aust. Geol. Geophys. Bull., 141, 163-172.

Aleva, G.J.J., E.H. Bon, J.J. Nossin, and W.J. Sluiter (1973b) A contribution to the geology of part of the Indonesian tinbelt: the sea areas between Singkep and Bangka Islands and around the Karimata Islands. Geol. Soc. Malaysia, Bull., 6, 257-271.

Anderson, R.J. (1972) Recent development in offshore mining. In: 4th Ann. Offshore Tech. Conf., Paper OTC, 1585, 1, 703-708.

Anonymous (1968) Ocean-bottom Minerals. Ocean Industry, 3, 61-73.

Anonymous (1971) COM to discontinue offshore diamond mining in South West Africa. Eng. Min. Jour., 172 (5), 30.

Aranyakanon, P. (1969) Tin deposits in Thailand. In: 2nd Technical Conf. on Tin. A. Fox (ed.), Intern. Tin Council, London, 81-104.

Barsukov, V.L. (1957) The geochemistry of tin. Geokhimiya, no. 1, 41-52; Geochemistry, Ann Arbor, no. 1, 41-52.

Bascom, W. (1964) Exploring the diamond coast. Geotimes, 9, 9-12.

Bateson, J.H. and E.A. Stephens (1967) An appraisal of diamond finds in peninsular Thailand. Sect. B, 76, B125-B126.

Bauer, M. (1968) Precious Stones. Trans. by L.J. Spencer. Dover Publications, New York, 621 pp. (Reprint of 1904 edition by Charles Griffin and Co., Ltd.).

Beckmann, W.C. (1975) Offshore alluvial mining at shallow depths. Oceanology International 75, Soc. Underwater Technol., London, 342-345.

Beiersdorf, H. (1972) Erkundung mariner Schwermineralvorkommen. Meerestech. Mar. Technol., 3 (6), 217-223.

Bon, E.H. (1979) Exploration techniques employed in the Pulau Tujuh tin discovery. Trans. Inst. Min. Metall., Sect. A, 88, A13-A22.

Borchers, D., C.G. Stocken, and A.E. Dall (1969) Beach mining at Consolidated Diamond Mines of South West Africa, Ltd. Exploitation of the area between the high and low-water marks. In: Mining and Petroleum Technology. M.J. Jones (ed.), Proc. 9th Commonw. Cong. Inst. Min. Metall. 1, 577-590.

Bouysse, P. (1972) La recherche minière sous-marine sur la marge continentale. In: Amenagement de la Marge Continentale Hommes et Robots. Coll. A.S.T.E.O., Paris, Cinq. Séance, C, 1-23.

Bowman, K.C., Jr. (1972) Evaluation of heavy mineral concentrations on the southern Oregon continental margin. 8th Ann. Conf. and Expos. Mar. Technol. Soc., 237-247.

Bradford, E.F. (1961) The occurrences of tin and tungsten in Malaya. Proc. 9th Pacific Sci. Cong., 12, 378-398.

Brown, G.A. (1971) Offshore mineral exploration in Australia. Underwater Jour., 3, 166-176.

Brown, G.A. and C. MacCullough (1970) Investigations for heavy minerals off the east coast of Australia. 6th Ann. Conf. Mar. Technol. Soc., 983-993.

Cobb, E.H. (1973) Placer deposits of Alaska. U.S. Geol. Surv. Bull., 1374, 200 pp.

Cruickshank, M.J. (1974) Mineral resources potential of continental margins. In: The Geology of Continental Margins. C.A. Burk and C.L. Drake (eds.), Springer, Heidelberg, 965-1000.

Daily, A.F. (1969) Off-the-ice placer prospecting for gold. 1st Ann. Offshore Technol. Conf., OTC 1029, I, 277-284.

Dieperink, F.J.H. and J.M. Donkers (1978) Offshore tin dredge for Indonesia. Trans. Inst. Min. Metall., Sect. A, 87, A39-A46.

Duane, D.B. (1976) Sedimentation and ocean engineering: placer mineral resources. In: Marine Sediment Transport and Environmental Management. D.J. Stanley and D.J.P. Swift (eds.), John Wiley and Sons, New York, 535-556.

Dunham, K.C. (1970) Gravel, sand, metallic placer and other mineral deposits on the East Atlantic continental margin. In: The Geology of the East Atlantic Continental Margin. ICSU/SCOR Working Party 31, Symp., Cambridge, 1970, Rep. No. 70/13, 79-85.

Dunham, K.C. and J.S. Sheppard (1970) Superficial and solid mineral deposits of the continental shelf around Britain. In: Mining and Petroleum Geology. M.J. Jones (ed.), Proc. 9th Commonw. Congr. Inst. Min. Metall., London, 2, 3-25.

Dunlop, A.C. and W.T. Meyer (1973) Influence of Late Miocene--Pliocene submergence on regional distribution of tin in stream sediments, southwest England. Trans. Inst. Min. Metall., Sect. B, 82, B62-B64.

Emery, K.O. and L.C. Noakes (1968) Economic placer deposits of the continental shelf. ECAFE, CCOP, Tech. Bull., 1, 95-111.

Fick, L.C. (1967) Offshore prospecting. A Technical Conference on Tin. Intern. Tin Council, London, 205-211.

Force, E.K. (1976) Metamorphic source rocks of titanium placer deposits--a geochemical cycle. U.S.G.S. Prof. Paper, 959-B, 13 pp.

Gardner, D.C. (1955) Beach sand heavy mineral deposits of Eastern Australia. Bur. Miner. Res. Aust. Bull., 28, 1-103.

Gilmore, G.A. (1971) Indonesian tin prospects. Mining Mag., 125, 331-343.

Griggs, A.B. (1945) Chromite-bearing sands of the southern part of the coast of Oregon. U. S. G. S. Bull., 945-E, 113-150.

Grubb, P.L.C. and P. Hannaford (1966) Magnetism in cassiterite. Miner. Dep. 1, 148-171.

Hails, J.R. (1972) The problem of recovering heavy minerals from the sea floor--an appraisal of depositional processes. 24th Int. Geol. Cong., Sect. 8, 157-164.

Hails, J.R. (1976) Placer deposits. In: Handbook of Stratabound and Stratiform Ore Deposits. K.H. Wolf (ed.), Elsevier, Amsterdam, 3, 213-244.

Hidalgo, I.C. (1973) Dredge mining for magnetite/iron ore in the Philippines. In: Ocean Mining Symposium: Wodcon Assn., San Pedro, Calif., 47-64.

Hidalgo, I.O. (1975) Mini-dredge (iron sand mining) in the Philippines. In: Proceedings of WODCON VI, Wodcon Assn., San Pedro, Calif. 401-406.

Hockin, H.W. (1957) Tantalum/niobium minerals in Malaya. Malayan Dep. Mines Bull., 2, 20 pp.

Hoffman, V.J. (1966) Thailand tin (offshore) U.C.C. Mining and Metals Division, unpublished m/s.

Hosking, K.F.G. (1969) Aspects of the geology of the tin fields of Southeast Asia. In: A 2nd Technical Conference on Tin. A. Fox (ed.), Intern. Tin Council, London, 41-79.

Hosking, K.F.G. (1970) The primary tin deposits of Southeast Asia. Min. Sci. Eng., 2, 24-50.

Hosking, K.F.G. (1971) The offshore tin deposits of Southeast Asia. ECAFE, CCOP, Tech. Bull., 5, 112-129.

Igrevskiy, V.I., N.P. Budnikov, and V.A. Levchenka (1972) Main tasks for marine exploration geology in U.S.S.R. Osnovnyye zadachi morskikn geologorozvedochnykn rabot VSSR, Sovetskaya Geologiya, 1972 n. 11, 9-24; trans. in Internat. Geol. Rev., 15, 1186-1196.

Isarangkron, P. (1973) Distribution of heavy minerals in the Pluket and Phangnga areas, southern Thailand. ECAFE, CCOP, Tech. Bull., 7, 11-21.

Kanjana-Vanit, R., P. Kham-Ourac and W.P.L. Champion (1969) Offshore mining of tin deposits in South Thailand. In: A 2nd Technical Conference on Tin. A. Fox (ed.), Intern. Tin Council, London, 689-722.

Kashcheyev, L.P., P.I. Kushnarev, and L.B. Khershberg (1972) Principal characteristics of geologic structure and variability of prospecting parameters of a coastal-marine placer: Vyssh. Ucheb. Zavedeniy Izv., Geologiya i Razvedka, no. 1, 76-82; trans. in Int. Geol. Rev., 15, 453-457.

Katili, J.A. (1975) III Geological environment of the Indonesian mineral deposits, a plate tectonic approach. ECAFE, CCOP, Tech. Bull., 9, 39-55.

Kennedy, J.S. (1971) The mineral industry of the territory of South West Africa. In: Minerals Yearbook, U.S.B.M., Washington, 3, 739-745.

Kogan, B.S., L.A. Naprasnikova and G.F. Ryabtseva (1974) Distribution and origin of local brach concentrates of gold as in one of the bays in South Primor'ye. Vyssh. Ucheb. Zavedeniy. Izv. Geologiya i Razvedka, no. 1, 54-60; trans. in Int. Geol. Rev., 17, 945-950.

Layton, W. (1966) Prospects of offshore mineral deposits on the eastern seaboard of Australia. Min. Mag., 115 (5), 344-351.

Lee, G.S. (1968) Prospecting for tin in the sands of St. Ives Bay Cornwall., Trans. Inst. Mining Metall., Sect. A., 77, A49-A64.

Libby, F. (1969) Searching for alluvial gold deposits off Nova Scotia: Ocean Industry, 4, 43-47.

McDonald, G.C.R. and W.K. Tong (1978) Exploration and development of a shallow coastal tin depsoit by suction dredging at Takua Pa, West Thailand, Trans. Inst. Min. Metall., Sect. A., 87, A29-A38.

McGuinness, W.T. and J.R. Hamilton (1972) Recovery of offshore Cornish tin sands. In: Oceanology International 72, Soc. Underwater Technol., London, 417-419.

McKellar, J.B. (1975) The eastern Australian rutile province. In: Economic Geology of Australia and Papua, New Guinea. Aust. Inst. Min. Metall. Monograph Series 5, 1, 1055-1061.

McLeod, I.R. (1965) Titanium and zirconium. In: Australian Mineral Industry: The Mineral Deposits. Bur. Miner. Res. Aust. Bull., 72, 623-634.

McManus, D.A., V. Kolla, D.M. Hopkins, and C.H. Nelson (1977) Distribution of bottom sediments on the continental shelf, Northern Bering Sea, U.S.G.S. Prof. Paper 759 C, 31 pp.

Mero, J.L. (1965) The Mineral Resources of the Sea. Elsevier, Amsterdam, 312 pp.

Mining Journal (1978) Mining Annual Review, London, p. 127.

Mitchell, A.H.G. and M.S. Garson (1972) Relationship of porphysy copper and circum Pacific tin deposits to paleo-Benioff Zons. Trans. Inst. Min. Metall., Sect. B, 81, B10-B25.

Murray, L.G. (1969) Exploration and sampling methods employed in the offshore diamond industry. In: Mining and Petroleum Geology. M.J. Jones (ed.), Proc. 9th Commonw. Cong., Inst. Min. Metall., London, 2, 71-94.

Murray, L.G., R.H. Joynt, D.O.C. O'Shea, R.W. Foster, and L. Kleinjan (1970) The geological environment of some diamond deposits off the coast of South West Africa. In: The Geology of the East Atlantic Continental Margin, ICSU/SCOR Working Party 31 Symposium, Cambridge, 1970. Rep. No. 70/13, 119-141.

Nelson, C.H. and D.M. Hopkins (1972) Sedimentary Processes and distribution of particulate gold in the Northern Bering Sea. U.S.G.S. Prof. Paper 689, 27 pp.

Nesbitt, A.C. (1967) Diamond mining at sea. Proceedings of WODCON, Wodcon Assn., Palos Verdes, 697-725.

Noakes, L.C. and H.A. Jones (1975) Mineral resources offshore. In: Economic Geology of Australia and Papua, New Guinea. Aust. Inst. Min. Metall., Victoria, 11, 1093-1104.

Osberger, R. (1967a) Prospecting tin placers in Indonesia. Min. Mag., 117, 97-105.

Osberger, R. (1967b) Dating Indonesian cassiterite placers. Min. Mag., 117, 260-264.

Osberger, R. (1968) Billiton tin placers: types, occurrences and how they were formed. World Mining, June, 22-28.

Overeem, A.J.A. van (1960a) The geology of the cassiterite placers of Billiton, Indonesia. Geol. Mijinbouw, 39, 444-457.

Overeem, A.J.A. van (1960b) Sonic underwater survesy to locate bedrock off the coasts of Billiton and Singkep, Indonesia. Geol. Mijinbouw, 39, 464-471.

Overeem, A.J.A. van (1970) Offshore tin exploration in Indonesia. Trans. Inst. Min. Metall., Sect. A, 79, 81-85.

Penhale, J. and C.T. Hollick (1968) Beneficiation testing of the St. Ives Bay, Cornwall, tin sands. Trans. Inst. Min. Metall., Sect. A, 77, 65-73.

Putzer, H. and U. von Stackelberg (1973) Exploration von Titanerzseifen vor Mocambique. In: Inter-Ocean '73, 1, 168-174.

Reimnitz, E. and G. Plafker (1976) Marine gold placers along the gulf of Alaska Margin. U.S.G.S. Bull., 1415, 16 pp.

Sainsbury, C.L. (1969) Tin resources of the world. U.S.G.S. Bull., 1301, 55 pp.

Sainsbury, C.L. and B.L. Reed (1973) Tin. U.S.G.S. Prof. Paper 820, 637-651.

Schott, W. (1976) Mineral (inorganic) resources of the oceans and ocean floors: a general review. Chap. 6. In: Handbook of Stratabound and stratiform ore deposits. K.H. Wolf (ed.), Elsevier, Amsterdam, 3, 245-294.

Sinlapajan, P. (1969) Tin dredging in Thailand. A 2nd Technical Conference on Tin. Intern. Tin Council, London, 673-683.

Sochneva, E.G. and O.V. Sukhadol'skaya (1974) Diamond sources of various ages for modern placers in the northeastern Siberian platform. An SSSR Izvestign, Ser. Geol., 1974, 32-35; Intern. Rev., 18, 341-344.

Stevens, C. (1975) The mineral industry of the territory of South West Africa. In: Minerals Yearbook U.S.B.M., Washington, 3, 875-884.

Straczek, J.A. (1968) Problems in offshore exploration for tin in Thailand: a case history. Proc. of a Symposium on Mineral Resources of the World Ocean, Univ. of R.I., occas. publ., no. 4, 66-79.

Tagg, A.R. and H.G. Greene (1973) High-resolution seismic survey of an offshore area near Nome, Alaska. U.S.G.S. Prof. Paper 759-A, 22 pp.

Taylor, J.T. (1969) Tin dredging off the coast of Cornwall. In: Mining and Petroleum Technology. M.J. Jones (ed.), Proc. 9th Commonw. Cong., Inst. Min. Metall., London, 1, 591-611.

Tooms, J.S. (1970) Some aspects of exploration for marine mineral deposits. In: Mining and Petroleum Geology. M.J. Jones (ed.), Proc. 9th Commonw. Min. Metall. Cong., London, 2, 285-296.

Ward, J. (1965) Heavy-mineral beach sands of Australia. In: Geology of Australia Ore Deposits. J. McAndrew (ed.), 8th Commonw. Min. Metall. Cong., Melbourne, 53-54.

Webb, B. (1965) Technology of sea diamond mining. In: Ocean Science and Engineering 1965, Mar. Technol. Soc., Washington, 1, 8-23.

Wilson, T.A. (1965) Offshore mining paves the way to ocean mineral wealth. Eng. Min. Jour., 166 (6), 124-132.

Winwood, K. (1974) Quaternary coastal sediments. In: The Mineral Deposits of New South Wales. N.L. Markham and H. Basden (eds.), Geol. Surv. N.S.W., 597-621.

Wopfner, H. and M. Schwarzbach (1976) Ore deposits in the light of paleoclimatology. Chapter 2 in: Handbook of Strata-bound and Stratiform Ore Deposits. K.H. Wolf (ed.), Elsevier, Amsterdam, 3, 43-92.

Zaalberg, P.H.A. (1970) Offshore tin dredging in Indonesia. Inst. Min. Metall. Trans., Sect. A, 79, A86-A95.